OXFORD STATISTICAL SCIENCE SERIES

Series Editors

A. C. ATKINSON R. J. CARROLL D. J. HAND
D. M. TITTERINGTON J.-L. WANG

OXFORD STATISTICAL SCIENCE SERIES

For a full list of titles please visit
http://www.oup.co.uk/academic/science/maths/series/osss/

10. J.K. Lindsey: *Models for Repeated Measurements*
11. N.T. Longford: *Random Coefficient Models*
12. P.J. Brown: *Measurement, Regression, and Calibration*
13. Peter J. Diggle, Kung-Yee Liang, and Scott L. Zeger: *Analysis of Longitudinal Data*
14. J.I. Ansell and M.J. Phillips: *Practical Methods for Reliability Data Analysis*
15. J.K. Lindsey: *Modelling Frequency and Count Data*
16. J.L. Jensen: *Saddlepoint Approximations*
17. Steffen L. Lauritzen: *Graphical Models*
18. A.W. Bowman and A. Azzalini: *Applied Smoothing Techniques for Data Analysis*
19. J.K. Lindsey: *Models for Repeated Measurements, Second Edition*
20. Michael Evans and Tim Swartz: *Approximating Integrals via Monte Carlo and Deterministic Methods*
21. D.F. Andrews and J.E. Stafford: *Symbolic Computation for Statistical Inference*
22. T.A. Severini: *Likelihood Methods in Statistics*
23. W.J. Krzanowski: *Principles of Multivariate Analysis: A User's Perspective, Revised Edition*
24. J. Durbin and S.J. Koopman: *Time Series Analysis by State Space Methods*
25. Peter J. Diggle, Patrick Heagerty, Kung-Yee Liang, and Scott L. Zeger: *Analysis of Longitudinal Data, Second Edition*
26. J.K. Lindsey: *Nonlinear Models in Medical Statistics*
27. Peter J. Green, Nils L. Hjort, and Sylvia Richardson: *Highly Structured Stochastic Systems*
28. Margaret Sullivan Pepe: *The Statistical Evaluation of Medical Tests for Classification and Prediction*
29. Christopher G. Small and Jinfang Wang: *Numerical Methods for Nonlinear Estimating Equations*
30. John C. Gower and Garmt B. Dijksterhuis: *Procrustes Problems*
31. Margaret Sullivan Pepe: *The Statistical Evaluation of Medical Tests for Classification and Prediction, Paperback*
32. Murray Aitkin, Brian Francis, and John Hinde: *Statistical Modelling in GLIM4, Second Edition*
33. Anthony C. Davison, Yadolah Dodge, N. Wermuth: *Celebrating Statistics: Papers in Honour of Sir David Cox on his 80th Birthday*
34. Anthony Atkinson, Alexander Donev, and Randall Tobias: *Optimum Experimental Designs, with SAS*
35. M. Aitkin, B. Francis, J. Hinde, and R. Darnell: *Statistical Modelling in R*
36. Ludwig Fahrmeir and Thomas Kneib: *Bayesian Smoothing and Regression for Longitudinal, Spatial and Event History Data*
37. Raymond L. Chambers and Robert G. Clark: *An Introduction to Model-Based Survey Sampling with Applications*

An Introduction to Model-Based Survey Sampling with Applications

Raymond L. Chambers
Centre for Statistical and Survey Methodology,
University of Wollongong, Australia

Robert G. Clark
Centre for Statistical and Survey Methodology,
University of Wollongong, Australia

OXFORD
UNIVERSITY PRESS

Great Clarendon Street, Oxford OX2 6DP

Oxford University Press is a department of the University of Oxford.
It furthers the University's objective of excellence in research, scholarship,
and education by publishing worldwide in

Oxford New York

Auckland Cape Town Dar es Salaam Hong Kong Karachi
Kuala Lumpur Madrid Melbourne Mexico City Nairobi
New Delhi Shanghai Taipei Toronto

With offices in

Argentina Austria Brazil Chile Czech Republic France Greece
Guatemala Hungary Italy Japan Poland Portugal Singapore
South Korea Switzerland Thailand Turkey Ukraine Vietnam

Oxford is a registered trade mark of Oxford University Press
in the UK and in certain other countries

Published in the United States
by Oxford University Press Inc., New York

© Raymond L. Chambers and Robert G. Clark 2012

The moral rights of the author have been asserted
Database right Oxford University Press (maker)

First published 2012

All rights reserved. No part of this publication may be reproduced,
stored in a retrieval system, or transmitted, in any form or by any means,
without the prior permission in writing of Oxford University Press,
or as expressly permitted by law, or under terms agreed with the appropriate
reprographics rights organization. Enquiries concerning reproduction
outside the scope of the above should be sent to the Rights Department,
Oxford University Press, at the address above

You must not circulate this book in any other binding or cover
and you must impose the same condition on any acquirer

British Library Cataloguing in Publication Data

Data available

Library of Congress Cataloguing in Publication Data

Data available

Typeset by SPI Publisher Services, Pondicherry, India
Printed and bound by
CPI Group (UK) Ltd, Croydon, CR0 4YY

ISBN 978-0-19-856662-5

1 3 5 7 9 10 8 6 4 2

Preface

The theory and methods of survey sampling are often glossed over in statistics education, with undergraduate programmes in statistics mainly concerned with introducing students to designs and procedures for choosing statistical models, checking model fit to available data, and estimating and making inferences about model parameters. Students may learn about models of considerable complexity, for example generalised linear models can be used for modelling the relationship of a range of explanatory variables to a response variable that can be continuous, binary or categorical. Increasingly, students are introduced to mixed models, time series models and models for spatial data, all of which are suitable for complex, correlated data sets. Non-parametric and semi-parametric methods based on kernel smoothing and spline smoothing are also increasingly important topics. In contrast, survey sampling is often only covered relatively briefly, and in contrast to these other topics, models either do not appear or are simple and are de-emphasised.

This is surprising because survey sampling is one of the most satisfying and useful fields of statistics:

- **The target of inference is satisfyingly solid and observable.** In classical modelling theory, the focus is on estimating model parameters that are intrinsically unobservable. In contrast, a primary aim in surveys is to estimate quantities defined on a finite population – quantities that can in principle be directly observed by carrying out a census of this population. For example, an aim in classical statistical modelling might be to estimate the expected value of income, assuming that the distribution of income can be characterised by a specified distributional family; in contrast, the aim in a survey could be to estimate the mean income for the population of working age citizens of a country at a certain point in time. This mean income actually exists, and so it is possible to check the performance of statistical procedures for estimating its value in specific populations, in a way that is not possible when estimating model parameters. The practicalities of running a survey are also considered by the statistician and the statistical researcher, perhaps more so than in other fields of statistics.
- **Survey sampling is a major field of application of statistics, and is one of the great success stories of mathematical statistics.** Before the mid-twentieth century, national statistics were based by and large on complete censuses of populations. This was enormously expensive and meant that only a limited range of data could be collected. Since then, the use of samples

has become widely accepted, due mainly to the leadership of mathematical statisticians in government statistical agencies, and to the rapid development of a body of theory and methods for probability sampling. Surveys remain a major area of application of statistics – probably **the** major area in terms of dollars spent. A high proportion of graduates from statistics programmes spend some of their career at organisations that conduct surveys.
- **The rich range of models and associated methods used in 'mainstream' statistics can also be used in survey sampling**. The key inferential objectives in survey sampling are fundamentally about prediction, and it is not difficult to transfer theoretical insights from mainstream statistics to survey sampling. Unfortunately, however, this remains a rather under-developed area because the use of models is often de-emphasised in undergraduate courses on survey sampling.

One of the reasons why modelling does not play much part when students are first taught sampling theory is that this theory has essentially evolved within the so-called design-based paradigm, with little or no reliance on models for inference. Instead, inference is based on the repeated sampling properties of estimators, where the repeated sampling is from a fixed finite population consisting of arbitrary data values. This is an attractive and logically consistent approach to inference, but is limiting because methods are required to work for virtually **any** population, and so cannot really exploit the properties of the **particular** population at hand. The model-assisted framework, which has existed in some form since the early 1970s, makes use of models for the population at hand, but in a limited way, so that the potential risks and benefits from modelling are likewise limited. This book is an introduction to the *model-based approach* to survey sampling, where estimators and inference are based on a model that is assumed to summarise the population of interest for a survey.

One way of presenting the model-based approach is to start with a very general linear model, or even generalised linear model, allowing also for any correlation structure in the population. Most of the methods in general use would then be special cases of this general model. We have instead chosen to start with simple models and build up from there, discussing the models suitable to different practical situations. With this aim in mind, this book is divided into three parts, with Part 1 focusing on estimating population totals under a range of models. Chapters 1 and 2 introduce survey sampling, and the model-based approach, respectively. Chapter 3 considers the simplest possible model, the homogenous population model. Chapter 4 extends this model to stratified populations. The stratified model is also quite simple, but nevertheless is very widely used in practice and is a good approximation to many populations of interest. Chapter 5 discusses linear regression models for populations with a single auxiliary variable, and Chapter 6 considers two level hierarchical populations made up of units grouped into clusters, with sampling carried out in two stages. Chapter 7 then integrates these results via the general linear population model. The approach in

Chapters 3 through 7 is to present a model and discuss its applicability, to derive efficient predictors of a population total, and then to explore sample design issues for these predictors.

Robustness to incorrectly specified models is of crucial importance in model-based survey sampling, particularly since much of the sample surveys canon has been model-free. Part 2 of this book therefore considers the properties of estimators based on incorrectly specified models. In practice, all statistical models are incorrect to a greater or lesser extent. To quote from Box and Draper (1987, page 74), 'all models are wrong; the practical question is how wrong do they have to be to not be useful'. Chapter 8 shows that robust sample designs exist, and that, under these designs, predictors of population totals will still be approximately unbiased (although perhaps less efficient), even if the assumed model is incorrect. Chapter 9 extends this exploration of robustness to the important problem of robustifying prediction variance estimators to model misspecification. Chapter 10 completes Part 2 of the book with an exploration of how survey sampling methods can be made robust to outliers (extreme observations not consistent with the assumed model), and also how flexible modelling methods like non-parametric regression can be used in survey sampling.

Parts 1 and 2 of this book are concerned with the estimation of population totals, and more generally with linear combinations of population values. This has historically been the primary objective of sample surveys, and still remains very important, but other quantities are becoming increasingly important. Part 3 therefore explores how model-based methods can be used in a variety of new problem areas of modern survey sampling. Chapter 11 discusses prediction of non-linear population quantities, including non-linear combinations of population totals, and population medians and quantiles. Prediction variance estimation for such complex statistics is the focus of Chapter 12, which discusses how subsampling methods can be used for this purpose. In practice, most surveys are designed to estimate a range of quantities, not just a single population total, and Chapter 13 considers issues in design and estimation for multipurpose surveys. Chapter 14 discusses prediction for domains, and Chapter 15 explores small area estimation methods, which are rapidly becoming important for many survey outputs. Finally, in Chapters 16 and 17 we consider efficient prediction of population distribution functions and the use of transformations in survey inference.

The book is designed to be accessible to undergraduate and graduate level students with a good grounding in statistics, including a course in the theory of linear regression. Matrix notation is not introduced until Chapter 7, and is avoided where possible to support readers less familiar with this notation. The book should also be a useful introduction to applied survey statisticians with some familiarity with surveys and statistics and who are looking for an introduction to the use of models in survey design and estimation.

Using models for survey sampling is a challenge, but a rewarding one. It can go wrong – if the model is not checked carefully against sample data, or if samples are chosen poorly, then estimates and inferences will be misleading. But if the model is chosen well and sampling is robust, then the rich body of knowledge that exists on modelling can be used to understand the population of interest, and to exploit this understanding through tailored sample designs and estimators. We hope that this book will help in this process.

Ray Chambers and Robert Clark

April 2011

Acknowledgements

This book owes its existence to the many people who have influenced our careers in statistics, and particularly our work in survey sampling. In this context, Ken Foreman stands out as the person whose inspiration and support in Ray's early years in the field set him on the path that eventually led to this book and to the model-based ideas that it promotes, while Ken Brewer and our many colleagues at the Australian Bureau of Statistics provided us with the theoretical and practical challenges necessary to ensure that these ideas were always grounded in reality.

Early in Ray's career he was enormously privileged to study under Richard Royall, who opened his eyes to the power of model-based ideas in survey sampling, and Alan Ross, who convinced him that it was just as necessary to ensure that these ideas were translated into practical advice for survey practitioners. To a large extent, the first part of this book is our attempt to achieve this aim. The book itself has its origin in a set of lectures that Ray presented to Eustat in Bilbao in 2003. Subsequently David Holmes was invaluable in providing advice on how these lectures should be organised into a book and with preparation of the exercises.

Robert also had the privilege to work with some great colleagues and mentors. Frank Yu of the Australian Bureau of Statistics encouraged Robert to undertake study and research into the use of models in survey sampling. David Steel's supervision of Robert's PhD developed his knowledge and interest in this area, as did a year in the stimulating environment of the University of Southampton, enriched by interaction with too many friends and colleagues to mention. Robert would also like to express his appreciation of his parents for their lifelong love and support, and for passing on their belief in education.

Many research colleagues have contributed over the years to the different applications that are described in this book, and we have tried to make sure that their inputs have been acknowledged in the text. However, special thanks are due to Hukum Chandra who helped considerably with the material presented in Chapter 15 on prediction for small areas and to Alan Dorfman whose long-standing collaboration on the use of transformations in sample survey inference eventually led to Chapter 17, and whose insightful and supportive comments on the first draft of the book resulted in it being significantly improved.

The book itself has been a long time in preparation, and we would like to thank the editorial team at Oxford University Press, and in particular Keith

Mansfield, Helen Eaton, Alison Jones and Elizabeth Hannon, for their patience and support in bringing it to a conclusion. Finally, we would express our sincere thanks to our wives, Pat and Linda, for freely giving us the time that we needed to develop the ideas set out in this book. Without their support this book would never have been written.

Contents

PART I BASICS OF MODEL-BASED SURVEY INFERENCE

1. **Introduction**   3
   1.1 Why Sample?   4
   1.2 Target Populations and Sampling Frames   5
   1.3 Notation   6
   1.4 Population Models and Non-Informative Sampling   9

2. **The Model-Based Approach**   14
   2.1 Optimal Prediction   16

3. **Homogeneous Populations**   18
   3.1 Random Sampling Models   19
   3.2 A Model for a Homogeneous Population   20
   3.3 Empirical Best Prediction and Best Linear Unbiased Prediction of the Population Total   21
   3.4 Variance Estimation and Confidence Intervals   23
   3.5 Predicting the Value of a Linear Population Parameter   24
   3.6 How Large a Sample?   24
   3.7 Selecting a Simple Random Sample   26
   3.8 A Generalisation of the Homogeneous Model   26

4. **Stratified Populations**   28
   4.1 The Homogeneous Strata Population Model   29
   4.2 Optimal Prediction Under Stratification   30
   4.3 Stratified Sample Design   31
   4.4 Proportional Allocation   31
   4.5 Optimal Allocation   34
   4.6 Allocation for Proportions   35
   4.7 How Large a Sample?   36
   4.8 Defining Stratum Boundaries   37
   4.9 Model-Based Stratification   40
   4.10 Equal Aggregate Size Stratification   42
   4.11 Multivariate Stratification   43
   4.12 How Many Strata?   45

5. **Populations with Regression Structure**   49
   5.1 Optimal Prediction Under a Proportional Relationship   49

	5.2 Optimal Prediction Under a Linear Relationship	52
	5.3 Sample Design and Inference Under the Ratio Population Model	53
	5.4 Sample Design and Inference Under the Linear Population Model	55
	5.5 Combining Regression and Stratification	56
6.	**Clustered Populations**	**61**
	6.1 Sampling from a Clustered Population	62
	6.2 Optimal Prediction for a Clustered Population	63
	6.3 Optimal Design for Fixed Sample Size	66
	6.4 Optimal Design for Fixed Cost	68
	6.5 Optimal Design for Fixed Cost including Listing	70
7.	**The General Linear Population Model**	**72**
	7.1 A General Linear Model for a Population	72
	7.2 The Correlated General Linear Model	74
	7.3 Special Cases of the General Linear Population Model	76
	7.4 Model Choice	79
	7.5 Optimal Sample Design	80
	7.6 Derivation of BLUP Weights	81

PART II ROBUST MODEL-BASED SURVEY METHODS

8.	**Robust Prediction Under Model Misspecification**	**85**
	8.1 Robustness and the Homogeneous Population Model	85
	8.2 Robustness and the Ratio Population Model	88
	8.3 Robustness and the Clustered Population Model	93
	8.4 Non-parametric Prediction	95
9.	**Robust Estimation of the Prediction Variance**	**101**
	9.1 Robust Variance Estimation for the Ratio Estimator	101
	9.2 Robust Variance Estimation for General Linear Estimators	103
	9.3 The Ultimate Cluster Variance Estimator	105
10.	**Outlier Robust Prediction**	**108**
	10.1 Strategies for Outlier Robust Prediction	108
	10.2 Robust Parametric Bias Correction	110
	10.3 Robust Non-parametric Bias Correction	113
	10.4 Outlier Robust Design	114
	10.5 Outlier Robust Ratio Estimation: Some Empirical Evidence	115
	10.6 Practical Problems with Outlier Robust Estimators	117

PART III APPLICATIONS OF MODEL-BASED SURVEY INFERENCE

11.	**Inference for Non-linear Population Parameters**	**121**
	11.1 Differentiable Functions of Population Means	121

11.2 Solutions of Estimating Equations	123
11.3 Population Medians	125

12. Survey Inference via Sub-Sampling — 129
12.1 Variance Estimation via Independent Sub-Samples — 130
12.2 Variance Estimation via Dependent Sub-Samples — 131
12.3 Variance and Interval Estimation via Bootstrapping — 135

13. Estimation for Multipurpose Surveys — 139
13.1 Calibrated Weighting via Linear Unbiased Weighting — 140
13.2 Calibration of Non-parametric Weights — 141
13.3 Problems Associated With Calibrated Weights — 143
13.4 A Simulation Analysis of Calibrated and Ridged Weighting — 145
13.5 The Interaction Between Sample Weighting and Sample Design — 151

14. Inference for Domains — 156
14.1 Unknown Domain Membership — 156
14.2 Using Information about Domain Membership — 158
14.3 The Weighted Domain Estimator — 159

15. Prediction for Small Areas — 161
15.1 Synthetic Methods — 162
15.2 Methods Based on Random Area Effects — 164
15.3 Estimation of the Prediction MSE of the EBLUP — 169
15.4 Direct Prediction for Small Areas — 173
15.5 Estimation of Conditional MSE for Small Area Predictors — 177
15.6 Simulation-Based Comparison of EBLUP and MBD Prediction — 180
15.7 Generalised Linear Mixed Models in Small Area Prediction — 184
15.8 Prediction of Small Area Unemployment — 185
15.9 Concluding Remarks — 192

16. Model-Based Inference for Distributions and Quantiles — 195
16.1 Distribution Inference for a Homogeneous Population — 195
16.2 Extension to a Stratified Population — 197
16.3 Distribution Function Estimation under a Linear Regression Model — 198
16.4 Use of Non-parametric Regression Methods for Distribution Function Estimation — 201
16.5 Imputation vs. Prediction for a Wages Distribution — 204
16.6 Distribution Inference for Clustered Populations — 209

17. Using Transformations in Sample Survey Inference — 214
17.1 Back Transformation Prediction — 214
17.2 Model Calibration Prediction — 215

17.3 Smearing Prediction 218
17.4 Outlier Robust Model Calibration and Smearing 219
17.5 Empirical Results I 221
17.6 Robustness to Model Misspecification 225
17.7 Empirical Results II 227
17.8 Efficient Sampling under Transformation and Balanced Weighting 229

Bibliography **233**

Exercises **241**

Index **261**

PART I
Basics of Model-Based Survey Inference

Statistical models for study populations were used in survey design and inference almost from the very first scientific applications of the sampling method in the late nineteenth century. Following publication of Neyman's influential paper (Neyman, 1934), however, randomisation or design-based methods became the dominant paradigm in scientific and official surveys in the mid-twentieth century, and models were effectively relegated to the secondary role of 'assisting' in the identification of efficient estimators for unknown population quantities. See Lohr (1999) for a development of sampling theory based on this approach. This situation has changed considerably over the last 30 years, with a resurgence of interest in the explicit use of models in finite population inference. Valliant *et al.* (2000) provide a comprehensive overview of the use of models in sample survey inference. To a large extent, this interest in the use of models is due to two mutually reinforcing trends in modern sample surveys. The first is the need to provide sample survey solutions for inferential problems that lie outside the domain of design-based theory, particularly situations where standard probability-based sampling methods are not possible. The second is the need for methods of survey inference that can efficiently integrate the increasing volume and complexity of data sources provided by modern information technology. In particular, it has been the capacity of the model-based paradigm to allow inference under a wider and more realistic set of sampling scenarios, as well as its capacity to efficiently integrate multiple sources of information about the population of interest, that has driven this resurgence.

This book aims to provide the reader with an introduction to the basic concepts of model-based sample survey inference as well as to illustrate how it is being used in practice. In particular, in Part 1 of this book we introduce the reader to model-based survey ideas via a focus on four basic model 'types' that are in wide use. These are models for homogeneous populations, stratified populations, populations with regression structure and clustered populations, as well as combinations of these basic structures.

1 Introduction

The standard method of scientific investigation is via controlled experimentation. That is, if a theory suggests some basic principles leading to outcomes that can be measured or observed, then if at all possible one attempts to validate it by carrying out an experiment where all extraneous effects influencing these outcomes are explicitly controlled or accounted for in the measurement process. By suitably modifying the conditions of the experiment, one can verify the theory by checking to see whether the observed responses are consistent with outcomes predicted under these conditions.

In many situations, however, such experimentation is impossible. All one can do is observe the behaviour of the objects of interest, and infer general principles from this behaviour. Astronomy and geology are two fields of study where this type of situation holds.

Another reason for not carrying out experimentation, especially in the biological sciences, is because ethical and legal structures prevent most kinds of deliberate experimentation, especially on human populations. In other cases, the fact of applying some sort of treatment alters the fundamental nature of the class of objects under study, and thereby precludes any useful analysis. This is often the case in the social and natural sciences.

Finally, there is a large class of problems for which controlled experimentation is not meaningful. These are problems relating to the description of study populations. For such problems what is usually needed is the calculation of some summary statistic defined in terms of the values of a characteristic of interest over the population. Problems of this type form the basis for most official statistical collections.

Collection of data by survey may be suitable in many of the instances where controlled experimentation is not possible. A survey can be defined as the planned observation of objects that are not themselves deliberately treated or controlled by the observer. In essence, 'nature' is assumed to have applied the treatments, and all the analyst can do is observe the consequences.

A survey can be of the complete population of interest, in which case it is often called a census, or can be of a subset of this population, in which case it is usually referred to as a sample survey. In this latter case the problem of inferring about some behaviour in the population given what has been observed on the sample must be considered. This is the domain of sample survey theory and the focus of this book.

Note that sample surveys are typically multipurpose in nature. A sample survey may be used to summarise characteristics of the study population as well to collect data for developing and/or evaluating theories about the mechanism underlying these characteristics. In this book we tackle the first of these objectives, that is the use of sample surveys for presenting summary population information, rather than their use as a data collection tool for research purposes. That is, we emphasise the enumerative aspects of sample surveys rather than their analytic aspects. By and large this reflects the way sample survey theory has evolved. In particular, the statistical theory for sample surveys that has been developed over the last 50 years has tended to concentrate largely on their enumerative use, and only recently has the problem of their analytic use received much attention. See Skinner *et al.* (1989) and Chambers and Skinner (2003) for recent developments on analysis of sample survey data.

1.1 Why Sample?

Why only look at a subset (i.e. a sample) when one could look at the complete set (i.e. the population)? Cochran (1977) lists four reasons:

(a) *Reduced cost* – samples, properly constructed, are usually much cheaper than censuses. This is especially true when the underlying population is very large, and where a sample that is only a small fraction of the population may still be large in terms of sample size and hence lead to highly precise sample estimates.

(b) *Greater speed* – data from a sample that is a relatively small fraction of a population can be collected, summarised and published much more quickly than comparable data from a census. Timeliness of output of survey results is usually of primary importance in official data collections.

(c) *Greater scope and flexibility* – the smaller size of sample survey operations means that greater effort can be invested in data collection for each sampled unit. Samples can thus be used to collect difficult to measure data that would be impracticable to collect via a census.

(d) *Greater accuracy* – the smaller scale of sample surveys means that greater effort can be put into ensuring personnel of higher quality can be employed and given more intensive training and supervision. More effort can also be put into quality control when the survey data are processed. The end result is that use of a sampling approach may actually produce higher quality data and more accurate results than would be possible under a census.

Of course there is a reverse side to these arguments. If accurate information is wanted for very rare population characteristics or for many very small groups in the population then the sample size needed to do the job may be so large that a census would be the most appropriate data collection vehicle anyway.

However, this type of situation is the exception rather than the rule, and the use of the survey sample as a cost efficient method for data collection has proliferated within the last 50 years. Censuses are still taken, but relatively infrequently, and in many cases as a method for benchmarking the much more frequent sample surveys that are the prime source of data for analysis.

1.2 Target Populations and Sampling Frames

The most basic concept of survey sampling theory is that of the underlying target population. Simply speaking, this is the aggregate of elements or units about which we wish to make an inference. Some examples of target populations for a particular country, in this case the United Kingdom (UK), are:

- all UK farming businesses in a particular year;
- all current adult residents of the UK;
- all transactions carried out by a UK business in a financial year;
- all long stay (greater than a week) patients admitted to UK public hospitals in a particular year;
- all animals of a particular species to be found in the UK county of Hampshire;
- all UK registered fishing boats operating in the English Channel.

Target populations can be finite or infinite. In this book we shall be concerned with finite populations, and all the examples above are of this type. However, target populations are not necessarily populations that can be surveyed. Often the units in a target population are only fuzzily defined. The actual population that is surveyed is the survey population. Thus, for the examples above the corresponding survey populations might be:

- all farm businesses that responded to the UK Agricultural Census in that year;
- all adults living in private dwellings and certain selected types of special dwelling (e.g. hotels, nursing homes, prisons, army barracks) in the UK on the night of 30 June;
- all records (e.g. computer records) of transactions by the business in that financial year;
- all hospital records showing more than seven days difference between first admission data and last discharge data for those patients discharged from UK public hospitals in that year;
- all 'visible' (i.e. excluding extremely young) members of a mobile species (e.g. a species of bird) that can be found in Hampshire;
- all current commercial fishing license holders for the UK Channel Fishery.

Ideally the survey and target populations should coincide. However, as the above examples show, this is hardly ever the case. To the extent that the two population definitions differ, the results from the survey will not truly reflect the full target population. It is vital, therefore, that at the planning stages of a survey all efforts are made to ensure that there is the strongest possible link between these two populations. In any event, the results of the survey should always explicitly define the survey population.

The next important concept in survey sampling is that of the sampling frame. This is the list, or series of lists, which enumerate the survey population and form the basis of the sample selection process. Again, referring to the examples above, some suitable frames might be:

- a computer list of unique identifiers associated with all businesses that responded to the UK Agricultural Census;
- a multiple level private dwelling list, with a first level consisting of a partition of the UK into small geographic areas (Enumeration Districts or EDs), a second level consisting of a list of all private dwellings within selected EDs, and a third level consisting of a list of all eligible adults within a selected sample of dwellings taken from the second level list; there may also be associated lists of special dwellings;
- a computer list of transactions;
- photocopies of hospital records for long stay patients, with name and address deleted to preserve confidentiality;
- impossible to construct a sampling frame;
- a computer printout of boat names, business addresses and license numbers for all commercial fishing license holders in the Fishery.

Note that for the animal population example above no sampling frame was available. Consequently, list or frame-based methods for survey sampling are inappropriate for this situation. The theory of survey sampling of natural populations, where frames are usually not available, is not covered in this book. Interested readers are referred to Thompson (1992).

Sampling frames can be complete or incomplete. The material presented in this book will assume a complete frame, that is one such that every element of the survey population is listed once and only once on the frame. Complete frames can be made up of a number of non-overlapping and exhaustive sub-frames, each covering a particular sub-class of elements in the population. An important ingredient of a complete frame is a unique identifier or label that can be associated with each element of the survey population and that enables the survey analyst to draw the specified sample from the frame. That is, the sample drawn is made up of a subset of the labels on the frame.

Incomplete frames, by definition, do not cover the whole survey population. Often, a survey analyst can have two or more incomplete and overlapping sub-frames, each covering an unknown proportion of the survey population, and such that the union of these sub-frames completely covers the population. Although procedures for handling incomplete frames can be based on the ideas described in this book, we do not explicitly consider this issue here.

1.3 Notation

Given a complete frame containing unique identifiers for elements of a finite survey population, we will denote this population by U and, without any loss of generality, associate each label in U with an integer in the set $\{1, 2, \ldots, N\}$, where N is the total survey population size.

Because of the one to one association between the uniquely defined labels on the framework and the set $\{1, 2, \ldots, N\}$, we can treat U as being indexed by the elements of this set, that is by $i = 1, 2, \ldots, N$. The sample s is then a subset of U and therefore a subset of these indices, that is a subset of the integers between 1 and N. The number of elements in s (the sample size) will be denoted by n, and the set of $N - n$ indices for the non-sampled elements, the complement of s, will be denoted r. Thus $s \cup r = U$ and $s \cap r = \emptyset$.

Surveys of human populations are usually targeted at attributes of the individuals making up the population. These can be demographic (e.g. age, gender, marital status, racial/ethnic group), socio-economic (e.g. income, employment status, education) or personal (e.g. political preference, health conditions and behaviours, time use). Surveys of economic populations tend to focus on the physical inputs and outputs for the economic entities making up the survey population, as well as financial performance measures like profit, debt and so on. We refer to any attribute measured in a survey as a variable, and use upper case to denote variables defined on a population and lower case to denote values specific to particular population units. Thus y_i denotes the value of a variable Y associated with the ith population element. Following standard convention, we do not distinguish between realisations and corresponding variables, and so write $E(y_i)$, $Var(y_i)$, $E(y_i|z_i)$ to represent the expected value, variance and conditional expected value of y_i. We also distinguish between two classes of variables. Survey variables correspond to measurements made in the survey whose population values exist but are only known on the sample. Auxiliary variables correspond to variables whose population values are known, although in practice it is often sufficient to know the sample values and the population totals of these variables. In this book, we will generally write survey variables as Y, and auxiliary variables as X or Z.

For example, in a survey of business incomes, the survey variables may include income, number of employees, profit and other financial variables from the most recent financial year. Auxiliary variables may include the number of employees or business income from an earlier period, perhaps obtained from taxation records. In a survey of employee satisfaction at a company, the survey variables would be various dimensions of satisfaction and morale. Auxiliary variables might consist of indicator variables summarising the age, gender, department and rank of each employee, available from personnel records. In a survey of the general population, auxiliary variables often consist of indicator variables summarising the age, gender and geographical region of each person in the population. Population totals for these indicator variables are then population counts by age, gender and region – these counts may be available from official population counts produced by a national statistics office.

Sample surveys are typically not concerned with the individual y_i values themselves, but with making inferences about suitable aggregates summarising the distribution of these values in the survey population. The population total of these values,

$$t_y = \sum_U y_i \qquad (1.1)$$

and the population mean,

$$\bar{y}_U = N^{-1} t_y \qquad (1.2)$$

are typically of interest. Sometimes the finite population distribution function defined by these values,

$$F_{Ny}(t) = N^{-1} \sum_U I(y_i \leq t) \qquad (1.3)$$

is also required. Here t is a dummy variable and $I(y_i \leq t)$ is the indicator function for the event $y_i \leq t$, that is it takes the value 1 when y_i is greater than or equal to t and the value zero when y_i is less than t. Associated with this distribution function are the finite population quantiles. These are values $Q_{Ny}(\alpha)$ such that

$$Q_{Ny}(\alpha) = \inf_t \{F_{Ny}(t) > \alpha\} \qquad (1.4)$$

where α is an index taking values between $1/N$ and $(N-1)/N$. That is, $Q_{Ny}(\alpha)$ is the smallest value of t for which at least $100\alpha\%$ of the population y_i values are less than or equal to that value. Note that $\alpha = 0.5$ defines the finite population median.

The quantities t_y, $\bar{y}_U$, $F_{Ny}(t)$ and $Q_{Ny}(\alpha)$ specified by (1.1)–(1.4) are called finite population parameters. In general, a finite population parameter is any well-defined function of the population values associated with one or more survey characteristics. Thus, for example, the ratio of the population totals (or averages) of two survey variables,

$$R_{yx} = \sum_U y_i \Big/ \sum_U x_i \qquad (1.5)$$

defines a finite population parameter, as does the average of the individual ratios of these characteristics,

$$\bar{r}_U = N^{-1} \sum_U y_i / x_i. \qquad (1.6)$$

A number of the arguments in this book are based on asymptotic considerations. That is, they relate to properties of statistics when the sample size n is large. Of course, since samples are always taken from populations, which are, by definition, finite, these arguments implicitly assume that the population size N is also large, in the sense that the difference $N - n$ is large. Since rigorous asymptotic arguments tend to be littered with technical conditions that are usually impossible to verify in any practical application, we avoid them in this book. Instead, we make free use of 'big oh' notation to indicate order of magnitude conditions that need to apply before results can be expected to hold. In particular, given two sequences of numbers $\{\alpha_n\}$ and $\{\beta_n\}$, we say that $\{\alpha_n\}$ is $O(\beta_n)$ if the sequence $\{\alpha_n/\beta_n\}$ remains bounded as n increases without limit. Thus, a statement of the form $Var(\hat{\theta}) = O(n^{-1})$ for some statistic $\hat{\theta}$ is just a shorthand way of saying that, as the sample size increases, the variance of this statistic decreases at the same rate as the inverse of the sample size.

An important aspect of asymptotic arguments is that they allow approximations. This is very useful when exact results about the distribution of a statistic are too complex to derive. In this book we will make frequent use of two approximations that are pervasive in statistics. The first is the central limit approximation, which essentially states that if sample sizes are large enough, the distributions of many statistics (and particularly linear statistics) are well approximated by normal distributions. The second is approximation of moments by leading terms in their Taylor series expansions (often also referred to as Taylor linearisation). A reader who is unsure about these concepts should refer to Serfling (1980) for a thorough examination of their application in statistics.

1.4 Population Models and Non-Informative Sampling

A finite population parameter is a quantity whose value would be known exactly if a census of the survey population were carried out. However, there is another type of statistical object that can be associated with the values that make up a finite population that is less well defined. This is a statistical model for these values, often referred to as a superpopulation model. In this book a statistical model for a population is defined broadly as a specification of the statistical properties of the population values of the survey variables of interest. In some cases this model may be tightly specified, in the sense that it explicitly identifies a stochastic process that generated these population values. More generally, such a model is usually rather weakly specified, in the sense that it only identifies some of the statistical properties (e.g. first and second order moments) of the distribution of the population values of the survey variables. In either case there will be parameters associated with the model specification (superpopulation parameters) whose values are unknown.

For example, it may be reasonable to postulate that the values $y_1, y_2, \ldots, y_N$ are in fact N independent and identically distributed realisations of a random variable with mean μ and variance σ^2. In this case μ and σ^2 are hypothetical constructs that could never be observed exactly even if a census of the survey population was carried out. Another example of a population model is a regression model. Given two survey variables with population values y_i and x_i, $i = 1, 2, \ldots, N$, it may be reasonable to assume that the conditional expectation of y_i given x_i is linear in x_i. That is, one can write $E(y_i|x_i) = \beta_0 + \beta_1 x_i$ where $E(.)$ denotes expectation relative to an underlying stochastic process that led to the population values y_i. The superpopulation parameters here are β_0 and β_1. Again, we note that these parameters are hypothetical. They only exist as a convenient way of characterising how the population values of Y tend to change as the corresponding population values of X change. The stochastic process that actually generated the population values of these variables is unspecified.

Given a model for a population, standard (infinite population) statistical theory provides various methods for efficient estimation of the parameters of this model. However, nearly all these methods assume the sample data are a random sample of realisations from the stochastic process defined by the population model. In practice this means that there is no systematic relationship between the values generated by the model and the method used to decide which of them are actually observed. With data obtained from survey samples this assumption may not be valid. Very often, the sample design used in the survey will favour observation of particular types of sample values. Ignoring this design information and analysing the survey data as if it had been obtained by some form of random sampling can lead to biased inference.

As noted earlier in this chapter, methods for making inferences about superpopulation parameters such as β_0 and β_1 are not the focus of this book. However, this does not mean that such parameters are unimportant for finite population inference (i.e. for inference about finite population parameters). In fact, this book is essentially about how our knowledge of these parameters (and the statistical models they characterise) can be used to develop efficient methods for finite population inference.

The key concept used to relate parametric models for populations to inference about finite population parameters is that of *non-informative sampling*. Broadly speaking, a method of sampling is non-informative for inference about the parameters of a superpopulation model for a variable if the same superpopulation model also holds for the sample values of this variable. That is, we can make valid inferences about these parameters on the basis of fitting the superpopulation model to the sample data. More formally, let θ be a finite population parameter defined by the population values of a (possibly multivariate) variable Y, let $\boldsymbol{X}_U$ consist of auxiliary information about the population, let $\boldsymbol{Y}_U$ be the values of Y for the population, let s be the set of n units selected using some sampling method, and let $\boldsymbol{Y}_s$ be the values of Y for the sampled units. We say that a method of sampling is non-informative for inference about θ given $\boldsymbol{X}_U$ if the joint conditional distribution of $\boldsymbol{Y}_s$ given $\boldsymbol{X}_U$ is the same as the joint conditional distribution of $\boldsymbol{Y}_U$ given $\boldsymbol{X}_U$ restricted to just those units in s. In effect, the method of sampling only influences inference about the parameters of the joint conditional population distribution of Y by determining which population units make up the sample. The outcome of the sampling process (the set s of sample labels) contains no further information about these parameters.

The Conditionality Principle (Cox and Hinkley, 1974, p. 38) states that one should always condition on ancillary variables in inference. An ancillary variable is one whose distribution depends on parameters that are distinct from those associated with the distribution of the variable of interest. As an ancillary statistic, s should therefore be treated as fixed in inference about the parameters of the joint distribution of the population values of Y given the auxiliary information.

Probability sampling methods form an important class of non-informative sampling methods. These are methods that use a probability mechanism to decide whether or not to include a particular population unit in sample, and where this mechanism only depends on the population values of an auxiliary variable Z (which can be vector valued). In this case, once we condition on the population values of Z, it is clear that the outcome of the selection process is independent of the values of any of the survey variables, and so the method of sampling is non-informative given Z. Note that simple random sampling, where every singleton, pair, triple and so on of population units has exactly the same chance of turning up in sample as any other singleton, pair, triple, and so on, is non-informative.

The importance of non-informative sampling to the model-based approach to finite population inference cannot be overstressed. This is because it allows valid inference for parameters of the conditional distribution of non-sampled population values of Y on the basis of models for the same conditional distribution of sampled values of Y. Thus, for example, we may have a model for a population that says that the regression of a variable Y on an auxiliary variable Z is linear. In effect the population values of these two variables satisfy

$$E(y_i|z_i) = \beta_0 + \beta_1 z_i. \tag{1.7}$$

If our method of sampling is non-informative given the population values of Z, we can then immediately say that (1.7) holds in both the sampled and non-sampled parts of the population. As a consequence, sample estimates of β_0 and β_1 can be validly used to estimate the regression of Y on Z in the non-sampled part of the population. As will become clear in the next chapter, this ability to use information about superpopulation parameters derived from the sample to make statements about the distribution of the non-sampled part of the population is critical for application of the model-based approach to survey inference.

A word of caution, however. Sampling methods are usually assumed to be non-informative conditional on Z. This is achieved by a combination of appropriate sample design, and the inclusion of relevant variables in Z, which can explain any differences between the sampled and non-sampled units. Very few sampling methods would be completely non-informative if Z was empty. However, there should always be some level of information about the outcome of the sampling process that allows us to distinguish the sampled population units from those that have not been sampled. Provided this information is included in Z, then we can safely ignore the sampling process in inference and treat sample and non-sample Y-values as drawn from the same distribution. To illustrate, suppose that (i) we have access to the population values of an auxiliary variable Z; (ii) we expect (1.7) to hold, with non-zero β_1; and (iii) we use a method of sampling such that the sample distribution of Z differs from its non-sample distribution. Then, under the model-based approach, we expect the marginal sample and non-sample distributions of Y to be different, and so we must condition on Z in our inference. If we did not condition on the auxiliary variable, then sampling would be informative, and our inferences would be invalid.

A classic example of this situation is where the probability of a particular population unit being included in sample depends on its value of Z, in the sense that units with larger values of Z tend to be included in sample more often than units with small values of Z. In such a case, we expect the sample and non-sample distributions of Z, and hence of Y, to be quite different. However, the conditional distribution of Y given Z is the same for both sets of units. That is, this method of sampling is non-informative given the population values of Z. We can then base our inference about a population characteristic of Y on this conditional distribution.

What about if the sampling method is informative? Here conditioning on Z is not sufficient to ensure that population and sample distributions of the variable Y are the same. In this case we have two options. Sometimes (typically not very often) we have sufficient information on the method of sampling to allow us to specify (and fit) a model for the distribution of the non-sample values of Y. Inference can then proceed on the basis of this model. An example is Sverchkov and Pfeffermann (2004). The other option is essentially our only choice when we do not have sufficient information to implement option one. This is to adopt robust methods of inference that allow for differences between the sample and non-sample distributions of Y. We discuss such robust methods of finite population inference later in this book. It should be noted, however, that such methods only work if the distribution of the non-sample values of Y is not too different from that of the sample values. No method of robust inference can protect against a total disconnect between the sample and non-sample distributions of the survey variable. In this context, it is advisable, if we suspect that an informative sampling method has been used, to collect enough additional information about the non-sampled part of the population to ensure that the sampling method then becomes non-informative, at least approximately.

Applying good survey practices can also be used to reduce the potential for sampling to be informative. Steps that can be taken include:

- *Selecting a sample using probability sampling or some other non-subjective method.* Designs where an expert chooses a set of units believed to be representative should be avoided, as in this case the sampling procedure will probably depend on variables other than Z, so that the sample distribution of $Y|Z$ could differ from the population distribution. If expert knowledge is available, it should be used to select which variables Z are likely to be relevant, rather than to select the actual sample. A sampling procedure based on Z should then be used. Extreme designs, for example where only the units with the largest values of Z are selected, should also be avoided, since assuming non-informativeness would then be equivalent to extrapolating the model for $Y|Z$ to an unobserved part of the domain of Z. Appropriate sample designs

reflecting this approach will be suggested for different situations throughout this book.
- *Achieving a high response rate.* When we say 'sample', we really mean the responding sample, that is those units who were selected, contacted, and agreed to participate in the survey. Of the initially selected sample, some units will be uncontactable or will decline to participate. The characteristics of the units who respond could well be different from those who do not. If response depends only on Z, then the sample will not be informative, but it may also depend on Y and other variables, leading to an informative sample. Achieving a high response rate will reduce the informativeness of the sampling process. This can be achieved by: using a sufficient number of callbacks when selected households do not answer in telephone or face-to-face interviewer surveys; well-designed questionnaires or interviews which do not overly burden the respondent; professional conduct, appearance and manner of interviewers; believable and justified assurances that respondents' data will be used only for statistical purposes and not for identifying individuals; a concise statement to the respondent of the value of the survey to the community; maintaining a public reputation for trustworthiness, professionalism and relevance; the use of pre-approach letters; and offering incentives for respondents to participate. Some national statistical offices also have the power to make surveys compulsory, which in conjunction with the other methods mentioned can lead to high response rates. For more information on survey methods, the reader is referred to Salant and Dillman (1994) and Groves *et al.* (2004).

2 The Model-Based Approach

In this chapter we develop the essentials of the model-based approach to sample survey design and estimation. In doing so, we focus on the population total $t_y = \sum_U y_i$ of a survey variable Y, and we denote an estimator of this quantity by $\hat{t}_y$.

Before we start our investigation of efficient estimators of this population total, it is useful to remind ourselves about which quantities are held fixed and which are allowed to be random under the model-based approach:

- Population values y_i are assumed to be generated by a stochastic model (the so-called superpopulation model) and are random. For example, (1.7) is a partial specification of such a model, giving the expected value of y_i.
- All expectations and variances are conditional on the outcome of the sample selection process. That is, the selected sample s is treated as a constant.
- The sample values of y_i are also random variables.
- The population total t_y is a sum of random variables and is therefore a random variable itself. Estimation of t_y is equivalent to prediction of the value of this random variable using the data available.
- Predictors $\hat{t}_y$ of t_y are functions of the sampled values $\{y_i, i \in s\}$ as well as of the auxiliary information $\{z_i, i = 1, \cdots, N\}$. The sampled values of Y are random variables, and so $\hat{t}_y$ is a random variable.
- Parameters of the model, such as β_0 and β_1 in (1.7), are assumed to be 'fixed but unknown' constants. In enumerative inference, we generally need to estimate model parameters, but only as a means to the end of predicting t_y and other finite population quantities.

It is important to be clear on what is treated as fixed or random, because other approaches to survey sampling do this differently. For example, the design-based (or randomization) (Cochran, 1977) and model-assisted (Särndal et al., 1992) approaches treat the population values of Y as unknown constants and the sample selected as the only source of randomness. The Bayesian approach (Ghosh, 2009; Ghosh and Meeden, 1997) treats all quantities as random variables, including model parameters. For a recent comparison of the model-based and design-based approaches, see Brewer and Gregoire (2009). A recent overview of the model-based approach is given in Valliant (2009); the two pioneering references are Brewer (1963) and Royall (1970).

In the following chapters we use the model-based approach to show how good predictors $\hat{t}_y$ of t_y can be constructed under some widely applicable models.

In particular, we show how the first two moments of $\hat{t}_y - t_y$ can be obtained under these models, and we then use this knowledge to design efficient sampling strategies for $\hat{t}_y$.

To start, we note that both t_y and $\hat{t}_y$ are realisations of random variables whose joint distribution is determined by two processes – the first one, assumed random, that led to the actual population values of Y, and the second the process (possibly random, possibly not) that was used to determine which population units were selected for the sample s, and which were not. The sample s will be defined throughout the book to contain those units which were selected for the survey and which fully responded. We assume a (superpopulation) model for the population-generating process. Typically, we do not model the sample selection (and non-response) process, assuming instead that this process is non-informative given the values of an auxiliary variable Z whose values are related to those of Y and are known for all units making up the population. As noted in the previous chapter, this means that the conditional distribution of Y given Z in the population is the same as that in the sample. This circumstance considerably simplifies our inference and will be taken as given unless specified otherwise.

Ideally, we want $\hat{t}_y$ to be close to t_y, or, equivalently, we want the sample error $\hat{t}_y - t_y$ to be close to zero. Of course, we do not know the value of this sample error, but under the model-based approach the statistical properties of $\hat{t}_y - t_y$ follow from the probability structure of the assumed model. In particular, the expected value and variance of $\hat{t}_y - t_y$ are of interest, in the sense that one would like the estimator $\hat{t}_y$ to lead to a small expected value and a small variance for $\hat{t}_y - t_y$. But which expected value and which variance? It turns out that, provided the method of sampling is non-informative given the population values of Z, then it is the mean and variance of $\hat{t}_y - t_y$ given these population Z-values that are relevant.

The first step in developing the model-based approach to prediction of t_y is to realise that this total can be decomposed as

$$t_y = \sum_s y_i + \sum_r y_i = t_{ys} + t_{yr}. \qquad (2.1)$$

That is, t_y is the sum of the sample total t_{ys} of the Y-values and the corresponding non-sample total t_{yr}. After the sample has been selected we obviously know t_{ys} so the basic problem is to predict t_{yr}. If we denote such a prediction by $\hat{t}_{yr}$, then the corresponding predictor of t_y satisfies $\hat{t}_y = t_{ys} + \hat{t}_{yr}$. In this context there are two basic questions one can ask:

- Given the assumed model, what is the 'best' predictor $\hat{t}_{yr}$ of t_{yr}?
- Given this model and this predictor, what is the best way to choose the sample s in order to 'minimise' the sample error $\hat{t}_y - t_y = \hat{t}_{yr} - t_{yr}$?

The answers to these questions will depend on our interpretation of 'best' and 'minimise' above. Here we use 'best' in the sense that

(i) $\hat{t}_y$ is a member of a class of 'acceptable' predictors of t_y; and

(ii) $\hat{t}_y$ generates the smallest value of $E(\hat{t}_y - t_y)^2$ within this class given the sample s,

where the expectation in (ii) above is with respect to the assumed model.

Furthermore, we seek to 'minimise' $\hat{t}_y - t_y$ by choosing s in order minimise $E(\hat{t}_y - t_y)^2$ over the set of all 'feasible' samples, that is those samples that practicality and resources constraints allow. The combination of an optimal predictor and the optimal sample s to choose given this estimator is then an optimal sampling strategy for t_y under the assumed model.

2.1 Optimal Prediction

As noted earlier, under the model-based approach the statistical properties of $\hat{t}_y$ as a predictor of t_y are defined by the distribution of the sample error $\hat{t}_y - t_y$ under the assumed model for the population. Thus, the prediction bias of $\hat{t}_y$ is the mean of this distribution, $E(\hat{t}_y - t_y)$, while the prediction variance of $\hat{t}_y$ is the variance of this distribution, $Var(\hat{t}_y - t_y)$. The prediction mean squared error of $\hat{t}_y$ is $E(\hat{t}_y - t_y)^2 = Var(\hat{t}_y - t_y) + \{E(\hat{t}_y - t_y)\}^2$. Recollect that both t_y and $\hat{t}_y$ are random variables here! The predictor $\hat{t}_y$ is said to be unbiased under the assumed model if its corresponding prediction bias $E(\hat{t}_y - t_y)$ is zero, in which case its prediction mean squared error is just its prediction variance, $Var(\hat{t}_y - t_y)$.

Our aim is to identify an optimal sampling strategy for t_y under the assumed model. The first step in this process is identification of an optimal predictor of t_{yr} for any given s. In order to do so, we use the following well-known statistical result.

Result 2.1 The minimum mean squared error predictor of a random variable W given the value of another random variable V is $E(W|V)$.

See exercise E.1 for proof of this result. We can immediately apply it to the problem of predicting t_{yr} (and hence prediction of t_y). We put W equal to t_y and V equal to our 'observed data', that is the sample Y-values and the population values of the auxiliary variable Z. The minimum mean squared error predictor of t_y is then

$$t_y^* = E(t_y|y_i, i \in s; z_i, i = 1, \cdots, N) = t_{ys} + E(t_{yr}|y_i, i \in s; z_i, i = 1, \cdots, N). \tag{2.2}$$

Clearly the conditional expectation in this result will depend on unknown parameters of the assumed model, so t_y^* is impossible to compute in practice. For example, if (1.7) holds, then (2.2) becomes

$$t_y^* = t_{ys} + \sum_r (\beta_0 + \beta_1 z_i),$$

which depends on β_0 and β_1. However, observe that these parameters will be those defining the conditional distribution of Y given Z, and our assumption of non-informative sampling given Z implies that we can estimate them efficiently using the sample values of Y and Z. Substituting these estimated parameter values for unknown true values and computing the conditional expectation on the right hand side of (2.2) then leads to a 'plug-in' approximation to t_y^*, which is sometimes referred to as an *empirical best* (EB) predictor $\hat{t}_y^{EB}$ of t_y.

In the following chapters we explore specifications for EB predictors of t_y under a number of widely used models for survey populations. We also consider specification of corresponding optimal model-based sampling strategies.

3 Homogeneous Populations

The first model for a survey population that we consider is the most basic that one might expect to encounter. This corresponds to a finite population where there are no auxiliary variables, or when it is clear a priori that any auxiliary variables are unrelated to Y. In this case, the distribution of $Y|Z$ is assumed not to depend on Z, so that the model for y_i is the same for every unit i in the population. We refer to this type of population as homogeneous.

This does not imply that the distribution of Y has to have low variance, or to follow a well-behaved distribution. The following examples illustrate when the homogenous model would be used:

- *A crate of oranges.* There is no sampling frame, and no auxiliary information. Oranges might be selected by physical sampling. Because there is no auxiliary information, the homogenous model is the only possible one, even though the oranges may vary considerably with respect to weight, colour, presence of mould and so on.
- *Children in a classroom.* The frame could be a class roll and might include date of birth, so age would be a potential auxiliary variable. Date of birth might be assumed to be unrelated to variables measured on the children, because all of the children would be of approximately the same age. In this case a homogenous model would be used, because there are no relevant auxiliary variables which would allow modelling of a different distribution for different children. There might be many characteristics which would be related to the Y values of the children, such as sex, racial/ethnic origin, home background and physical limitations (e.g. short sight). However, these variables are not available for the population, and so do not form part of Z. The homogenous model applies because the distribution of $y_i|z_i$ is the same for every child in the classroom, due to the paucity of information available in Z. (If it was thought that age might be a relevant variable, then a model other than the homogenous model would be used. Some non-homogenous models will be discussed in Chapters 4 through 7.)
- *Items on an assembly line.* Y might be the weight of the item. If there were no auxiliary variables, then the homogenous model would apply. If the assembly line is dedicated to the production of a single item, then Y might be expected to follow a fairly well-behaved distribution, as production and legal standards are usually such that the items produced have to be as alike as possible. In a multi-item assembly line, the homogeneity model would still apply as $\{y_i|z_i\}$ would follow the same distribution for all i because z_i is empty.

However, the distribution would probably be an inconvenient one, with high variance and multiple modes. The theory in this chapter would still be applicable, but the large sample confidence intervals of Section 3.4 would perhaps require a larger sample size to be reliable than would otherwise be the case. (If the type of item was available for all items in the population, then type could be used as an auxiliary variable. The stratified model, which will be described in Chapter 4, would probably be the most appropriate.)

The common thread in these examples is that z_i does not contain any information which would allow $(y_i|z_i)$ to be different for different i. Whenever this is the case, the homogenous model applies.

3.1 Random Sampling Models

This lack of information in the sample labels means that all samples of the same size are equally informative. There is no reason for the survey designer to prefer any one sample to any other. For this reason, a random sampling method that gives equal probability of selection to all possible samples of the same size seems an intuitively sensible way of sampling from a homogeneous population. In addition, there are a number of strong, but essentially pragmatic, arguments for adopting such an approach that will be discussed later in this book. Such a random sampling method will be referred to as simple random sampling, or just SRS, in what follows.

An additional argument for utilising a random approach to sample selection in this situation is that it sometimes makes it straightforward to derive a probability model directly from the probability sampling method. For example, consider a typical urn problem.

An urn is known to contain N balls, some of which are white and some are black. It is required to estimate the proportion of white balls in the urn. Beyond knowing that the urn contains N balls, and that these are either black or white, nothing else is known about the distribution of the balls in the urn.

However, by the simple expedient of vigorously stirring the balls in the urn, and then selecting a sample of n distinct balls 'at random' to observe, one can immediately generate a known distribution for the random variable corresponding to the number of white balls observed in the sample. This is the hypergeometric distribution with parameters N = the total number of balls in the urn (known), W = the total number of white balls in the urn (unknown), and n = the sample number of balls taken from the urn (known). Under this model the probability that w white balls turn up in the sample is:

$$p(w) = \binom{W}{w}\binom{N-W}{n-w} \Big/ \binom{N}{n}.$$

An unbiased predictor of W under this model is $\hat{W} = N(w/n)$. Derivation of the variance of $\hat{W}$ under this model is left to exercise E.2.

Note that the use of the random selection procedure above is sufficient to guarantee a hypergeometric distribution for the sample data. However, this does not mean that such a random selection procedure is necessary to guarantee such a distribution. We may know of other factors which imply that the urn is already 'randomly mixed'. If so, selection of any sample, not necessarily a random one, still allows use of the hypergeometric model. For example, one could select the n balls at the 'top' of the urn. However, even in this case, using a random selection procedure seems a wise precaution, in the event that our knowledge of these other factors may be imperfect, for example white balls might be slightly lighter than black balls and so tend to congregate more at the top of the urn.

Application of this seemingly trivial model is widespread in survey sample practice. For example, it forms the basis of sample inference in opinion polls. Here the urns correspond to selected polling booths, the balls correspond to votes cast at these booths, the colours correspond to the candidates (or political parties) endorsed by these votes, and randomisation is necessary because of possible trends in the sequence in which the votes for particular candidates are cast at the booth.

3.2 A Model for a Homogeneous Population

A general model for a homogeneous population starts with the concept of exchangeability. The random variables whose realisations are the population values y_i of Y are said to be exchangeable up to order K if the joint distribution of $\{y_i; i \in A\}$ is the same for any permutation A of $k = 1, 2, \ldots, K$ distinct labels from the population.

It is easy to see that in an exchangeable population all moments of products of population Y-values up to order K are the same. In particular, if K is greater than or equal to two, then all units in the population have Y-values with the same mean and the same variance, and all pairs of distinct units in the population have Y-values with the same covariance. Such a population will be referred to as second order homogeneous (or just homogeneous) in what follows. We will assume for now that values from different units are independent so that all covariances are zero, although Section 3.8 will remove this restriction.

The second order *homogeneous population model* represents the basic 'building block' for more complex models which we will describe in later chapters that can be used to represent real world variability. The population Y-values under the homogeneous model satisfy

$$E(y_i) = \mu \qquad (3.1a)$$

$$Var(y_i) = \sigma^2 \qquad (3.1b)$$

$$y_i \text{ and } y_j \text{ independent when } i \neq j. \qquad (3.1c)$$

3.3 Empirical Best Prediction and Best Linear Unbiased Prediction of the Population Total

For the homogeneous population model, there is no auxiliary information. That is, we cannot identify a variable Z whose population values vary and are all known and is such that the conditional distribution of Y given Z (and in particular $E(Y|Z)$) varies with Z. From (2.2) we know that the minimum mean squared error predictor of t_y is $t_y^* = t_{ys} + E\left[t_{yr}|y_i, i \in s\right]$. For model (3.1), this is given by $t_y^* = t_{ys} + (N-n)\mu$. Of course, we do not know μ and so must replace this parameter by an estimate in order to define the EB predictor. Intuitively, in an exchangeable population all sample values provide the same information about μ, and so it seems sensible to use the sample mean $\bar{y}_s$ of Y as our estimator of μ. This leads to the predictor:

$$t_y^E = t_{ys} + (N-n)\hat{\mu} = t_{ys} + (N-n)\bar{y}_s = \frac{N}{n}t_{ys}. \qquad (3.2)$$

The predictor defined by (3.2) above is commonly called the *expansion estimator*.

Note that the EB predictor is not necessarily unique, as there may be several possible estimators of unknown parameters such as μ. However, it provides a simple way to construct predictors, and will be statistically efficient if a sensible method of parameter estimation is used. An alternative, more complex approach, called Best Linear Unbiased Prediction, can also be used. This method does yield a unique best predictor, called the Best Linear Unbiased Predictor, or BLUP. In many cases, including model (3.1) and estimator (3.2), the BLUP is also an EB predictor.

To define the BLUP, we will first define linear predictors to be those that can be written as a linear combination of the values of Y associated with sample units. Linear predictors are used extensively in survey sampling, mainly because of their simplicity of use. The BLUP $\hat{t}_y^{BLUP}$ of t_y under a specified model satisfies three conditions:

- It is a linear predictor; that is it can be written in the form $\hat{t}_y^{BLUP} = \sum_s w_i y_i$, where the w_i are weights that have to be determined. Note that there is no restriction on these weights, except that they must not depend on any values of Y. In particular, they can, and often do, depend on the population units that make up the sample s, and the auxiliary variable Z in cases where the model includes Z.
- It is unbiased for t_y, that is its sample error has an expectation of zero, $E\left(\hat{t}_y^{BLUP} - t_y\right) = 0$.
- For any sample s its sample error has minimum variance among the sample errors of all unbiased linear predictors of t_y, that is $Var\left(\hat{t}_y^{BLUP} - t_y\right) \leq Var(\hat{t}_y - t_y)$ where $\hat{t}_y$ is any other unbiased linear predictor of t_y.

It turns out that to derive the BLUP, we only need to assume that different observations are uncorrelated, rather than the stronger assumption of independence in (3.1c). We first note that for any linear predictor of t_y we have the decomposition

$$\hat{t}_y = \sum_s w_i y_i = \sum_s y_i + \sum_s (w_i - 1) y_i = t_{ys} + \sum_s u_i y_i$$

where $u_i = w_i - 1$. Consequently the sample error can be expressed as

$$\hat{t}_y - t_y = \sum_s u_i y_i - \sum_r y_i.$$

We can think of u_i as essentially defining the 'prediction weight' of unit i, that is the weight attached to its Y-value when predicting the non-sample total of Y.

Clearly, in order to define the BLUP, all we need to do is work out the weights w_i, or equivalently the prediction weights u_i, that define this predictor. By definition there are two restrictions on these weights – they should lead to an unbiased predictor, and they should lead to the smallest possible prediction variance for such an unbiased predictor. To start, we focus on unbiasedness. This condition is equivalent to saying that for any linear predictor, including the BLUP, we have

$$Bias(\hat{t}_y) = E(\hat{t}_y - t_y) = \mu \left[\sum_s u_i - (N-n) \right] = 0$$

which is true only if

$$\sum_s u_i - (N-n) = 0. \qquad (3.3)$$

Next, we seek to minimise the prediction variance. From standard statistical manipulations, we obtain

$$Var(\hat{t}_y - t_y) = Var(\hat{t}_{yr} - t_{yr}) = Var(\hat{t}_{yr}) - 2Cov(\hat{t}_{yr}, t_{yr}) + Var(t_{yr})$$

where

$$Var(\hat{t}_{yr}) = \sigma^2 \sum_s u_i^2 \qquad (3.4a)$$

$$Var(t_{yr}) = (N-n)\sigma^2 \qquad (3.4b)$$

$$Cov(\hat{t}_{yr}, t_{yr}) = 0. \qquad (3.4c)$$

Note that the last result (3.4c) makes use of the fact that the sample and non-sample values of Y are uncorrelated under model (3.1). Since $Var(t_{yr})$ and $Cov(\hat{t}_{yr}, t_{yr})$ are not functions of the u_i, it follows that $Var(\hat{t}_y - t_y)$ will be minimised with respect to these weights when $Var(\hat{t}_{yr})$ is minimised. That is, optimal values of u_i (and hence w_i) are obtained by minimising $Var(\hat{t}_{yr})$ defined by (3.4a), or equivalently $\sum_s u_i^2$, subject to the unbiasedness constraint (3.3). In order to do so, we form the Lagrangian L for this minimisation problem:

$$L = \sum_s u_i^2 - 2\lambda \left(\sum_s u_i - (N-n) \right).$$

Differentiating L with respect to u_i and equating to zero we obtain

$$u_i = \lambda$$

Substituting this expression into the unbiasedness constraint (3.3) and solving for λ leads to

$$\lambda = \frac{N-n}{n},$$

which implies $u_i = \frac{N-n}{n}$ and hence $w_i = \frac{N}{n}$. That is, the BLUP $\hat{t}_y^{BLUP}$ of t_y under the homogeneous population model (3.1) is the expansion estimator (3.2).

3.4 Variance Estimation and Confidence Intervals

Substituting the optimal prediction weights $u_i = \frac{N-n}{n}$ in (3.4a) leads to

$$\operatorname{Var}\left(\hat{t}_{ry}^E\right) = \sigma^2 \frac{(N-n)^2}{n}$$

and hence

$$\begin{aligned}
\operatorname{Var}\left(\hat{t}_y^E - t_y\right) &= \operatorname{Var}\left(\hat{t}_{yr}^E\right) + \operatorname{Var}(t_{yr}) - 2\operatorname{Cov}\left(\hat{t}_{yr}^E, t_{yr}\right) \\
&= \sigma^2\left[\frac{(N-n)^2}{n} + (N-n)\right] \\
&= \sigma^2 (N-n)\left(\frac{N}{n}\right).
\end{aligned}$$

That is, the prediction variance of the expansion estimator (3.2) under the homogeneous population model (3.1) is

$$\operatorname{Var}\left(\hat{t}_y^E - t_y\right) = \frac{N^2}{n}\left(1 - \frac{n}{N}\right)\sigma^2. \tag{3.5}$$

In order to create confidence intervals for t_y based on (3.2), we need to be able to estimate (3.5). An unbiased estimator of σ^2 in (3.1) is the sample variance of Y,

$$s_y^2 = \frac{1}{n-1}\sum_s (y_i - \bar{y}_s)^2.$$

This implies that an unbiased estimator of the prediction variance (3.5) is

$$\hat{V}\left(\hat{t}_y^E\right) = \frac{N^2}{n}\left(1 - \frac{n}{N}\right)s_y^2. \tag{3.6}$$

See exercise E.3 for a proof of the unbiasedness of (3.6). For large sample sizes, standard central limit theory implies that the distribution of the z statistic

$$z = \left(\hat{t}_y^E - t_y\right) \Big/ \sqrt{\hat{V}\left(\hat{t}_y^E\right)}$$

is (approximately) normal with zero mean and unit standard deviation. Consequently, an approximate $100(1-\alpha)\%$ confidence interval for t_y is

$$\hat{t}_y^E \pm q_{\alpha/2}\sqrt{\hat{V}\left(\hat{t}_y^E\right)}$$

where $q_{\alpha/2}$ is the $(1-\alpha/2)$-quantile of an $N(0,1)$ distribution. Since t_y is a random variable, such an interval is often referred to as a prediction interval.

3.5 Predicting the Value of a Linear Population Parameter

Suppose that we are interested in predicting the value of

$$A = \sum_{i=1}^N a_i y_i = \sum_U a_i y_i$$

where $a_1, \ldots, a_N$ is a set of N known constants. For example, A could be the mean $\bar{y}_U$ of the population Y-values in which case $a_i = N^{-1}$. Often, A corresponds to a mean of Y for some identifiable subgroup of size M of the population, in which case $a_i = M^{-1}$ when unit i is in the subgroup and is zero otherwise. The BLUP for A under the homogeneous population model is then

$$\hat{A} = \sum_s a_i y_i + \bar{y}_s \sum_r a_i. \qquad (3.7)$$

It can be shown that the prediction variance of (3.7) is

$$\mathrm{Var}(\hat{A}-A) = \sigma^2\left[n^{-1}\left(\sum_r a_i\right)^2 + \sum_r a_i^2\right]$$

which has the unbiased estimator

$$\hat{V}(\hat{A}) = \left[n^{-1}\left(\sum_r a_i\right)^2 + \sum_r a_i^2\right](n-1)^{-1}\sum_s (y_i - \bar{y}_s)^2.$$

We may sometimes be interested in predicting other population parameters such as ratios of population means or totals. For example, the economic indicator 'average weekly earnings' is often defined as the sum of the 'wages paid' variable for a population of businesses, divided by the population sum of 'number of employees'. See Section 11.1 for extensions of the above results to estimation of ratios.

3.6 How Large a Sample?

As noted earlier, simple random sampling (SRS) is an obvious way of selecting a sample from a homogeneous population. Certainly, this method of sampling is one of the simplest probability-based methods of sample selection. There are two basic design questions that need to be answered before a sample can be selected via this method. These are:

- How big a sample should we take?
- How do we go about selecting a sample via SRS?

The answer to the first question depends on the constraints imposed on the sample design process. For example, suppose that it is required to select a sample of sufficient size so as to ensure that the expansion estimator $\hat{t}_y^E$ has a relative standard error (RSE) of A percent. The RSE of a predictor (also known as its coefficient of variation, or CV) is the square root of its prediction variance, expressed as a percentage of the value of the population quantity being predicted. So

$$RSE\left(\hat{t}_y^E\right) = \left[\sqrt{Var\left(\hat{t}_y^E - t_y\right)}/t_y\right] \times 100.$$

By substituting $Var\left(\hat{t}_y^E - t_y\right)$ from (3.5) above, then setting this expression equal to A, and solving for n, we obtain

$$n = \left(N^{-1} + (A^2/10^4)\bar{y}_U^2/\sigma^2\right)^{-1} \qquad (3.8)$$

where $\bar{y}_U$ is the mean of the population Y-values. Thus, the required sample size to meet the RSE objective depends on the value of the ratio

$$C = \sigma^2/\bar{y}_U^2 \qquad (3.9)$$

and hence on the population Y-values. Typically, this ratio can be estimated directly from data obtained in a pilot study preceding the main survey, or, if such an assumption seems reasonable, by setting it equal to the relative variance of another population variable, say Z, whose values are known for all population elements (i.e. Z is an auxiliary variable). In many cases, these values are the historical values of Y from a past census of the population.

Alternatively, if the survey is a continuing one, then an estimate of the RSE of $\hat{t}_y^E$ based on data from the immediately preceding survey can be calculated and, assuming the relative variance of Y has remained unchanged, substituted into (3.8) to give the sample size required for the target RSE of the present survey.

To illustrate this situation, consider the case where in a past survey of the same population (or one very much like it), with sample size m, say, an estimated RSE equal to B was obtained. Assuming that the relative variance of Y is the same in both populations, we can then estimate the ratio C in (3.9) by

$$\hat{C} = \frac{B^2/100^2}{m^{-1} - N^{-1}}.$$

Substituting this estimate in the expression for n in (3.8) above, and discarding lower order terms leads to
$$n = (B^2/A^2)m.$$
That is, the ratio of the estimated relative standard errors from both surveys is the inverse of the ratio of the square roots of the respective sample sizes for the surveys.

3.7 Selecting a Simple Random Sample

How do we go about selecting a simple random sample? The simplest way, given the population and sample sizes involved are small, is to use a table of random numbers. For example, Fisher and Yates (1963) provide a list of two digit random numbers, together with instructions on how to use them to select a simple random sample.

When sample sizes and populations are large, it is usually most convenient to use a computer to select the sample. Most computer packages include a pseudo-random number generator, which can be used in this regard. A simple way of selecting a SRS of size n, using a computer-based random number generator, is to randomly order the population units on the sampling frame, then take the first n of these randomly ordered units to be the sample. This random ordering is easily accomplished by independently assigning a pseudo-uniform [0,1] random variate to each of the N units on the sampling frame. We then re-order these N units according to these random values.

The above so-called 'shuffle algorithm' has the disadvantage that it requires a pass through the population (to allocate the random numbers), then another pass to re-order the population. This can be expensive (in computer time) in very large populations. Another procedure, therefore, is to generate random numbers between 1 and N until n distinct numbers are generated. These then define the labels of the selected sample units. Vitter (1984) contains a discussion of several efficient algorithms for computer-based selection of a simple random sample by one sequential pass through a computer list of the population labels.

Implicit in the preceding discussion about methods of selecting a simple random sample of size n is that the same population unit cannot occur more than once in any particular selected sample. In other words any sample we select must contain n distinct population units and all samples of the same size must be equally likely. This method of sampling is typically referred to as *Simple Random Sampling Without Replacement* (SRSWOR).

3.8 A Generalisation of the Homogeneous Model

Model (3.1) implies that the values of Y are uncorrelated for distinct units. Suppose that we generalise the model, by allowing a uniform correlation ρ between every pair of values:

A Generalisation of the Homogeneous Model

$$E(y_i) = \mu \qquad (3.10a)$$
$$Var(y_i) = \sigma^2 \qquad (3.10b)$$
$$Cov(y_i, y_j) = \rho\sigma^2 \text{ when } i \neq j. \qquad (3.10c)$$

It turns out that the expansion estimator (3.2) is still the BLUP for t_y under this more general homogeneous model. Furthermore, (3.6) is still unbiased for the prediction variance of $\hat{t}_y^E$. For proof, see exercise E.3.

We have focused on the more restrictive model (3.1) rather than (3.10) in this chapter because a uniform correlation between all pairs of units in a population does not have a sensible interpretation for most populations in the real world. However, correlations between pairs of units will be relevant in Chapter 6 where we consider sampling from a population made up of units grouped into clusters, because we then use a superpopulation model where values for units from the same cluster satisfy (3.10).

4 Stratified Populations

The reality of sample survey practice is that target populations, and especially the large populations of interest in social and economic data collections, are almost never homogeneous. In many cases, these target populations can be modelled as being made up of a number of distinct and non-overlapping groups of units, each one of which could be considered to be internally homogeneous, but which may differ considerably from one another. These groups, each one of which is usually referred to as a stratum, and collectively as strata, are often large in size with the average value of Y varying significantly across the strata. As a result, every stratum is sampled, since information about the distribution of Y obtained from units in the sampled strata tells us very little about the distribution of Y in the non-sampled strata.

In many cases, strata are 'naturally defined'. For example, if units are businesses, then strata might be industries, and if units are people, strata might be states or provinces. In other cases, there may be information on the population frame that allows a choice of how the population can be stratified. Typically, this information consists of the known values, listed on the frame, of one or more auxiliary or benchmark variables defined on the population. It is known that there is systematic variation in the survey variables associated with the variation of the benchmark variables on the frame. By judicious choice of appropriate ranges of values for these benchmarks, the survey analyst can define strata within which one can assume that the survey variables have small systematic variation relative to their variation across the population as a whole.

Stratification of target populations is extremely common in survey sampling. Typically, samples are then selected independently from each stratum, referred to as stratified sampling. Aside from the statistical reason of stratifying in order to control for systematic heterogeneity in the target population, there are many practical reasons why stratified sampling is adopted as a sample survey technique. Cochran (1977) lists three sensible reasons for the use of this technique:

- *Domains of interest.* The subpopulations defining the strata can be of interest in themselves, and estimates of known precision may be required for them. For example, states or provinces are often considered important output categories in national household surveys.
- *Efficient survey management.* In many situations the target population is spread across a wide geographic area and administrative convenience may

dictate the use of stratification; for example, the agency conducting the survey may have field offices, each one of which can supervise the survey for part of the population.
- *Different methods of sampling.* Sampling problems may differ markedly in different parts of the population. With human populations, people living in institutions (e.g. hospitals, army bases, prisons) are often placed in a different stratum from people living in ordinary homes because a different approach to sampling respondents is appropriate for the two situations. In sampling businesses we may place the largest firms in a separate stratum because the level of detail in the data we require from these firms may be quite different from the data we require from smaller firms.

Some examples of stratified populations are:

- *Children in a school system.* Children can be stratified on the basis of the level of school (primary/secondary/college), the type of school (government/private/other), and class levels within each school. Depending on the information available to the survey analyst, stratification on the basis of gender could also be considered.
- *Households in a city.* Households can be stratified on the basis of the wards making up the city; special strata can also be constructed for special dwelling types like caravans, hotels, armed forces bases and institutional dwelling arrangements (e.g. hospitals, prisons).
- *Businesses in a sector of a country's economy.* In this case stratification is usually on the basis of the industries (or groups of industries) to which businesses making up the sector belong, the physical locations of the businesses themselves (based on an appropriate geographic identifier) and their sizes (measured in some appropriate way).
- *Books in a library, or files in an archive.* This type of example comes up in surveys for auditing purposes. Stratification would often involve physical location, for example books could be stratified by room and shelf, and files could be stratified by location, filing cabinet and drawer.

4.1 The Homogeneous Strata Population Model

A model for a stratified population follows naturally from our definition of strata as made up of population elements that are homogeneous with respect to other elements of the same stratum and heterogeneous with respect to elements of other strata. We will use the following model for the distribution of population Y-values across strata $h = 1, \ldots, H$:

$$E(y_i | i \in h) = \mu_h \qquad (4.1\text{a})$$

$$Var(y_i | i \in h) = \sigma_h^2 \qquad (4.1\text{b})$$

$$y_i \text{ and } y_j \text{ are independent when } i \neq j. \qquad (4.1\text{c})$$

Here $i \in h$ indicates that population unit i is in stratum h. We refer to this model as the *homogeneous strata population model* in what follows. The assumption (4.1c) means that distinct population units are independent as far as their Y-values are concerned. We will see that (4.1c) is necessary to derive an EB predictor, while a weaker assumption of zero covariance is sufficient to derive the BLUP.

4.2 Optimal Prediction Under Stratification

As in Section 3.3, we develop an EB predictor for this case. Since each stratum constitutes a separate homogeneous population, following model (3.1), the sample mean of Y within each of the strata is an EB predictor of the corresponding stratum population mean. The predictor $\hat{t}_y^{EB}$ of the overall population total t_y is then the sum of the individual stratum level expansion estimators $\hat{t}_{yh}^E$, since these are EB predictors of the stratum population totals t_{yh} of Y, that is

$$\hat{t}_y^{EB} = \hat{t}_y^S = \sum_h \hat{t}_{yh}^E = \sum_h N_h \bar{y}_{sh}. \tag{4.2}$$

Here h indexes the strata, N_h is the stratum population size, n_h is the stratum sample size and $\bar{y}_{sh}$ is the sample mean of Y in stratum h. The predictor $\hat{t}_y^S$ defined by (4.2) above is usually called the *stratified expansion estimator*.

Using the same derivation as in Section 3.3, it is straightforward to show that the stratified expansion estimator is also the BLUP under model (4.1). It is sufficient to assume that covariances are zero between different units, rather than the stronger assumption of independence in (4.1c).

Note that if some strata are not represented in the sample, then it is not possible to use the stratified expansion estimator, since $\bar{y}_{sh}$ will not be defined for those strata. In this case, there is no unbiased estimator of t_y. Ideally, we should select our sample in such a way that every stratum is represented, so that this problem does not arise.

In order to compute the prediction variance of $\hat{t}_y^S$, and hence develop an estimator for it, we observe that since distinct population units are mutually uncorrelated the prediction variance of $\hat{t}_y^S$ under the stratified population model (4.1) is the sum of the individual prediction variances of the stratum specific BLUPs $\hat{t}_{yh}^E$, and each of these is given by (3.5) with the addition of a stratum subscript, that is

$$\text{Var}\left(\hat{t}_y^S - t_y\right) = \sum_h \text{Var}\left(\hat{t}_{yh}^E - t_{yh}\right) = \sum_h \left(N_h^2/n_h\right)(1 - n_h/N_h)\sigma_h^2. \tag{4.3}$$

Unbiased estimation of (4.3) is straightforward. One just sums unbiased stratum level estimators of the prediction variances of the $\hat{t}_{yh}^E$ (see (3.6)) to get

$$\hat{V}\left(\hat{t}_y^S\right) = \sum_h \hat{V}\left(\hat{t}_{yh}^E\right) = \sum_h \left(N_h^2/n_h\right)(1 - n_h/N_h)s_{yh}^2. \tag{4.4}$$

where $s_{yh}^2 = \frac{1}{n_h-1} \sum_{s_h} (y_i - \bar{y}_{sh})^2$ denotes the unbiased estimator of the variance σ_h^2 of Y-values in stratum h. Here s_h denotes the sample units in stratum h.

Provided the strata population and sample sizes are large enough, the Central Limit Theorem applies within each stratum, and so applies overall, allowing us to write:

$$\left(\hat{t}_y^S - t_y\right) / \sqrt{\hat{V}\left(\hat{t}_y^S\right)} \sim N(0,1).$$

Confidence intervals for t_y follow directly: an approximate $100(1-\alpha)\%$ confidence interval for t_y is

$$\hat{t}_y^S \pm q_{\alpha/2} \sqrt{\hat{V}\left(\hat{t}_y^S\right)}.$$

4.3 Stratified Sample Design

The stratified expansion estimator (4.2) can be used whenever model (4.1) is a reasonable approximation to the population. We do not necessarily have to use the strata in the sample design and selection. However, we want to ensure that every stratum is represented in the sample. Also, given that the strata are the most important feature of the population, it makes sense to build this into the design. We will see that we can develop efficient stratified designs that lead to low variance for the stratified expansion estimator when the model is true.

In some surveys, it is not feasible to base selection on strata. The most common reason for this is that strata are not known in advance of sampling for every unit in the population. The stratified expansion estimator (4.2) can still be used (as long as we have at least one unit in sample from each stratum). To calculate its value, all we need to know are the population stratum sizes, and the stratum memberships of sampled units—this information is sometimes available even if the stratum membership of every population unit is not. The estimator (4.2) is sometimes called a post-stratified estimator in this scenario, as the strata are only formed after the sample is selected.

The remainder of this chapter is concerned with how to design the sample when strata are known in advance. Some questions which we will consider are: how many units to select from each stratum; how to decide on the total sample size; how to form strata by categorising a continuous variable; how many such strata should be used; and how to construct strata when there are multiple auxiliary variables.

4.4 Proportional Allocation

An intuitive method of allocating the sample to the strata is via proportional allocation, where the stratum sample proportion $f_h = n_h/n$ is equal to the stratum population proportion $F_h = N_h/N$. This implies a stratum h sample size $n_h = nF_h$. Under proportional allocation the stratified expansion estimator (4.2) reduces to the simple expansion estimator. Note, however, that the prediction

Stratified Populations

Table 4.1 Population counts of 64 cities (in 1000s) in 1920 and 1930. Note that cities are arranged in the same order in both years.

| \multicolumn{4}{c}{Z = 1920 population count} | \multicolumn{4}{c}{Y = 1930 population count} |

Z = 1920 population count				Y = 1930 population count			
$h=1$		$h=2$		$h=1$		$h=2$	
797	314	172	121	900	***364***	***209***	113
773	298	172	120	***822***	317	183	115
748	296	163	119	781	328	163	123
734	258	162	118	805	302	253	154
588	256	161	118	***670***	288	232	***140***
577	243	159	116	1238	291	260	***119***
507	238	153	116	573	253	201	***130***
507	237	144	113	***634***	***291***	147	127
457	235	138	113	***578***	308	***292***	100
438	235	138	110	487	272	164	107
415	216	138	110	***442***	284	143	***114***
401	208	138	108	451	255	169	111
387	201	136	106	459	***270***	139	163
381	192	132	104	464	214	***170***	116
324	180	130	101	400	195	150	***122***
315	179	126	100	366	260	143	134

variance of the stratified estimator is still based on (4.3) and is not the prediction variance (3.5) of the simple expansion estimator under a homogeneous population model.

For example, consider the following population, taken from Cochran (1977, page 94). This population consists of 64 cities in the USA, with the variable of interest, Y, being their 1930 population counts (in 1000s). The total of these counts (which would, in practice, not be known by the sampler) is $t_y = 19,568$. It is assumed that the sampler knows the corresponding 1920 population counts for these cities, which we denote by Z, and can use this information for stratifying the population. The total of these 'auxiliary' counts is 16,290. Values of the 1920 and 1930 counts for the 64 cities are shown in Table 4.1.

Consider two different approaches to stratification of this population, both resulting in two strata:

1. Put the 16 cities with the largest values of Z (1920 population counts) into one stratum and the remaining 48 cities into another. Call this *size stratification*. The two size strata are shown in Table 4.1, with stratum $h = 1$ containing the cities with the largest 1920 population counts, and stratum $h = 2$ containing the remainder.
2. Randomly allocate 16 of the 64 cities to stratum 1 and the remaining 48 to stratum 2. The Y-values (1930 population counts) of the 16 randomly chosen

cities making up stratum 1 are shown in italic boldface in Table 4.1. Call this *random stratification*.

Let $\sum_{U_h}$ denote summation over all the population units in stratum h. We assume the homogenous strata model, (4.1). The strata variance parameters σ_h^2 are approximately equal to

$$\hat{\sigma}_h^2 = S_{yh}^2 = (N_h - 1)^{-1} \sum_{U_h} (y_i - \bar{y}_h)^2$$

where $\bar{y}_h$ denotes the average value of Y in stratum h.

Under size stratification, $S_{y1}^2 = 53,843$, while $S_{y2}^2 = 5,581$. On the other hand, under random stratification, $S_{y1}^2 = 52,144$ with $S_{y2}^2 = 53,262$. Since the population variance S_y^2 is 52,448, it is clear that the strata defined via size stratification are internally less variable (or at least stratum 2 is) than the overall population. This is not the case for the strata defined by random stratification, where we see no real reduction in variability within either stratum compared to that of the population as a whole.

How can we assess the impact of reduced within-strata variability brought about by size stratification? Consider taking a sample of $n = 16$ cities from this population for the purpose of estimating the total 1930 count of the 64 cities. Suppose we assume that these cities are homogeneous with respect to their 1930 counts, take a simple random sample from all 64 (i.e. ignore stratification), and use the simple expansion estimator (3.2) to generate our estimate. The theory developed in Section 3.4 can then be applied, with the population variance S_y^2 substituted for the corresponding Y-variance σ^2, to show that the prediction variance (3.5) of the simple expansion estimator based on such a sample is approximately:

$$\text{Var}\left(\hat{t}_y^E - t_y\right) \approx \frac{N^2}{n}\left(1 - \frac{n}{N}\right)S_y^2 = \frac{64^2}{16}\left(1 - \frac{16}{64}\right) \times 52,448 = 10,070,010$$

under the homogenous population model (3.1).

Now, suppose that instead we use stratified sampling with our two size strata, with proportional allocation. The population sizes of the two strata are $N_1 = 16$ and $N_2 = 48$, so we obtain the following sample sizes:

$$n_1 = nN_1/N = 16 \times 16/64 = 4$$
$$n_2 = nN_2/N = 16 \times 48/64 = 12$$

So, the variance of the stratified expansion estimator would be approximately equal to

$$\text{Var}\left(\hat{t}_y^S - t_y\right) \approx \sum_h \left(N_h^2/n_h\right)(1 - n_h/N_h)S_h^2$$
$$= \left(16^2/4\right)(1 - 4/16) \times 53,843 + \left(48^2/12\right)(1 - 12/16) \times 5,581$$
$$= 3,388,113$$

assuming the homogenous strata population model (4.1) holds, where strata are given by our two size strata.

Alternatively, we might consider stratified sampling with our two random strata, with proportional allocation. The population sizes of the two strata are still $N_1 = 16$ and $N_2 = 48$, so we again obtain $n_1 = 4$ and $n_2 = 12$. The variance of our estimator can then be approximated by:

$$\begin{aligned} Var\left(\hat{t}_y^S - t_y\right) &\approx \sum_h \left(N_h^2/n_h\right)(1 - n_h/N_h)S_h^2 \\ &= \left(16^2/4\right)(1 - 4/16) \times 52{,}144 + \left(48^2/12\right)(1 - 12/16) \times 53{,}262 \\ &= 10{,}172{,}572 \end{aligned}$$

assuming the homogenous strata population model (4.1) holds, where strata are now given by our two random strata.

We can compare the three variance approximations we have obtained to give an indication of the relative efficiencies of the three designs. This comparison suggests that for the Cities' population and a sample size of $n = 16$, size stratification with two strata and proportional allocation is about three times as efficient as random stratification with two strata and the same allocation. In fact, the latter sample design is virtually equivalent (in terms of prediction variance) to not stratifying at all. (We should note that these three variances are not strictly comparable as they are based on different models. Ideally we should have decided on a best model, and compared the relative performance of the three designs and estimators based on this model, but this would have been much more complex.)

It follows that there are considerable gains to be had in stratifying so that the resulting strata are more homogeneous than the original population.

4.5 Optimal Allocation

Can we do better than size stratification and proportional allocation? If our aim is to minimise the prediction variance of the stratified expansion estimator (4.2) subject to an overall sample size of n, that is $\sum_h n_h = n$, the answer is yes.

To see how, consider again the formula for the prediction variance (4.3) of this estimator. We see that it can be decomposed into two terms,

$$Var\left(\hat{t}_y^S - t_y\right) = \sum_h N_h^2 \sigma_h^2 / n_h - \sum_h N_h \sigma_h^2$$

Only the first term depends on the n_h, and minimising $Var\left(\hat{t}_y^S - t_y\right)$ is therefore equivalent to choosing n_h in order to minimise $\sum_h N_h^2 \sigma_h^2 / n_h$ subject to the restriction $\sum_h n_h = n$. It can be shown (see exercise E.6) that this minimum occurs when $n_h \propto N_h \sigma_h$, which implies

$$n_h = nN_h\sigma_h / \sum_g N_g\sigma_g. \quad (4.5)$$

This optimal method of allocation is often referred to as Neyman Allocation, after Neyman (1934), whose fundamental paper gave the method wide prominence.

Notice that for two strata of the same size, (4.5) allocates a greater sample size to the more variable stratum. If the σ_h^2 is the same (or approximately the same) in each stratum then this method of allocation is equivalent to proportional allocation.

Applying the method to the Cities' population, and again substituting stratum variances for Y-variances, we see that for size stratification $N_1 S_{y1} = 3717$ and $N_2 S_{y2} = 3586$, and hence an optimal allocation corresponds to $n_1 = 8.1$ and $n_2 = 7.9$ which would be rounded to $n_1 = n_2 = 8$.

On the other hand, for random stratification $N_1 S_{y1} = 3654$ and $N_2 S_{y2} = 11078$, leading to an optimal allocation defined by $n_1 = 4$ and $n_2 = 12$, that is proportional allocation. This result is hardly surprising given that the stratum variances in the two random strata are approximately the same and equal to the overall population variance.

Under optimal allocation, it can easily be calculated that the prediction variance (4.3) of the stratified expansion estimator (with σ_h^2 replaced by S_{yh}^2) for the Cities' population is 2,200,908 under size stratification and 10,172,572 under random stratification. Comparing these figures with those for proportional allocation, we note a further improvement in precision under size stratification but not under random stratification. Both these results are consistent with the theory developed above.

The formula (4.5) is actually a special case of a more general optimal allocation formula, which minimises the prediction variance (4.3) subject to a fixed survey budget rather than a fixed sample size, where the cost is assumed to be a linear combination of the stratum sample sizes. See exercise E.8. In the case of (4.5) there is the implicit assumption that there is no cost differential when sampling in different strata.

True optimal allocation assumes knowledge of the variances σ_h^2. In practice, of course, these quantities will not be known. However, estimates of them can often be obtained either from a preliminary pilot study of the population, or, since it is just the relative sizes of these variances between the strata that are needed, we can assume that these are unchanged from past studies of the same population, or are the same as the relative sizes of the stratum variances of another variable whose values are known for all population elements.

For example, in the case of the Cities' population, we know the 1920 counts for all 64 cities, and can base an 'optimal' allocation on these counts. See exercise E.7 for an examination of the efficiency of such an approach.

4.6 Allocation for Proportions

An important type of survey variable is one that takes the value one or zero depending on whether the corresponding population element has, or does not have, a particular characteristic. That is, Y can be modelled as a Bernoulli variable. The population mean of such a variable is the proportion of population elements with this characteristic, and its distribution satisfies

$$E(y_i|i \in h) = \Pr(y_i = 1|i \in h) = \pi_h \quad (4.6a)$$

$$Var(y_i|i \in h) = E\left(y_i^2|i \in h\right) - E^2(y_i|i \in h) = \pi_h(1 - \pi_h) \quad (4.6b)$$

$$y_i \text{ and } y_j \text{ are independent when } i \neq j. \quad (4.6c)$$

which is a special case of (4.1). Note that $\sigma_h^2 = \pi_h(1 - \pi_h)$ here, and since the gains from stratified sampling and optimal allocation (relative to simple random sampling) increase with the difference between the largest and smallest values of the σ_h^2, it follows that in the case of estimating proportions optimal allocation offers significant gains over simple random sampling only if the values of $\pi_h(1 - \pi_h)$ vary considerably between the strata. However, this will usually not be the case, since the function $\pi(1 - \pi)$ remains relatively constant for values of π between 0.25 and 0.75.

The gains from optimal allocation relative to proportional allocation in the case of estimating proportions are even less. If we ignore lower order terms, it can be shown that the ratio of the prediction variance of the stratified expansion estimator $\hat{t}_y^S$ under optimal allocation to the same variance under proportional allocation is

$$\frac{Var\left(\hat{t}_y^S - t_y|\text{optimal allocation}\right)}{Var\left(\hat{t}_y^S - t_y|\text{proportional allocation}\right)} \approx \frac{\left[\sum_h F_h \sqrt{\pi_h(1 - \pi_h)}\right]^2}{\sum_h F_h \pi_h(1 - \pi_h)}. \quad (4.7)$$

Cochran (1977, page 109) shows empirically that for the special case of two equal-sized strata the smallest value this ratio can take (and hence the greatest gain in efficiency from the use of optimal allocation) is 0.94 for values $0.1 \leq \pi_h \leq 0.9$. These results have been theoretically confirmed and extended to the case of an arbitrary number of (non-equal) strata by Sadooghi-Alvandi (1988). It therefore follows that optimal allocation is hardly ever worthwhile when estimating proportions, with proportional allocation being the preferred approach in these cases.

The implications of this result are greatest for multipurpose surveys, where the aim is not just to predict population totals and means for specified variables, but also to estimate the distributions of these variables across the population. These estimated distributions typically correspond to estimates of proportions (see Chapter 16), and so the above recommendations are directly applicable. In such cases, a compromise allocation somewhere between proportional and optimal allocation is usually employed.

4.7 How Large a Sample?

In the previous two sections we discussed the allocation of the sample between the strata. But what should the total sample size n be under stratified sampling? As we observed in Section 3.6, sample size calculation depends on the practical constraints (e.g. available funds to carry out the survey) as well as the statistical constraints (e.g. a maximum RSE for the estimate) that are applicable. As in that section, we shall assume that a sample of sufficient size to ensure a RSE of A percent for the stratified expansion estimator (4.2) is required. If optimal allocation is used, the prediction variance (4.3) of this estimator can be obtained by substituting (4.5) into (4.3), which gives:

$$Var\left(\hat{t}_y^S - t_y\right) = n^{-1}\left(\sum_h N_h \sigma_h\right)^2 - \sum_h N_h \sigma_h^2.$$

The magnitudes of the terms on the right hand side are $O(N^2/n)$ and $O(N)$ respectively, so provided the sampling fraction n/N is small, the second term can be ignored:

$$Var\left(\hat{t}_y^S - t_y\right) \approx n^{-1}\left(\sum_h N_h \sigma_h\right)^2 \quad (4.8)$$

The required sample size is obtained by equating the right hand side in the above expression to $(At_y/100)^2$ and solving for n. This leads to

$$n = (10^4/A^2)\left(\sum_h N_h \sigma_h \Big/ \sum_h N_h \bar{y}_h\right)^2. \quad (4.9)$$

In practice, we calculate (4.9) by substituting preliminary estimates for the stratum standard deviations σ_h and the stratum means $\bar{y}_h$. Alternatively, if size stratification (based on a size measure Z) is being used, we can substitute $\bar{z}_h$ for $\bar{y}_h$ and the stratum variance S_{zh}^2 of Z for σ_h^2. See exercise E.9 for development of comparable sample size formulae under proportional and equal allocation.

4.8 Defining Stratum Boundaries

So far we have not addressed the important issues of how many strata there should be, or exactly how these strata should be defined. Sample frames often contain one or more auxiliary variables whose values can be used to form strata. For example, a business survey frame will typically contain variables that characterise the industrial classification of each business on the frame as well as variables that measure the size of each business in terms of number of employees or amount of tax paid. Similarly, an area-based frame for a social survey will contain information about the size of each area in terms of numbers of people or households at the last census and so on. In either case, we have access to one or more continuous auxiliary size variables whose values are known for the entire population and which are typically correlated with the survey variable Y of interest. This correlation usually means that the mean and variance of Y change with the values of these variables, and so it makes sense to partition the population by defining strata on the basis of these auxiliaries in order to control the variation of Y-values in our sample by sampling independently from the different strata. The question then becomes one of deciding how these size strata should be defined.

To start, we assume we have just one continuous auxiliary variable, Z, to use in defining the strata. We have seen that strata should be defined to be as internally homogeneous as possible, and as externally (i.e. between different strata) heterogeneous as possible. Ideally, the elements of a stratum should all be identical, with each stratum corresponding to a different 'type' in the population. Of course, in reality such perfection in discrimination is impossible to attain. However, there are some fairly practical 'rules of thumb' for setting size stratum boundaries (i.e. defining size strata) that are in common use and enable one to reduce the within-stratum variability of Y as much as possible given the auxiliary information available.

Possibly the most widely used method of determining size strata boundaries based on a single continuous auxiliary size variable Z is that of Dalenius and Hodges (1959). This approach makes the following assumptions:

- The number of strata H is given.
- The aim is to optimally predict the population total t_z of Z (although of course the real aim is to predict the population total t_y of Y).
- Optimal (Neyman) allocation for estimating t_z is used.
- There exists a continuous function $f(z)$ that closely approximates the frequency of population values of Z at any point z in the closed interval $a \leq z \leq b$.

Suppose that the ordered values of Z in the population are denoted by $z_{(1)}, z_{(2)}, \ldots, z_{(N)}$. The smallest and largest values of Z in the population are therefore $z_{(1)}$ and $z_{(N)}$ respectively. In order to define H strata based on these population values of Z, we need to define $H-1$ strata boundaries $c_1, \ldots, c_{H-1}$. For $h = 1, \ldots, H$, let

$$A_h = \int_{c_{h-1}}^{c_h} \sqrt{f(z)}\, dz$$

with $c_0 = a$ and $c_H = b$. Assuming that there are a sufficient number of strata, the within-stratum distribution of Z can be considered to be approximately uniform, with the within-stratum variance given by $\sigma_{zh}^2 = (c_h - c_{h-1})^2/12$. Also, the fraction $F_h = N^{-1} N_h$ of the population falling in stratum h is approximately $F_h \approx d_h \times (c_h - c_{h-1})$, where $f(z)$ is assumed constant, with value d_h, in stratum h. Furthermore, under the uniform assumption, $A_h \approx \sqrt{d_h} \times (c_h - c_{h-1})$.

From (4.8), we know that the prediction variance of the stratified expansion estimator $\hat{t}_y^S$ is minimised when $\sum_h F_h \sigma_h$ is minimised. Substituting σ_{zh} for σ_h, replacing F_h by its approximation, and simplifying leads to

$$\sum_h F_h \sigma_h \cong \sum_h (d_h(c_h - c_{h-1})) \frac{1}{\sqrt{12}}(c_h - c_{h-1})$$

$$\cong \frac{1}{\sqrt{12}} \sum_h d_h (c_h - c_{h-1})^2$$

$$\cong \frac{1}{\sqrt{12}} \sum_h A_h^2.$$

So, minimizing the prediction variance is equivalent to minimising $\sum_h A_h^2$. Note that $\sum_h A_h = \int_a^b \sqrt{f(z)} \, dz$ is equal to a constant, K say, independently of the choice of A_h. It immediately follows that $\sum_h A_h^2$ is then minimised when A_h is a constant for all values of h, that is $A_h = H^{-1} K$, $h = 1, \ldots, H$.

In other words, the optimal points of stratification (assuming optimal allocation) are obtained by taking equal intervals on the cumulative of $\sqrt{f(z)}$, that is each stratum accounts for $1/H$ of the cumulative of $\sqrt{f(z)}$. This is often referred to as the cumulative square root rule for forming strata.

In practice, we can approximate $\sqrt{f(z)}$ using a histogram based on the N population values of Z. The procedure is to construct a histogram based on J equal width classes, where J is much larger than the number of strata H, then calculate $\sqrt{f_j}$ where f_j is the number of Z values in the jth interval ($j = 1, \ldots, J$). Then starting at the left tail of the histogram, neighbouring classes are joined together to form strata in such a way that each stratum contributes approximately $1/H$ of the accumulated scores $\sqrt{f_j}$.

To illustrate this, consider the application of the Dalenius–Hodges method to the Cities' population, with Z equal to the 1920 population count. We take $J = 14$, and a class width of 50. Table 4.2 below sets out the details of the calculations for the case of $H = 2$. Note that the method requires the population be first ordered by increasing values of Z.

The cumulative total of $\sqrt{f_j}$ for this population is 24.526. The optimal stratum boundary (under $H = 2$, there is only one) is then determined by partitioning this total into two equal parts, giving a boundary for cum($\sqrt{f_j}$) of $24.526/2 = 12.263$. This corresponds to stratum 1 (large units) being defined by $Z > 300$ and stratum 2 (small units) being defined by $Z \leq 300$. As it turns out, these strata are very similar to the strata defined earlier under size stratification where the boundary was set at $Z = 314$.

How does the Dalenius–Hodges method perform in terms of the prediction variance of the stratified expansion estimator for the total 1930 population? We have $N_1 = 17$ and $N_2 = 47$, with the corresponding stratum variances being $S_{y1}^2 = 54{,}620$ and $S_{y2}^2 = 5{,}090$. Substituting these variances for the actual variances σ_h^2, the optimal allocation of a sample of size $n = 16$ based on this stratification is then $n_1 = 9$ and $n_2 = 7$, and the prediction variance of the stratified expansion estimator is equal to 2,192,277. Note that this value is very similar to the prediction variance of this estimator under equal allocation and the size stratification rule defined earlier, because we happen to have chosen a stratum boundary close to the Dalenius–Hodges optimal boundary.

Table 4.2 Calculation of stratum boundaries for the Cities' population using the Dalenius–Hodges cumulative $\sqrt{f}$ rule applied to the 1920 population count.

Class index	Z range	f_j	$\sqrt{f_j}$	cum($\sqrt{f_j}$)
1	100–150	25	5.000	5.000
2	151–200	10	3.162	8.162
3	201–250	8	2.828	10.991
4	251–300	4	2.000	12.991
5	301–350	3	1.732	14.723
6	351–400	2	1.414	16.137
7	401–450	3	1.732	17.869
8	451–500	1	1.000	18.869
9	501–550	2	1.414	20.283
10	551–600	2	1.414	21.697
11	601–650	0	0.000	21.697
12	651–700	0	0.000	21.697
13	701–750	2	1.414	23.112
14	751–800	2	1.414	24.526

4.9 Model-Based Stratification

Dalenius–Hodges stratification is aimed at optimising stratified expansion estimation of t_z, but in reality, we want to stratify by the auxiliary variable Z in order to estimate t_y. What can we say about the appropriate stratification in this case? The answer to this depends on what model we believe represents the relationship between Y and Z. Many of the populations encountered in survey practice have Z-distributions that are intrinsically positive and skewed to the right, with long tails. The 'local' Y-variability of such a population tends to increase the further out into the tail of the Z-distribution one goes. We might then suppose that the variance of Y is approximately proportional to Z or Z^2, or more generally, to $Z^{2\gamma}$, for some value of γ in the range $0 \leq \gamma \leq 1$. (We will consider models of this type in more detail in Chapter 5.) We might then approximate this model using the stratified model (4.1), with the stratum variances given by

$$\sigma_h^2 = \sigma^2 \frac{1}{N_h} \sum_{U_h} z_{hi}^{2\gamma} \qquad (4.10)$$

where σ is an unknown scale coefficient. If the strata consist of reasonably narrow ranges of Z, so that Z varies little within each stratum, then (4.10) can be further approximated by:

$$\sigma_h^2 = \sigma^2 \left(\bar{z}_h\right)^{2\gamma} \qquad (4.11)$$

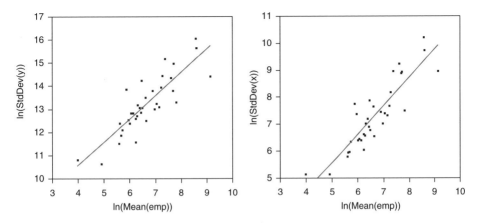

Fig. 4.1 Plots of $\log(\sigma_h)$ vs. $\log(\bar{z}_h)$ for WSS industry strata.

or equivalently

$$\ln(\sigma_h) = \ln(\sigma) + \gamma \ln(\bar{z}_h).$$

Some empirical evidence for the model (4.11) is provided by the plots shown in Figure 4.1 below, based on data extracted from a Wages and Salaries Survey (WSS). For the 35 industry strata used in the design of this survey, the plots show the relationship between the logarithms of the (estimated) standard deviations of two important survey variables (left = wages, right = actual employment) and the logarithm of the mean value of the survey auxiliary variable (Z = register employment, that is employment as indicated on the register of businesses from which the sample was selected). The slopes of the least squares lines fitted to these plots (also shown) are 1.0065 (wages) and 1.0517 (employment). This provides some evidence that $\gamma = 1$ for these variables and this population.

Each point shown in the plots in Fig. 4.1 corresponds to a different industry represented in the WSS sample. However, the type of stratification characterised by the model (4.11) is at a lower level of aggregation, where the industry strata are themselves partitioned on the basis of the auxiliary variable Z. Consequently in Fig. 4.2 we take two of these industry strata and break them down further into so-called 'size' strata (defined by ranges of values of Z), showing the same relationships as in Fig. 4.1. There are now two lines in each plot, corresponding to the two different industries. The slope coefficients in this case vary from 0.9464 to 1.2323. This is consistent with an $\gamma = 1$ assumption in the variance model (4.11).

Suppose now we are prepared to accept (4.11) as being a reasonable assumption for our population, and because our population has a fairly long tail we

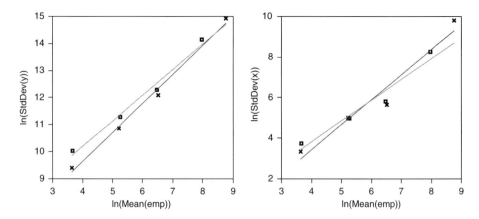

Fig. 4.2 Plots of $\log(\sigma_h)$ vs. $\log(\bar{z}_h)$ for size strata in two WSS industries (indicated by different plotting symbols).

decide to fix $\gamma = 1$. Substituting into the approximate expression (4.8) for the prediction variance of the stratified expansion estimator, we obtain

$$\mathrm{Var}\left(\hat{t}_y^S - t_y\right) \approx n^{-1}\left(\sum_h N_h \sigma \bar{z}_h\right)^2 = n^{-1}\left(\sum_h \sigma t_{Zh}\right)^2 = n^{-1}\sigma^2 t_z^2 \quad (4.12)$$

where t_{hz} is the total of Z in stratum h and t_z is the population total of Z over all strata. We can see that (4.12) does not depend on the method used to stratify the population. Any stratification method leads to the same (approximate) prediction variance for the stratified expansion estimator. We should note, however, that this finding depends on an assumption that Z varies little within strata, so that (4.10) simplifies to (4.11). We still need to choose strata in a sensible way, for example using the Dalenius–Hodges approach, or using one of the 'equal aggregate' rules which we will now describe.

4.10 Equal Aggregate Size Stratification

Another approach that has been proposed for defining size stratum boundaries is an equal aggregate size rule. The population units are sorted from low to high based on the value of a strictly positive variable Z, and then strata formed so that each contains about the same total of $Z^{\alpha/2}$ or a monotone transformation of it. Equalising the stratum totals $\sum_{U_h} z_i^{\alpha/2}$ leads to several common stratification rules used in practice (see Table 4.3). References are Valliant et al. (2000; section 6.5.2) and Särndal et al. (1992; sections 12.4 and 12.5).

The equal aggregate size rule with $\alpha = 2$ can lead to a simpler optimal allocation. If model (4.11) applies with $\gamma = 1$, so that $\sigma_h^2 = \sigma^2 \bar{z}_h^2$, then the optimal allocation is given by substituting into (4.5):

$$n_h = nN_h\sigma_h/\sum_g N_g\sigma_g = nN_h\bar{z}_h/\sum_g N_g\bar{z}_g = nt_{zh}/\sum_g t_{zg} = n/H.$$

So the optimal allocation in this case is to have equal sample sizes in every stratum. The result of this allocation for the Cities' population with $H = 2$ was described at the end of Section 4.5, where we noted that it lead to a prediction variance of 2,200,908 for the stratified expansion estimator.

On the other hand, applying the equal aggregate $\sqrt{Z}$ stratification method to the Cities' population with $H = 2$ we see that the stratum boundary is determined by dividing the cumulative total of $\sqrt{z_i}$ by 2, that is $970.15/2 = 485.075$. This corresponds to stratum 1 (large units) being defined by $Z \geq 237 (N_1 = 24)$ and stratum 2 (small units) being defined by $Z \leq 235$ ($N_2 = 40$). The corresponding stratum variances are $S_1^2 = 59,955$ and $S_2^2 = 3,747$, and so the optimal allocation of a sample of size $n = 16$ is then $n_1 = 11$ and $n_2 = 5$. Substituting these values back into the expression for the prediction variance of the stratified expansion estimator we see that using the equal aggregate $\sqrt{Z}$ rule and optimal allocation leads to a prediction variance of 2,749,664. Note that this value is somewhat greater than that previously obtained using the Dalenius–Hodges stratification method or the equal aggregate Z rule.

4.11 Multivariate Stratification

So far our discussion has been in terms of stratifying on the basis of a single positive-valued continuous auxiliary ('size') variable Z. However, this is not always the case. Sometimes the auxiliary information available to the survey analyst also includes a number of other categorical measurements on the population elements, together with one or more other continuous valued measurements. Stratifying on Z as well as each one of these variables individually, and then cross-classifying the strata, would lead to such a proliferation of strata that most of them would either be empty, or would have unworkably small sample sizes.

One approach to getting around this problem, and one that makes use of all the available auxiliary information, is to construct the strata using a clustering algorithm. An algorithm that appears to perform well in this regard is the binary segmentation method used in AID (Automatic Interaction Detection – Sonquist et al., 1971). This is very similar to the CART algorithm (Classification and Regression Trees – Breiman et al., 1984).

Table 4.3 Some equal aggregate size stratification rules.

$\alpha = 0$ (equal size)	$N_h = H^{-1}N$
$\alpha = 1$ (equal aggregate $\sqrt{Z}$)	$\sum_{U_h} z_i^{1/2} = H^{-1}\sum_U z_i^{1/2}$
$\alpha = 2$ (equal aggregate Z)	$\sum_{U_h} z_i = H^{-1}\sum_U z_i$

We briefly describe the binary segmentation algorithm below. This assumes that the set of auxiliary variables to be used in clustering the population are either categorical in nature, or have been categorised for this purpose.

The basic idea in binary segmentation is to cluster the population by a succession of binary splits. Each split creates two 'child' clusters from one 'parent' cluster by placing, in one of the child clusters, all those elements from the parent cluster with a specified range of values of one of the auxiliary variables. The remaining elements from the parent cluster are placed in the other child cluster.

The decision on the cluster to split, and the auxiliary variable and associated set of values determining the split, is such as to minimise the within-cluster variation (and maximise the between-cluster variation) of a suitable criterion variable over all possible alternative binary splits that can be applied to the clusters that have been created in previous splits of the population.

The result of all these binary splits is an inverted 'tree' with branches determined by split points or 'nodes' each defined in terms of a split based on the categories associated with one of the auxiliary variables. At each level of this tree the population can therefore be decomposed into a number of relatively homogeneous clusters (or strata), with the homogeneity of these clusters increasing, but their size decreasing, the further down the tree levels one goes.

Ideally, we would like to use the survey variable, Y, as the criterion variable in constructing this tree. Unfortunately, we do not usually have access to this variable at the survey design stage, otherwise we would not need to conduct a survey at all. Alternatively, we might have access to data from a pilot survey, or a past survey collecting the same or a similar variable, and use this data set instead.

The following example serves to illustrate the use of the AID algorithm for stratification. The population in this case corresponds to fishing boats licensed to operate in Australia's Northern Prawn Fishery. The criterion variable (Z) used was total boat catch of prawns. The stratifying variables were:

- total hours fished (HOUR – 33 categories: few = 1, ..., many = 33)
- age of boat (AGE – 29 categories: old = 1, ..., new = 29)
- underdeck volume of boat (SIZE – 9 categories: small = 1, ..., big = 9)
- engine power (ENG – 22 categories: low = 1, ..., high = 22)
- type of boat ownership (COY – private = 1, fleet = 2)
- main species targeted (BANSPECT: banana = 1, else = 0; TIGSPECT: tiger = 1, else = 0; EKSPECT: endeavour/king = 1, else = 0)
- main species in management area catch (BANSPECM: banana = 1, else = 0; TIGSPECM: tiger = 1, else = 0; EKSPECM: endeavour-king = 1, else = 0)
- species catch mainly from management area (REGBAN: banana = 1, else = 0; REGTIG: tiger = 1, else = 0; REGEK: endeavour-king = 1, else = 0)

The aim in stratifying this population was to end up with 20–25 strata defined in terms of the above variables, and then to select one sample boat from each stratum from which to obtain data for constructing an economic model of the Fishery. Applying the AID binary segmentation algorithm to these data resulted in the sequence of splits detailed in Table 4.4.

4.12 How Many Strata?

Finally, we briefly consider the choice of the number of strata. In some cases the strata are determined by output categories of interest, or operational convenience (e.g. geographic strata) and so there is very little flexibility for the survey analyst in how they are defined. However, when stratifying on the basis of a 'size' variable, we usually have the option of choosing between a minimum of 2 strata and a maximum of $[n/k]$ strata, where k denotes a minimum acceptable stratum sample size, and $[.]$ denotes integer part. The minimum stratum sample size, k, is usually set to at least 2 to allow variances to be estimated, and more commonly to 5 or 6.

Cochran (1977, pages 132–34) considered this question, and, on the basis of some empirical analysis, as well as on the basis of a linear model for the survey data, recommended that the number of strata be kept to around 6 or 7. Beyond this number the gains from further stratification are small and are usually outweighed by the extra administrative cost associated with designing the sample to take account of the extra strata.

A similar conclusion follows from an analysis based on the variance model (4.10). Under this model with $\gamma = 1$, equal aggregate stratification on Z, and equal allocation, the prediction variance of the stratified expansion estimator is approximately equal to

$$V = n^{-1} \sum_h F_h \sigma_h$$

$$= n^{-1} \sum_h F_h \sqrt{N_h^{-1} \sum_{U_h} z_{hi}^2}$$

$$= n^{-1} \sum_h F_h \sqrt{N_h^{-1} \sum_k \sum_{U_k} z_{ki}^2 / H}$$

$$= n^{-1} \sum_h F_h N_h^{-1/2} H^{-1/2} \sqrt{\sum_k \sum_{U_k} z_{ki}^2}$$

which is proportional to $D(H) = H^{-1/2} \sum_h F_h^{1/2}$.

We will review the values of $D(H)$ for a simple case to see how the benefit from size stratification depends on H. In practice, most auxiliary variables used

Table 4.4 AID-based analysis of Northern Prawn Fishery using Z = Total Boat Catch (of prawns). The rows with values of S_z correspond to the nodes that define the 21 strata eventually used for this Fishery.

Split	Child	Parent	Defined by	N	$\bar{z}$	S_z
1	2	1	HOUR = 0–14	91	24,019	
	3		HOUR = 15–28	94	37,498	
2	4	2	BANSPECT = 1	35	34,342	
	5		BANSPECT = 0	56	17,568	
3	6	3	BANSPECM = 1	10	60,169	
	7		BANSPECM = 0	84	34,799	
4	8	4	AGE = 7–24	24	29,079	
	9		AGE = 25–29	11	45,824	
5	10	7	ENG = 5–10	38	30,289	
	11		ENG = 11–21	46	38,526	
6	12	9	SIZE = 6	6	39,318	9,924
	13		SIZE = 7	5	53,630	25,415
7	14	5	HOUR = 0–10	21	11,123	
	15		HOUR = 11–14	35	21,435	
8	16	11	HOUR = 15–20	40	37,090	
	17		HOUR = 21–24	6	48,100	6,234
9	18	10	HOUR = 15–21	33	28,322	
	19		HOUR = 22–28	5	43,263	5,693
10	20	8	HOUR = 6–9	9	21,185	6,326
	21		HOUR = 10–14	15	33,815	
11	22	16	TIGSPECT = 1	27	34,508	
	23		TIGSPECT = 2	13	42,452	
12	24	6	ENG = 9–13	5	51,341	9,979
	25		ENG = 15–16	5	68,997	11,077
13	26	15	SIZE = 1–2	6	11,746	1,563
	27		SIZE = 3–7	29	23,440	
14	28	18	EKSPECT = 1	16	25,025	5,072
	29		EKSPECT = 2	17	31,426	
15	30	14	HOUR = 0–8	11	8,179	6,023
	31		HOUR = 9–10	10	15,047	4,613
16	32	23	COY = 1	5	38,922	10,263
	33		COY = 2	8	44,659	5,299
17	34	21	AGE = 7–22	5	27,796	2,749
	35		AGE = 23–24	10	36,825	6,961
18	36	27	SIZE = 3–5	16	20,988	3,900
	37		SIZE = 6–7	13	26,458	4,074
19	38	22	ENG = 12–13	14	32,130	3,352
	39		ENG = 14–21	13	37,068	4,200
20	40	29	AGE = 13–23	11	29,308	3,803
	41		AGE = 24–29	6	36,510	5,397

to define strata are non-negative, and follow a distribution such that smaller values are more common. For example, this is typically the case when Z is the number of employees, for a population of businesses. Suppose for the purposes of illustration that Z come from an exponential distribution. Without loss of generality, suppose that $E(Z) = 1$. The probability density function of Z for $z \geq 0$ is then $f(z) = e^{-z}$. We assume that equal aggregate stratification is to be used, so that $t_{zh} = t_z/H$. Under our assumed distribution,

$$t_{zh} \approx N \int_{c_{h-1}}^{c_h} zf(z)\,dz = N \int_{c_{h-1}}^{c_h} ze^{-z}\,dz = N\left\{e^{-c_{h-1}}(1+c_{h-1}) - e^{-c_h}(1+c_h)\right\}$$

and

$$t_z \approx N \int_{c_0}^{c_H} zf(z)\,dz = N \int_0^\infty ze^{-z}\,dz = N$$

Equating t_{zh} to t_z/H for each h leads to the following equations, which define the stratum cutoffs:

$$-e^{-c_h}(c_h + 1) = -e^{-c_{h-1}}(c_{h-1} + 1) + H^{-1} (h = 1, \ldots, H-1)$$

Table 4.5 Efficiency of size stratification for different numbers of strata where the size variable is exponentially distributed.

Number of Size Strata (H)	$D(H)$
1	1.000
2	0.943
3	0.926
4	0.917
5	0.911
6	0.908
7	0.905
8	0.903
9	0.902
10	0.900
20	0.894
50	0.890
200	0.888
1000	0.887
∞	$\Gamma(1.5) = 0.886$

with $c_0 = 0$ and $c_H = \infty$. Also note that F_h is approximately equal to:

$$F_h \approx \int_{c_{h-1}}^{c_h} f(z)\,dz = e^{-c_{h-1}} - e^{-c_h}$$

Solving the above equations numerically, some values of $D(H)$ are given in Table 4.5. Here $\Gamma(1.5)$ denotes the value of the Gamma function at 1.5, which can be shown to be the limiting value of $D(H)$ in this case.

There is little gain in going beyond 5–10 strata, with only 2.8% further gain possible beyond $H = 5$, and only 1.6% beyond $H = 10$. Moreover, in practice the strata containing the very largest population units would often be completely enumerated, with $n_h = N_h$, so that the prediction variance V would not be exactly proportional to $D(H)$. This probably means that even fewer size strata are needed before further improvements become negligible.

5 Populations with Regression Structure

Stratification on the basis of a size variable is not the only way to model heterogeneity in a population. Another way of accounting for this heterogeneity is via the correlation between the survey variable and the size variable. This corresponds to using a linear regression model linking the survey variable and the size variable to predict the non-sample values of the survey variable. We continue to assume that the method of sampling is non-informative, so that the regression model that applies in the population also applies in the sample.

5.1 Optimal Prediction Under a Proportional Relationship

It is often reasonable to assume a proportionality relationship between Y and the size variable, Z, with the conditional variability of Y increasing with Z. One could then use a linear regression model of the form

$$E(y_i|z_i) = \beta z_i \tag{5.1a}$$

$$Var(y_i|z_i) = \sigma^2 z_i^{2\gamma} \tag{5.1b}$$

$$y_i \text{ and } y_j \text{ independent when } i \neq j \tag{5.1c}$$

This model is sometimes called the *gamma population model*. The parameter γ controls how much the variance depends on Z, and generally lies between 0 and 1.

In many cases $\gamma = 0.5$ is a reasonable choice, and we will refer to the gamma model in this case as the *ratio population model*:

$$E(y_i|z_i) = \beta z_i \tag{5.2a}$$

$$Var(y_i|z_i) = \sigma^2 z_i \tag{5.2b}$$

$$y_i \text{ and } y_j \text{ independent when } i \neq j. \tag{5.2c}$$

The ratio population model is particularly appropriate for many business survey populations, as the scatterplots in Fig. 5.1 below attest. These plot the values of four important economic variables, *Receipts*, *Costs*, *Profit* (defined as *Receipts* - *Costs*), all measured in A$, and *Harvest* (measured in tonnes) against the values

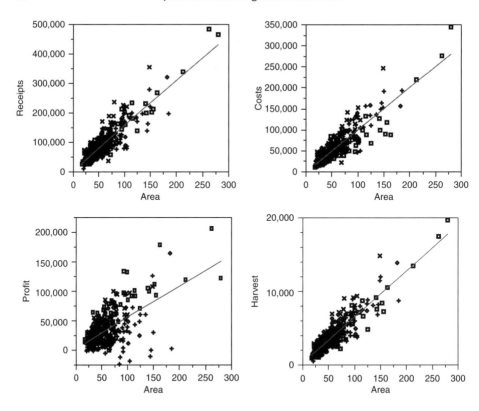

Fig. 5.1 Sample data scatterplots for sugarcane farm sample ($n = 338$).

of an auxiliary variable *Area* (measured in hectares) for a sample of $n = 338$ farms growing sugarcane in Australia in the 1980s. Here *Area* denotes the area allocated for growing sugarcane by each of the farms, with values of this variable known for all farms making up the population of interest. The straight line shown in each plot is the least squares fit to these data, and indicates that the proportionality assumption (5.2a) is reasonable.

Recollect that the population total t_y of a survey variable Y can always be decomposed as $t_y = t_{ys} + t_{yr}$ where t_{ys} is the sample sum of Y and t_{yr} is the non-sample sum of Y. Our best predictor of t_y (see Result 2.1) is $t_y^* = t_{ys} + E[t_{yr}|y_i, i \in s]$, as in Chapter 3. For model (5.1), this is equal to $t_y^* = t_{ys} + \sum_r \beta z_i = t_{ys} + \beta t_{zr}$. An EB predictor is obtained by substituting an efficient estimator for β. An efficient sample-based estimate of this quantity is given by the *best linear unbiased estimator* (BLUE) for β under the gamma population model (5.1),

$$b = \sum_s y_i z_i^{1-2\gamma} / \sum_s z_i^{2-2\gamma}.$$

A proof of this assertion is not difficult to write down: If $b = \sum_s c_i y_i$ then under (5.1) $E(b) = \sum_s c_i \beta z_i = \beta \sum_s c_i z_i$ and $Var(b) = \sum_s c_i^2 \sigma^2 z_i^{2\gamma} = \sigma^2 \sum_s c_i^2 z_i^{2\gamma}$. For b to be unbiased for any β, we therefore require $\sum_s c_i z_i = 1$. The value of c_i that minimises $\sum_s c_i^2 z_i^{2\gamma}$ subject to this constraint is $c_i \propto z_i^{1-2\gamma}$ - this can be shown using Lagrange multipliers or by applying the Cauchy–Schwarz Inequality.

Substituting b for β, our EB predictor of the population total t_y of Y under the gamma population model is then

$$\hat{t}_y = t_{ys} + bt_{zr} = t_{ys} + t_{zr} \sum_s y_i z_i^{1-2\gamma} / \sum_s z_i^{2-2\gamma} \qquad (5.3)$$

Using the same argument as that used to show that the expansion estimator $\hat{t}_y^E$ is the BLUP of t_y under the homogeneous population model (3.1), we can show that (5.3) is also the BLUP of t_y under the gamma population model, (5.1). See exercise E.14. As with models (3.1) and (4.1) in the previous two chapters, it turns out that the derivation of the BLUP only requires zero correlation, rather than the stronger statement of independence, in (5.1c).

We are particularly interested in the ratio population model (5.2), and in this case the BLUE of β simplifies to:

$$b = \bar{y}_s / \bar{x}_s$$

and the corresponding EB predictor of t_y is

$$\hat{t}_y^R = t_{ys} + bt_{zr} = \left(\sum_s y_i / \sum_s z_i\right) \sum_U z_j = (\bar{y}_s / \bar{z}_s) t_z. \qquad (5.4)$$

This predictor is usually called the *ratio estimator* of t_y. It is a linear predictor of t_y, as it can be expressed as $\hat{t}_y^R = \sum_s w_i y_i$ with weights defined by

$$w_i = \sum_U z_j / \sum_s z_i = \frac{N \bar{z}_U}{n \bar{z}_s}.$$

Use of the ratio estimator is clearly justifiable in situations where:

- the ratio population model itself is plausible –that is there is a proportionality relationship between the population values of the size variable Z and the survey variable Y;
- the values of Z are known for all sampled elements;
- (at least) the population total of Z is known;
- (preferably) the individual non-sample values of Z are known.

In practice, these conditions tend to hold for economic and social surveys based on frames that contain good quality auxiliary information strongly related to the variables being measured. Official economic surveys are often of this type, with auxiliary size information based on economic performance data collected in previous censuses and stored on a central, integrated register of business units. Furthermore, there is often good reason to expect a proportional relationship between this size information and the survey variables (the amount of stocks held by a company is typically roughly proportional to its past sales, for example). However, the approach can in principle be used with any collection.

A ratio population model like (5.2) provides a crude but convenient way of describing what tends to be observed in data collected in these surveys – that is, sample Y-values that increase proportionately with corresponding sample Z-values, in such a way that their variation also tends to increase with Z.

5.2 Optimal Prediction Under a Linear Relationship

Although the ratio population model described in the previous section suffices to describe a large number of economic and social populations observed in practice, there is another popular linear regression model that is often applied to 'natural' populations, where the proportionality assumption (i.e. the assumption of a zero intercept) is usually not tenable, and where the residual variance remains (relatively) constant over the range of Z values in the population. This is the *linear population model*, where for $i = 1, \ldots, N$,

$$E(y_i|z_i) = \alpha + \beta z_i \tag{5.5a}$$

$$Var(y_i|z_i) = \sigma^2 \tag{5.5b}$$

$$y_i \text{ and } y_j \text{ independent when } i \neq j. \tag{5.5c}$$

For example, consider again the Cities' population introduced in Section 4.4. There we used the correlation between the 1920 count and the 1930 count to justify stratifying the population on the basis of the (known) 1920 count as an efficient substitute for stratification of the (unknown) 1930 count.

Another way of using this relationship would be to fit model (5.5). Fig. 5.2 shows the scatterplot of the 1930 counts (Y) against the 1920 counts (Z). With one notable exception, there is (as one would expect) a very strong linear relationship between these counts, with their correlation $r = 0.9397$, increasing to 0.9871 if the outlier city is excluded. The spread of the values of Y appears to be fairly constant across the values of Z if one excludes the outlier city, suggesting that model (5.5) may be more appropriate than the Ratio Population Model (5.2) for characterising the 'non-outliers' in this population.

It is well known that *ordinary least squares* (OLS) is an optimal parameter estimation technique under the linear population model, with the OLS estimators of α and β given by

$$a_L = \bar{y}_s - b_L \bar{z}_s$$

and

$$b_L = \sum_s (y_i - \bar{y}_s)(z_i - \bar{z}_s) / \sum_s (z_i - \bar{z}_s)^2$$

respectively.

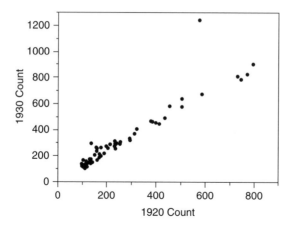

Fig. 5.2 Scatterplot of 1930 count against 1920 count for the Cities' population.

The EB predictor of t_y under (5.5) is obtained in the usual way. Since under this model

$$t_{ys} + E(t_{yr}|z_i; i \in r) = t_{ys} + \sum_r (\alpha + \beta z_i),$$

we can substitute the OLS estimators above for the unknown values of α and β to obtain the corresponding EB predictor under (5.5), usually referred to as the *regression estimator* in the literature. This is

$$\hat{t}_y^L = t_{sy} + \sum_r (a_L + b_L z_i) = N[\bar{y}_s + b_L(\bar{z}_U - \bar{z}_s)] = N(a_L + b_L \bar{z}_U). \quad (5.6)$$

Like the ratio estimator, the regression estimator $\hat{t}_y^L$ has dual optimality properties. First, it is an EB predictor of t_y under the linear population model (5.5). It is also the BLUP of t_y under this model, with weights defined by

$$w_i = \frac{N}{n}\left(1 + \frac{(\bar{z}_U - \bar{z}_s)(z_i - \bar{z}_s)}{(1 - n^{-1})s_z^2}\right)$$

where $s_z^2 = (n-1)^{-1}\sum_s (z_i - \bar{z}_s)^2$ is the sample variance of Z. The derivation of the BLUP requires only that covariances between values of Y from different units are zero, rather than the stronger assumption of independence in (5.5c).

5.3 Sample Design and Inference Under the Ratio Population Model

The assumption of within-stratum homogeneity for the homogeneous strata population model considered in the previous chapter leads logically to the population labels being regarded as informative only in the sense of identifying the stratum to which an element belongs. This in turn implies that simple random sampling within each stratum is an appropriate method of sample selection.

With a regression model for the population, however, these labels become much more informative. When we observe a sample s, we also observe the Z-values associated with this sample. In general, these values will all be distinct, and so will generally define a unique or 'recognisable' subset in the class of all subsets of n different Z-values that can be drawn from the (assumed known) population of N different Z-values. Since the variability of a Y-value and hence the 'information' obtained when observing this value, is in general related to its associated Z-value, it follows that different samples are no longer interchangeable in terms of the information they provide – depending on their 'spread' of Z-values, some samples provide more information than other samples, and so simple random sampling is not the obvious sample selection method. How do we go about selecting the sample now? In what follows we explore optimal sampling when it does matter which population units are included in sample.

To start, we consider populations where the ratio population model (5.2), is appropriate. Here the ratio estimator (5.4) is optimal. It is straightforward to see that the prediction variance of the ratio estimator is just the prediction variance of its implied predictor of the non-sample total t_{yr} of Y. That is,

$$Var\left(\hat{t}_y^R - t_y\right) = Var\left(b\sum_r z_i - \sum_r y_i\right)$$

$$= \frac{\sigma^2}{\sum_s z_i}\left(\sum_r z_i\right)^2 + \sigma^2 \sum_r z_i$$

$$= \sigma^2 N \bar{z}_U \left(\frac{\sum_r z_i}{\sum_s z_i}\right) \quad (5.7a)$$

$$= \sigma^2 \left(\frac{N^2 \bar{z}_U^2}{n \bar{z}_s} - N\bar{z}_U\right).$$

An equivalent expression which is sometimes convenient is

$$Var\left(\hat{t}_y^R - t_y\right) = \sigma^2(N^2/n)(1 - n/N)\bar{z}_r \bar{z}_U/\bar{z}_s. \quad (5.7b)$$

It is easy to see that (5.7a) is minimised when the sample s contains those n units in the population with largest values of Z. Consequently, if we believe that (5.2) represents the 'true' model for our population, then use of the ratio estimator (5.4) together with this extreme sample represents an optimal sampling strategy for estimating t_y. However, this design has the disadvantage that only large values of Z are observed in the sample, but the model is then used to predict all of the unobserved Y, even those with small Z values. This extrapolation would be acceptable if the model is certain to be correct, but it does not allow model checking over the whole domain of Z, and so is particularly susceptible to incorrect model specification. In Chapter 8 we will explore alternative, more robust, strategies that can be used for sample design under (5.2). If in fact (5.2) does hold, then these strategies will be less efficient than this extreme sample choice. However, they will also have the advantage of not being as risky.

Estimation of the prediction variance (5.7) is straightforward once we estimate the scale parameter σ^2. Under (5.2) we can show that an unbiased estimator of this parameter is

$$\hat{\sigma}^2 = n^{-1} \sum_s \left\{ z_i \left(1 - \frac{z_i}{n\bar{z}_s}\right)\right\}^{-1} (y_i - bz_i)^2$$

leading to an unbiased estimator of (5.7),

$$\hat{V}\left(\hat{t}_y^R\right) = \hat{\sigma}^2 N \bar{z}_U \left(\frac{\sum_r z_i}{\sum_s z_i}\right). \qquad (5.8)$$

Assuming (5.2) holds, we can then invoke the central limit theorem, which allows us to construct large sample prediction intervals $\hat{t}_y^R \pm q_{\alpha/2} \sqrt{\hat{V}\left(\hat{t}_y^R\right)}$ for t_y based on the ratio estimator and (5.8).

5.4 Sample Design and Inference Under the Linear Population Model

The prediction variance of the regression estimator $\hat{t}_y^L$ under the linear population model (5.5) is

$$\text{Var}\left(\hat{t}_y^L - t_y\right) = \frac{N^2}{n} \sigma^2 \left[\left(1 - \frac{n}{N}\right) + \frac{(\bar{z}_U - \bar{z}_s)^2}{(1 - n^{-1})s_z^2}\right]. \qquad (5.9)$$

The proof of this result is left for the reader (exercise E.15). Inspection of (5.9) immediately shows that this prediction variance is minimised by choosing the sample such that the sample Z-mean and the population Z-mean coincide, that is $\bar{z}_s = \bar{z}_U$. This type of sample is said to be *first-order balanced* on Z. That is, under the linear population model (5.5), choice of such a balanced sample and use of the regression estimator represents an optimal strategy for predicting the population total t_y. Surprisingly, the second order and other properties of the sample do not matter. The sample can be selected so that all of the sample Z values are close to $\bar{z}_U$, or so that the sample Z values are very dispersed and include the smallest and largest values of Z. As long as $\bar{z}_s = \bar{z}_U$, (5.9) will take on its minimum value. In practice, however, it may not be possible to achieve $\bar{z}_s = \bar{z}_U$ exactly. A design which has all of the sample values close to $\bar{z}_U$ would be dangerous because both $(\bar{z}_U - \bar{z}_s)^2$ and s_z^2 would be small, which could possibly result in (5.9) being large. A more sensible strategy would be to select a spread of values of Z in the sample, so that $\bar{z}_s \approx \bar{z}_U$ and s_z^2 is large.

Observe that under first-order balanced sampling $\hat{t}_y^L$ reduces to the expansion estimator $\hat{t}_y^E$, which makes it particularly simple to compute. Also, in this case (5.9) reduces to

$$\text{Var}\left(\hat{t}_y^L - t_y\right) = \frac{N^2}{n}\left(1 - \frac{n}{N}\right)\sigma^2.$$

A note of warning, however – this simplification only holds in balanced samples. In any other sample these estimators can be quite different. In fact, in all other samples the expansion estimator is biased for t_y under the linear population model (5.5). Furthermore, this does *not* mean that assuming model (3.1) will give the same results as assuming model (5.5) when a balanced sample is selected. The prediction variance of $\hat{t}_y^E$ is given by (3.5), which looks formally identical to the expression above for the prediction variance of $\hat{t}_y^L$ given a balanced sample. However, the σ^2 parameter in (3.5) is quite different from the σ^2 parameter in (5.5). In (3.5) this parameter is the marginal variance of Y, which will usually be much larger than the value of σ^2 in (5.5), where this parameter corresponds to the conditional variance of Y given Z.

From (5.9) we see that the only unknown quantity in the prediction variance of the regression estimator under the linear population model is the parameter σ^2 in (5.5). From standard linear regression theory we know that an unbiased estimator of this regression error variance is the estimated mean squared error

$$\hat{\sigma}^2 = (n-2)^{-1} \sum_s (y_i - a_L - b_L z_i)^2. \qquad (5.10)$$

The corresponding unbiased estimator of the prediction variance of the regression estimator under the linear population model (5.5) is then

$$\hat{V}\left(\hat{t}_y^L\right) = \frac{N^2}{n}\hat{\sigma}^2 \left[\left(1 - \frac{n}{N}\right) + \frac{(\bar{z}_U - \bar{z}_s)^2}{(1 - n^{-1})s_z^2}\right]. \qquad (5.11)$$

Under first-order balanced sampling, the second term in the square brackets on the right hand side of (5.11) disappears. However, $\hat{\sigma}^2$ is still calculated using (5.10). Confidence interval estimation follows directly.

The preceding analysis clearly shows that the structure of the assumed model plays a crucial role in determining an optimal sample. The extreme sample that is optimal for the ratio estimator under (5.2) is quite different from the balanced sample that is optimal for the regression estimator under (5.5). Obviously, we need to be quite sure of our model if we wish to pursue an optimal sampling strategy to use with either estimator. However, as we shall see in Chapter 8, balanced samples also play an important role when we are unsure of the exact nature of the linear dependence between the survey variable Y and the auxiliary variable Z.

5.5 Combining Regression and Stratification

In most populations of interest, the relationship between Y and Z will be more complex than that implied by either the ratio population model or the linear population model. However, we can accommodate this complexity by stratifying the population so that separate versions of these models hold in different strata. For example, if different versions of (5.2) hold in the different strata, with parameters β_h and σ_h, then we refer to the overall model as the *separate ratio population model*. Under this model

$$E(y_i|z_i, i \in h) = \beta_h z_i \qquad (5.12a)$$

$$Var(y_i|z_i, i \in h) = \sigma_h^2 z_i \qquad (5.12b)$$

$$y_i \text{ and } y_j \text{ independent when } i \neq j \qquad (5.12c)$$

holds for all units in the population.

It is straightforward to see that the EB predictor of the population total t_y under (5.12) is then obtained by aggregating the individual stratum level ratio estimators of the stratum population totals. This leads to the *separate ratio estimator*

$$\hat{t}_y^{SR} = \sum_h \hat{t}_{yh}^R = \sum_h \frac{\bar{y}_{sh}}{\bar{z}_{sh}} t_{zh}. \qquad (5.13)$$

Here a subscript of h denotes a stratum h specific quantity. Clearly, the separate ratio estimator is also the BLUP for t_y under the model defined by (5.12). If the parameters β_h and σ_h are actually the same for the different strata (i.e. (5.2) actually holds across the entire population), then $\hat{t}_y^{SR}$ is an inefficient estimator compared with the standard ratio estimator $\hat{t}_y^R$. This loss in efficiency is the price we must pay for an estimator that is optimal across a much wider, and more likely, range of models for the populations that are observed in survey practice.

To illustrate, we return to the sugarcane farms example that we used to motivate (5.2). However, this time we note that the sample of $n = 338$ farms were actually drawn from four sugarcane growing regions, with different climatic conditions, water availability and soil structure. None of these factors are reflected in the auxiliary size variable *Area*, but one would certainly expect them to be reflected in the distribution of values for the four survey variables. Figure 5.3 shows that this is the case. Here, in addition to the actual data values, we show the straight line least squares fit in each region (four lines per plot). Clearly, (5.2) is still a reasonable assumption within each region, but, equally clearly, β_h (and most likely σ_h) varies considerably between regions.

Exactly the same type of argument as used to motivate the separate ratio estimator (5.13) above can be used when combining stratification information with the linear population model (5.5). This leads to the *separate linear population model*, defined by

$$E(y_i|z_i, i \in h) = \alpha_h + \beta_h z_i \qquad (5.14a)$$

$$Var(y_i|z_i, i \in h) = \sigma_h^2 \qquad (5.14b)$$

$$y_i \text{ and } y_j \text{ independent when } i \neq j. \qquad (5.14c)$$

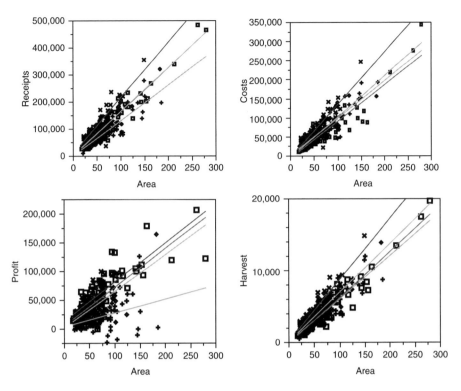

Fig. 5.3 Sample data scatterplots for sugarcane farm sample showing regional fits.

The corresponding EB predictor and BLUP for t_y is then the *separate regression estimator*

$$\hat{t}_y^{SL} = \sum_h \hat{t}_{yh}^L = \sum_h N_h \left[\bar{y}_{sh} + b_{Lh}(\bar{z}_h - \bar{z}_{sh}) \right] \quad (5.15)$$

where

$$b_{Lh} = \sum_{s_h} (y_i - \bar{y}_{sh})(z_i - \bar{z}_{sh}) / \sum_{s_h} (z_i - \bar{z}_{sh})^2.$$

The more statistically knowledgeable reader will no doubt have realised that both the separate ratio estimator (5.13) and the separate regression estimator (5.15) are actually special cases of estimators based on multivariate (multiple) regression models for the survey variable Y. In both cases we have two sources of variation that are being used to explain variation in Y – variation in the size variable Z and variation between the strata. Development of the theory for efficient prediction of population totals of survey variables based on such multiple regression models requires no new concepts, but does require the use of matrix notation and some matrix algebra. We therefore defer this until Chapter 7.

What about sample design under the separate ratio population model (5.12) and the separate linear population model (5.14)? Strata are sometimes based on 'natural' categories, for example the industry strata in the population underpinning the Wages and Salaries Survey sample in Section 4.9 and the region strata in the sugar farms population above. Alternatively, strata may created by subdividing a population on the basis of an auxiliary size variable (possibly within industry or region). If size strata are used, an equal aggregate Z rule, such as that discussed in Chapter 4, would be a sensible approach to determining the boundaries of these strata.

The main sample design problems are those of deciding on a total sample size and then allocating the sample across the various strata. Sample selection within a stratum depends on the confidence the survey analyst has in the assumed within-stratum regression model. If this model appears reasonable, then it would be sensible to select the optimal sample within the stratum – that is, the stratum sample that minimises the prediction variance of the BLUP of the stratum total t_{yh}.

Thus, if (5.12) is assumed to hold, and the separate ratio estimator (5.13) is to be used, then selection of the n_h elements in each stratum with largest values of Z represents an optimal sampling strategy. On the other hand, if (5.14) is assumed to hold, and the separate regression estimator (5.15) is to be used, then the optimal sampling strategy is to select a *stratified balanced sample*, where each stratum sample is first-order balanced with respect to Z, so that $\bar{z}_{sh} = \bar{z}_h$, where $\bar{z}_h$ denotes the stratum h average value of Z.

Consider allocation of the sample across the strata when the separate ratio estimator $\hat{t}_y^{SR}$ is to be used. Under (5.12), its prediction variance follows from (5.7) and is given by

$$\text{Var}\left(\hat{t}_y^{SR} - t_y\right) = \sum_h \sigma_h^2 \left(\frac{N_h^2 \bar{z}_h^2}{n_h \bar{z}_{sh}} - N_h \bar{z}_h\right). \tag{5.16}$$

The optimal sample allocation between the strata is obtained by choosing the n_h so that this expression is minimised, subject to the constraint that these stratum sample sizes sum to the overall sample size n. If we select the largest n_h units in each stratum, then the stratum sample mean, $\bar{z}_{sh}$, of Z also depends implicitly on n_h, and there is no explicit solution to this minimisation problem in general. As we shall show in Chapter 8, it may be preferable to use stratified balanced sampling when assuming (5.12) since this will provide some protection against this model not being properly specified. If this is done, an explicit formula for the optimal stratum allocation is easily obtained. This is

$$n_h = n \left[\frac{\sigma_h N_h \sqrt{\bar{z}_h}}{\sum_g \sigma_g N_g \sqrt{\bar{z}_g}}\right].$$

Under this allocation, and still assuming stratified balanced sampling,

$$\text{Var}\left(\hat{t}_y^{SR} - t_y\right) = n^{-1} \left[\sum_h \sigma_h N_h \sqrt{\bar{z}_h}\right]^2 - \sum_h \sigma_h^2 N_h \bar{z}_h. \tag{5.17}$$

Given an overall RSE objective of α percent for this estimator and sample design, the total sample size n required to achieve this objective is obtained by setting the variance expression in (5.17) equal to $\alpha^2 t_y^2 \times 10^{-4}$ and solving for n. As usual, the result depends on the (unknown) values of the σ_h^2, as well as the population total t_y. These must be replaced by suitable preliminary estimates obtained either from similar, known, populations (e.g. the population of Z-values) or from a pilot test of the survey.

Turning now to sample allocation for the separate regression estimator, we note that under the separate linear population model (5.14), the prediction variance of the separate regression estimator is

$$\mathrm{Var}\left(\hat{t}_y^{SL} - t_y\right) = \sum_h \frac{N_h^2}{n_h} \sigma_h^2 \left[\left(1 - \frac{n_h}{N_h}\right) + \frac{(\bar{z}_h - \bar{z}_{sh})^2}{\left(1 - n_h^{-1}\right) s_{zh}^2}\right]. \tag{5.18}$$

Clearly, (5.18) is minimised under stratified balanced sampling, in which case it reduces to

$$\mathrm{Var}\left(\hat{t}_y^{SL} - t_y\right) = \sum_h \left(N_h^2/n_h\right)\left(1 - n_h/N_h\right)\sigma_h^2.$$

Optimal stratum sample allocation (for fixed n) in this case is given by

$$n_h = n\left(\frac{N_h \sigma_h}{\sum_g N_g \sigma_g}\right). \tag{5.19}$$

Note however that, unlike the case of the optimal allocation (4.5) for the stratified expansion estimator, the σ_h in (5.19) above refers to the residual standard deviation in stratum h. With this allocation,

$$\mathrm{Var}\left(\hat{t}_y^{SL} - t_y\right) = n^{-1}\left[\sum_h N_h \sigma_h\right]^2 - \sum_h N_h \sigma_h^2. \tag{5.20}$$

The overall sample size (n) is finally determined by setting (5.20) equal to $\alpha^2 t_y^2 \times 10^{-4}$ and solving for n.

6 Clustered Populations

Most human and 'natural' populations are clustered, either geographically, socially, economically or (in the case of natural populations) genetically. In many cases, clustering occurs on all these dimensions simultaneously. For example, the individuals making up the population of a country can be grouped into families, each made up of individuals who are similar in terms of where they live (geographic clustering), have similar ethnic backgrounds (social clustering), similar wealth characteristics (economic clustering) and, obviously, similar genetic characteristics. If the survey variable Y is related to any of these characteristics, then it too will be clustered in its distribution across the population, and this clustering will need to be allowed for when modelling this variable.

Clustering is typically also multi-level. Thus, the population of individuals referred to in the previous paragraph is usually grouped into households or families. The population whose units consist of these households is then grouped into a further population whose units consist of villages and suburbs. Contiguous villages or suburbs are finally grouped into a population of units defined by compact geographical areas. This hierarchical grouping structure tends to result in population units that are all in a group at one level in this hierarchy (e.g. the individuals within a household or the households in a suburb or village) being more alike than those from different groups at this level of the hierarchy. Modelling this clustering effect is especially important when there is little auxiliary information available to explain why some units are more alike than others.

We consider the simplest version of this situation, a two level population, where the population elements are grouped into clusters. We let $g = 1, 2, \ldots, Q$ index the Q clusters in the population, and $i = 1, 2, \ldots, M_g$ index the M_g elements within cluster g. It immediately follows that the overall population size satisfies $N = \sum_g M_g$. Beyond knowing the cluster affiliations of the population elements, we shall assume there is no auxiliary information, and so the simple model of a common mean and variance for the population Y-values is appropriate. Each cluster is taken as defining a 'micro-realisation' of this population, and the association between different elements in the same cluster is then accounted for by assuming a constant correlation between the values of Y for elements in the same cluster. This leads to the *clustered population model*:

$$E(y_i | i \in g) = \mu \quad \text{(6.1a)}$$

$$Var(y_i | i \in g) = \sigma^2 \quad \text{(6.1b)}$$

$$Cov(y_i, y_j | i \neq j, i \in g, j \in f) = \rho \sigma^2 I(g = f) \quad \text{(6.1c)}$$

where $I(g = f)$ takes the value 1 if $g = f$ and is zero otherwise. Notice that (6.1) is the generalisation of (3.1) where there can be a non-zero correlation within clusters. It is also a generalisation of the homogenous model in Section 3.8, with (3.10) being the special case of (6.1) where the population consists of a single cluster.

In practice, it would be highly desirable to make use of auxiliary variables Z in the model, as well as modelling correlations within clusters. Section 7.2 in the next chapter will present a generalisation of (6.1) that allows for auxiliary variables.

6.1 Sampling from a Clustered Population

How do we sample from a clustered population? Given that we believe that our hierarchy can be modelled by (6.1), it makes sense to sample the clusters. Some or all of the elements would then be selected within each selected cluster. This approach is called *two-stage sampling*, and has several practical advantages:

- There might be no list available of the population elements, while there may be a list of clusters. For example, in a survey of employees, there might be a list of all employers but no list of employees. In this case a sample of businesses could be selected, and a list of employees in selected businesses could be compiled and used to select the final sample.
- Sampling clusters and then sampling elements within selected clusters may lead to reduced travel costs for a survey, if elements within a cluster tend to be close together. An obvious example is a population of dwellings, which can be clustered by grouping neighbouring dwellings together, for example on the basis of street blocks in a city or other physical boundaries (roads, watercourses, etc.) in non-urban areas.

Given that population elements within a cluster are exchangeable under (6.1) (at least in terms of their first two moments), it is clear that sample selection within each cluster can be carried out at random. For a selected cluster g, therefore, let m_g denote the sample size within the cluster. Typically the total number Q of clusters in the population is large, so only $q < Q$ of them can be selected. We denote this first stage sample of clusters by s. The non-sampled clusters are denoted by r. The second stage sample is then made up of the actual population elements of interest, made up of the m_g elements selected from each cluster in s. The total sample size is therefore $n = \sum_s m_g$. Remember that, unlike the previous development, the sample s here is a sample of clusters (the primary sampling units) so summation over s (and r) below is over clusters, and **not** elements. We denote the sample of elements from sampled cluster g by s_g. Population elements are indexed by i, so $i \in s_g$ ($i \in r_g$) denotes the sampled (non-sampled) elements in sampled cluster g.

There are two basic sample design problems that arise with two-stage sampling:

- How many (q) and which (s) clusters to select?
- How many (m_g) population elements to select in each sampled cluster?

We address these problems below, after first developing an efficient estimator of t_y given two-stage sample data obtained from a clustered population.

In some cases, the clusters will be dictated by practical considerations, such as the availability of a frame of these units. In other cases, the sample designer may be free to decide how best to group elements into clusters, for example this would sometimes be the case if neighbouring dwellings are to be grouped into areas. If so, this adds two more design considerations:

- How many (M_g) population elements to group in forming each cluster?
- How many (Q) clusters to form in the population?

Consideration of these two additional aspects of two-stage sample design is beyond the scope of this book. See Cochran (1977, Chapter 9) and Kish (1965, Chapter 9). We assume that Q and the values M_g are known characteristics of the population.

6.2 Optimal Prediction for a Clustered Population

We will now derive the BLUP of the population total t_y under (6.1). We will not derive an EB estimator, because for correlated data, $E(t_{yr}|\{y_i, i \in s_g, g \in s\})$ cannot be derived without a much more complete specification of the dependency structure than (6.1). This BLUP was derived under a generalisation of (6.1) by Royall (1976). Let $\hat{t}_y = \sum_s w_i y_i$, then:

$$Var\left(\hat{t}_y - t_y\right) = Var\left[\sum_{g \in s} \sum_{i \in s_g} w_i y_i - \sum_{g \in U} \sum_{i \in U_g} y_i\right]$$

$$= Var\left[\sum_{g \in s}\left\{\sum_{i \in s_g}(w_i - 1)y_i - \sum_{i \in r_g} y_i\right\} - \sum_{g \in r}\sum_{i \in U_g} y_i\right]$$

$$= \sum_{g \in s}\left[Var\left\{\sum_{i \in s_g}(w_i - 1)y_i\right\} + Var\left(\sum_{i \in r_g} y_i\right)\right.$$

$$\left. -2Cov\left\{\sum_{i \in s_g}(w_i - 1)y_i, \sum_{i \in r_g} y_i\right\}\right] + \sum_{g \in r} Var\left(\sum_{i \in U_g} y_i\right)$$

$$= \sigma^2 \sum_{g \in s}\left\{\sum_{i \in s_g}(w_i - 1)^2 + \sum_{i \neq j \in s_g} \rho(w_i - 1)(w_j - 1) + (M_g - m_g)\right.$$

$$\left. + \sum_{i \neq j \in r_g} \rho - 2\sum_{i \in s_g} \rho(w_i - 1)(M_g - m_g)\right\} + \sigma^2 \sum_{g \in r}\left(M_g + \sum_{i \neq j \in U_g} \rho\right).$$

With some elementary manipulation, this becomes

$$\operatorname{Var}(\hat{t}_y - t_y) = \sigma^2 (1-\rho) \sum_{g \in s} \sum_{i \in s_g} (w_i - 1)^2 + \sigma^2 \rho \sum_{g \in s} \left\{ \sum_{s_g} (w_i - 1) \right\}^2$$
$$- 2\sigma^2 \rho \sum_{g \in s} (M_g - m_g) \sum_{i \in s_g} (w_i - 1) \qquad (6.2)$$
$$+ \sigma^2 \sum_{g \in s} (M_g - m_g) \{1 + (M_g - m_g - 1)\rho\}$$
$$+ \sigma^2 \sum_{g \in r} M_g \{1 + (M_g - 1)\rho\}.$$

The bias of $\hat{t}_y$ under (6.1) is

$$E(\hat{t}_y - t_y) = E\left[\sum_{g \in s} \sum_{i \in s_g} w_i y_i - \sum_{g \in U} \sum_{i \in U_g} y_i\right]$$
$$= \mu \left\{ \sum_{g \in s} \sum_{i \in s_g} (w_i - 1) - (N - n) \right\}.$$

We therefore need to minimise (6.2) subject to $\sum_{g \in s} \sum_{i \in s_g} (w_i - 1) = (N - n)$. Taking the derivative with respect to w_i, for $i \in s_g$, using Lagrange multipliers, and dropping σ^2, we obtain

$$0 = 2(1-\rho)(w_i - 1) - 2\rho(M_g - m_g) + 2\rho \sum_{j \in s_g} (w_j - 1) - 2\lambda. \qquad (6.3)$$

It is clear from (6.3) that the weights w_i are equal to the same value, say w_g, within a cluster g. Substituting $w_i = w_g$ into (6.3) and expressing in terms of w_g gives the BLUP weight for a sample element in cluster g:

$$w_g = 1 + \frac{\lambda + \rho(M_g - m_g)}{1 - \rho + \rho m_g}. \qquad (6.4)$$

The unbiasedness constraint determines λ, and we end up with

$$\lambda = \frac{N - n - \rho \sum_s m_g (M_g - m_g)(1 - \rho + \rho m_g)^{-1}}{\sum_s m_g (1 - \rho + \rho m_g)^{-1}}. \qquad (6.5)$$

Computation of (6.4) using (6.5) assumes that we know the intracluster correlation ρ. This is extremely unlikely. Three options we might therefore consider are:

- Put $\rho = 0$. In this case the weights become $w_i = N/n$, that is the simple expansion weight.
- Put $\rho = 1$. In this case, the weights become

$$w_g = \frac{M_g}{m_g} + \frac{N - \sum_s M_g}{q}.$$

- Typically ρ lies somewhere between 0 and 1. When this is the case, (6.1) is a special case of a two-level random intercepts model, which in turn is a special case of a multi-level model. See, for example, Goldstein (2003) for a text on this very important class of models. Most statistical software packages have procedures to fit this kind of model, and to produce an estimate of ρ, which can then be substituted into (6.4) and (6.5). The resulting predictor $\hat{t}_y$ of t_y is called an *empirical* BLUP (or EBLUP), because it is calculated by substituting an empirical estimate of unknown parameters (in this case ρ) into the formula for the BLUP.

Which of the options we decide to use depends very much on how we intend to process the survey data. If our intention is to use the same sample weight for a number of different Y variables, then the third option is inappropriate since it results in a different sample weight for each survey Y variable since each can lead to a different estimate of ρ. In this case the first option above is typically the option of choice since in reality ρ is usually closer to 0 than to 1 for most variables of interest in multi-stage surveys. On the other hand, option 3 will clearly lead to a more accurate estimator of the intracluster correlation ρ (and hence a better predictor of t_y) provided we have a large enough sample of clusters, and elements within selected clusters, to estimate this correlation accurately.

However, in an important special case, this choice becomes redundant. This is where the number of population and sampled elements in each cluster is constant. Let $M_g = M$ and $m_g = m$, so the overall sample size becomes $n = mq$. Then (6.4) and (6.5) reduce to

$$w_i = w_g = N/n$$

which is the simple expansion weight, regardless of the value of ρ.

For the purpose of understanding the behaviour of the weights (6.4), and also for motivating the sample designs considered in the next section, some approximations would be helpful. We will firstly derive an approximation to the prediction variance (6.2) of $\hat{t}_y$, and then derive approximately optimal weights based on this approximation. If the sampling fraction q/Q at the first stage is small, then the first two terms of (6.2) will dominate the remaining terms, allowing us to write:

$$\text{Var}\left(\hat{t}_y - t_y\right) \cong \sigma^2 (1-\rho) \sum_{g \in s} \sum_{i \in s_g} (w_i - 1)^2 + \sigma^2 \rho \sum_{g \in s} \left\{ \sum_{s_g} (w_i - 1) \right\}^2.$$

Substituting $w_i = w_g$ gives

$$\text{Var}\left(\hat{t}_y - t_y\right) \cong \sigma^2 \sum_{g \in s} (w_g - 1)^2 m_g (1 - \rho + m_g \rho). \tag{6.6}$$

We now minimise this approximation to the prediction variance subject to the unbiasedness constraint $N - n = \sum_{g \in s} \sum_{i \in s_g} (w_i - 1) = \sum_{g \in s} m_g (w_g - 1)$. Using Lagrange multipliers and differentiating leads to:

$$2(w_g - 1) m_g (1 - \rho + m_g \rho) - 2\lambda m_g = 0$$

Table 6.1 Approximate inefficiency from using BLUP based on $\rho = 0$.

True ρ	Range of Cluster Sample Sizes				
	20–20	19–21	18–22	15–25	10–30
0.000	1.000	1.000	1.000	1.000	1.000
0.010	1.000	1.000	1.000	1.001	1.002
0.025	1.000	1.000	1.001	1.003	1.009
0.050	1.000	1.000	1.001	1.006	1.022
0.100	1.000	1.001	1.002	1.012	1.040
0.250	1.000	1.001	1.004	1.019	1.066
0.500	1.000	1.002	1.005	1.022	1.080
1.000	1.000	1.002	1.005	1.025	1.090

or

$$w_g = 1 + \lambda \left(1 - \rho + m_g \rho\right)^{-1}.$$

The unbiasedness constraint determines λ, and results in the approximate BLUP weights:

$$w_i = w_g = 1 + (N - n) \frac{(1 - \rho + m_g \rho)^{-1}}{\sum_{f \in s} m_f (1 - \rho + m_f \rho)^{-1}}. \qquad (6.7)$$

Substituting back into (6.6) then gives a first order approximation to the prediction variance of the BLUP:

$$\text{Var}\left(\hat{t}_y^{BLUP} - t_y\right) \cong \sigma^2 (N - n)^2 \left\{\sum_{g \in s} m_g (1 - \rho + m_g \rho)^{-1}\right\}^{-1}. \qquad (6.8)$$

Do we lose much if we assume $\rho = 0$ if the true ρ is non-zero? We can answer this using the approximation (6.8). We have already shown that we do not lose anything if clusters are of equal size. Table 6.1 shows the approximate variance of the BLUP based on $\rho = 0$ divided by the approximate variance of the BLUP based on the correct ρ, for various values of ρ and a range of cluster sample sizes. The calculation is based on an average sample size of 20 elements per cluster, but with actual cluster sample sizes distributed uniformly on the integers in a specified range around 20. The table shows that the inefficiency is negligible unless the true ρ is quite large (0.25 or higher) and the sample sizes within clusters vary substantially.

6.3 Optimal Design for Fixed Sample Size

First, we consider the optimal choice of m_g and q when the aim is to minimise the prediction variance of the BLUP under (6.1) subject to fixed total sample size of n elements. We will assume that q is much smaller than Q, so that approximation (6.8) to the prediction variance can be used. In this case, the aim is to identify

q, a set s of q clusters to sample, and within cluster sample sizes $\{m_g : g \in s\}$, so that we minimise (6.8) subject to $\sum_{g \in s} m_g = n$. To do this, we first minimise (6.8) with respect to $\{m_g : g \in s\}$ subject to this constraint with s and q fixed.

Note that (6.8) can be re-written as

$$\mathrm{Var}\left(\hat{t}_y^{BLUP} - t_y\right) \cong \sigma^2(N-n)^2 \left\{ \sum_{g \in s} \frac{\rho^{-1}(1 - \rho + m_g \rho) - \rho^{-1}(1-\rho)}{1 - \rho + m_g \rho} \right\}^{-1}$$

$$= \sigma^2(N-n)^2 \left\{ q - \rho^{-1}(1-\rho) \sum_{g \in s} (1 - \rho + m_g \rho)^{-1} \right\}^{-1}$$

so that minimising $\mathrm{Var}\left(\hat{t}_y^{BLUP} - t_y\right)$ with respect to $\{m_g : g \in s\}$ subject to $\sum_{g \in s} m_g = n$ is equivalent to minimising

$$\sum_{g \in s} (1 - \rho + m_g \rho)^{-1} \qquad (6.9)$$

(provided $\rho > 0$, which would almost always be the case in practice). Differentiating and using Lagrange multipliers gives us

$$0 = (1 - \rho + m_g \rho)^{-2} \rho + \lambda$$

which means that the values of $\{m_g\}$ are all equal to $m = n/q$.

We now substitute $m_g = m$ into (6.8) and optimise with respect to s. In this case

$$\mathrm{Var}\left(\hat{t}_y^{BLUP} - t_y\right) \cong \sigma^2(N-n)^2 \left\{ \sum_{g \in s} m_g (1 - \rho + m_g \rho)^{-1} \right\}^{-1}$$

$$= \sigma^2(N-n)^2 \left\{ qm(1 - \rho + m\rho)^{-1} \right\}^{-1}$$

$$= \sigma^2(N-n)^2 n^{-1} (1 - \rho + m\rho).$$

So, apart from the value of m (or equivalently of q, since $mq = n$), which sample of clusters s we take does not change our approximation (6.8) to the prediction variance of the BLUP. This prediction variance is clearly minimised by making m as small as possible (or equivalently making q as large as possible), which can be achieved by setting $m = 1$ and $q = n$. However, even though this is the best design for fixed total sample size, it might be prohibitively expensive to implement, as the sample is maximally dispersed, with only one element sampled per cluster.

Given that the actual sample of clusters chosen does not affect the minimum value of the approximate prediction variance (6.8), it makes sense to take a simple random sample of clusters, or something similar (e.g. a systematic size-ordered sample or a sample of clusters balanced on their sizes). In contrast, Royall (1976) shows that the actual prediction variance of the BLUP is minimised by taking the largest clusters. Based on our approximation (6.8), it is likely that the increase in precision from using this extreme sample is likely to be small under (6.1).

6.4 Optimal Design for Fixed Cost

In most practical two-stage sample design situations the design constraint is fixed overall cost, rather than fixed sample size of elements. The cost depends on the number of clusters selected, the size of the selected clusters and the second stage allocation in the selected clusters. Cochran (1977, page 313) gives a simple cost function that incorporates these features,

$$C = c_0 + c_1 q + c_2 \sum_s m_g + c_3 \sum_s M_g. \qquad (6.10)$$

The first term of C reflects costs not related to the sample size or sample design, such as questionnaire development. These do not affect the optimal design so we will henceforth assume $c_0 = 0$. The second term relates to costs associated with the number of clusters in sample, such as travel between clusters, which is often a significant component of surveys utilising face-to-face interviewing. The third term consists of costs associated with the number of elements in the sample, for example interviewing time or time spent processing each form. The final term is the cost associated with listing all elements in selected clusters, prior to sampling these elements. This may consist, for example, of an interviewer driving around the cluster and compiling a list of addresses.

This linear cost model is obviously only a rough approximation to how a survey's cost really depends on the sample design used. For example, travel costs might be expected to be a non-linear function of the number of clusters selected, and cost models with square root and quadratic terms in q have been proposed. Or increasing m_g beyond a certain point might mean that an interviewer needs to return to a cluster for a second day, rather than being able to complete all interviews in one visit; this would mean a discontinuity in the cost as a function of m_g. The linear cost model is still useful, however, and gives sensible designs which reflect the fact that both the number of clusters and the number of elements in sample affect the cost. Even if the true cost model is non-linear, it might be expected to behave linearly when relatively small changes to a design are being considered. Another argument in favour of linear cost models rather than more complex non-linear models is that data on survey costs is often not rich enough or reliable enough to enable the parameters of these more sophisticated models to be estimated.

In some cases, listing costs do not apply or are relatively minor, so in this section we will consider the best design when $c_3 = 0$. The next section will discuss the optimal design when this parameter is non-zero. When both $c_0 = 0$ and $c_3 = 0$, the cost model (6.10) becomes

$$C = c_1 q + c_2 \sum_s m_g. \tag{6.11}$$

Our aim therefore is to identify a sample size, q, a set s of q clusters, and associated within cluster sample numbers $\{m_g : g \in s\}$, in order to minimise the approximation (6.8) for the prediction variance of the BLUP under (6.1), subject to a fixed value for C in (6.11). As in Section 6.3, we start by minimising with respect to $\{m_g : g \in s\}$ for given s and q. As shown in Section 6.3, this is equivalent to minimising (6.9) subject to this constraint. Differentiating and using Lagrange multipliers gives us

$$0 = (1 - \rho + m_g \rho)^{-2} \rho + \lambda c_2$$

which, as in Section 6.3, implies that $m_g = m$. The cost constraint is $C = c_1 q + c_2 q m$, which implies that

$$q = C (c_1 + c_2 m)^{-1}.$$

On substitution back into (6.8), and also assuming that $n << N$, we obtain

$$\begin{aligned}
\text{Var}\left(\hat{t}_y^{BLUP} - t_y\right) &\cong N^2 \sigma^2 \left\{ \sum_{g \in s} m_g (1 - \rho + m_g \rho)^{-1} \right\}^{-1} \\
&= N^2 \sigma^2 \left\{ qm (1 - \rho + m\rho)^{-1} \right\}^{-1} \\
&= N^2 \sigma^2 C^{-1} (c_1 + c_2 m) m^{-1} (1 - \rho + m\rho) \\
&= N^2 \sigma^2 C^{-1} \left\{ c_1 (1 - \rho) m^{-1} + c_2 \rho m + c_2 (1 - \rho) + c_1 \rho \right\}.
\end{aligned}$$

So minimising this prediction variance approximation is equivalent to minimising $c_1(1 - \rho)m^{-1} + c_2 \rho m$. Note that this expression does not depend on s, which means that it makes no difference which m clusters are selected. Differentiating it with respect to m and setting to zero finally gives

$$m_{opt} = \sqrt{\frac{c_1 (1 - \rho)}{c_2 \rho}}$$

with q then determined by the cost constraint. It is interesting to note that:

- The optimal within-cluster sample sizes are all equal, even when clusters vary in size.
- The optimal within-cluster sample size m_{opt} does not depend on the budget C.
- A high value of ρ leads to a less clustered design (i.e. smaller m_{opt}) – this makes sense, since we would expect that sampling more elements from the same cluster will add less information when ρ is large.
- If the costs associated with numbers of clusters (c_1) are high compared to the costs associated with the number of elements (c_2), then the design will be

more clustered, in the sense that a smaller number of clusters will be selected, and more elements will be sampled in the sampled clusters.

This result for m_{opt} was originally derived in a design-based framework. Royall (1976) re-derived m_{opt} in a model-based framework, and also found that it was efficient to select a sample of large clusters of nearly equal size, if possible. However, the approximate variance (6.8) does not depend on which clusters are selected, so the improvement from selecting large clusters is most likely small. Moreover, selecting the largest clusters could result in sensitivity to the model (6.1) being incorrectly specified. Robustness to model breakdown will be explored in Chapter 8, including a robust approach to selecting clusters in Section 8.3.

6.5 Optimal Design for Fixed Cost including Listing

We finally consider optimal design for fixed cost when listing costs are **not** negligible. As before, we base this design on the prediction variance approximation (6.9), in particular the optimal design is obtained by choosing q, s and m_g to minimise (6.9) subject to a fixed cost given by (6.10). We will again assume that c_0 is zero.

First, we note that given q, and a set of q values of m_g, the optimal choice of s consists of the q clusters with the **smallest** values of M_g. To see this, suppose that a set of q clusters which does **not** consist of the smallest clusters is chosen. The cost of this design is given by (6.10). Suppose that we replace the largest sampled cluster by a smaller cluster from the non-sampled clusters. Let K be the amount by which the new cluster is smaller than the old. Call this new sample of q clusters s_{new}, with element sample sizes $m_{g(new)}$. We increase one or more of the values of $m_{g(new)}$ relative to m_g, such that $\sum_s m_{g(new)} = \sum_s m_g + Kc_3/c_2$. The cost of this new sample is:

$$\begin{aligned} C_{new} &= c_1 q + c_2 \sum_{s_{new}} m_g + c_3 \sum_{s_{new}} M_g \\ &= c_1 q + c_2 \left(\sum_s m_g + Kc_3/c_2 \right) + c_3 \left(\sum_s M_g - K \right) \\ &= c_1 q + c_2 \sum_s m_g + c_3 \sum_s M_g \\ &= C \end{aligned}$$

so the new sample has the same cost as the original sample, and therefore meets the cost constraint. However, (6.9) is clearly smaller for the new sample than our original sample, since q is unchanged, and one or more of the values of m_g have increased while none have decreased. Hence it is possible to improve on any sample in this way, unless it consists of the q smallest clusters.

Optimal Design for Fixed Cost including Listing

Since the smallest clusters are to be selected, our problem becomes to minimise (6.9) subject to

$$C = c_1 q + c_2 \sum_s m_g + c_3 \sum_{g=1}^{q} M_{(g)}$$

where $M_{(1)}, M_{(2)}, \ldots, M_{(Q)}$ are the population cluster sizes in ascending order. Given the sample s, it is straightforward to show that the optimal values of m_g are all equal to a constant value m, using the same arguments as Section 6.4. The cost constraint implies that

$$m = c_2^{-1} q^{-1} \left(C - c_1 q - c_3 \sum_{g=1}^{q} M_{(g)} \right).$$

Substituting this expression for m into the equation preceding (6.9), and assuming $n << N$, gives

$$\operatorname{Var}\left(\hat{t}_y^{BLUP} - t_y\right) \cong \sigma^2 (N-n)^2 \left\{ q - \rho^{-1}(1-\rho) \sum_{g \in s} (1 - \rho + m_g \rho)^{-1} \right\}^{-1}$$

$$= \sigma^2 N^2 \left[q - \rho^{-1}(1-\rho) q \left\{ 1 - \rho + c_2^{-1} q^{-1} \left(C - c_1 q - c_3 \sum_{g=1}^{q} M_{(g)} \right) \rho \right\}^{-1} \right]^{-1}$$

This expression can be minimised with respect to q to give the optimal design. No algebraic solution is available, so the solution would have to be calculated numerically for the population of interest.

Note that this approach of selecting the smallest clusters is only optimal under the clustered population model (6.1). In particular, selecting the smallest clusters could be sensitive to incorrect specification of model (6.1), in the same way as discussed at the end of Section 6.4.

7 The General Linear Population Model

So far in this book, we have introduced a number of useful models for populations often encountered in survey practice, and developed optimal methods for predicting a linear parameter (e.g. t_y) using samples drawn from these populations. In this chapter we show that all of these population models are in fact special cases of a very general population model that is widely used in statistics. This is the *general linear population model*. Specification of this model, and derivation of optimal predictors based on it, requires, however, that we abandon the scalar notation used so far and use vector and matrix notation instead. For readers unfamiliar with this notation, and with the general principles of matrix arithmetic and matrix algebra, we recommend reading Appendix A of Anderson (1984). Throughout the following development, vectors are column vectors.

7.1 A General Linear Model for a Population

To start, we let $\mathbf{y}_U$ denote the vector containing the N population values of the survey variable Y. Similarly, let $\mathbf{z}_{1U}, \mathbf{z}_{2U}, \ldots, \mathbf{z}_{pU}$ denote the vectors containing the N population values of p auxiliary variables $Z_1, Z_2, \ldots, Z_p$. We define the matrix $\mathbf{Z}_U$ as the concatenation of these auxiliary variable vectors. That is, $\mathbf{Z}_U$ is a $N \times p$ matrix whose k^{th} column is $\mathbf{z}_{kU}$. In a slight abuse of notation, we let $\mathbf{z}_i$ denote the vector of dimension p defined by the ith row of $\mathbf{Z}_U$. The random vector $\mathbf{y}_U$ is then said to follow the general linear population model defined by $\mathbf{Z}_U$ when we can write

$$E(y_i|\mathbf{z}_i) = \mathbf{z}'_i \beta \tag{7.1a}$$

$$Var(y_i|\mathbf{z}_i) = \sigma^2 \tag{7.1b}$$

$$y_i \text{ and } y_j \text{ independent conditional on } \mathbf{Z}_U \text{ when } i \neq j \tag{7.1c}$$

where β is an unknown vector of regression parameters of dimension p. Once again, we note that (7.1c) is necessary to derive an EB predictor, but the weaker assumption of zero covariances is sufficient for the BLUP.

An EB predictor of the population total t_y of Y under (7.1) is easily specified. Observe that, under this model, the expected value of t_y given the sample data is

$$t_{ys} + E(t_{yr}|\mathbf{y}_s, \mathbf{Z}_U) = t_{ys} + \sum_r \mathbf{z}'_i \beta = t_{ys} + \mathbf{t}'_{zr}\beta \tag{7.2}$$

A General Linear Model for a Population

where $\mathbf{y}_s$ denotes the sample component of $\mathbf{y}_U$ and $\mathbf{t}_{zr}$ denotes the p-vector of non-sample totals of the p auxiliary variables. Furthermore, standard statistical theory tells us that the best linear unbiased estimator of β is then the ordinary least squares regression estimator,

$$\hat{\beta}_{ols} = (\mathbf{Z}_s'\mathbf{Z}_s)^{-1}\mathbf{Z}_s'\mathbf{y}_s = \left(\sum_s \mathbf{z}_i\mathbf{z}_i'\right)^{-1}\sum_s \mathbf{z}_i y_i$$

where $\mathbf{Z}_s$ is the $n \times p$ sub-matrix of $\mathbf{Z}_U$ defined by the sampled population elements. The EB predictor is then defined by substituting $\hat{\beta}_{ols}$ for β in (7.2). This turns out to also be the BLUP of t_y assuming (7.1), and is equal to:

$$\hat{t}_y^{BLUP} = t_{ys} + \mathbf{t}_{zr}'\hat{\beta}_{ols}. \tag{7.3}$$

We can write $\hat{t}_y^{BLUP}$ in weighted form as

$$\hat{t}_y^{BLUP} = \sum_s w_i y_i$$

where the weights are given by

$$w_i = 1 + \mathbf{t}_{zr}'\left(\sum_s \mathbf{z}_j\mathbf{z}_j'\right)^{-1}\mathbf{z}_i.$$

Notice that the weights depend on the sample and non-sample values of the auxiliary variables, but not on any of the values of Y. This is convenient, because the same weights can be used for different Y variables, provided that each follows a model of the same form as (7.1), though not necessarily with the same β or σ^2.

The prediction variance of $\hat{t}_y^{BLUP}$ under (7.1) follows directly. Since $Var\left(\hat{\beta}_{ols}\right) = \sigma^2 \left(\mathbf{Z}_s'\mathbf{Z}_s\right)^{-1}$,

$$Var\left(\hat{t}_y^{BLUP} - t_y\right) = Var\left(\mathbf{t}_{zr}'\hat{\beta}_{ols} - t_{yr}\right) = Var\left(t_{yr}\right) + \mathbf{t}_{zr}' Var\left(\hat{\beta}_{ols}\right)\mathbf{t}_{zr}$$

$$= \sigma^2 \left\{(N-n) + \mathbf{t}_{zr}'\left(\mathbf{Z}_s'\mathbf{Z}_s\right)^{-1}\mathbf{t}_{zr}\right\}.$$

Similarly, since

$$\hat{\sigma}^2 = (n-p)^{-1}\sum_s \left(y_i - \mathbf{z}_i'\hat{\beta}_{ols}\right)^2$$

is an unbiased estimator of σ^2, we can estimate the prediction variance of $\hat{t}_y^{BLUP}$ under (7.1) using

$$\hat{V}\left(\hat{t}_y^{BLUP}\right) = \hat{\sigma}^2 \left\{(N-n) + \mathbf{t}_{zr}'\left(\mathbf{Z}_s'\mathbf{Z}_s\right)^{-1}\mathbf{t}_{zr}\right\}. \tag{7.4}$$

7.2 The Correlated General Linear Model

We now look at a generalisation of (7.1) where the values of Y for different units can be correlated. In this case it turns out to be more convenient to use matrix and vector notation to define this model. In particular, we shall assume that

$$\mathbf{y}_U = \mathbf{Z}_U \beta + \mathbf{e}_U, \tag{7.5}$$

where $\mathbf{e}_U$ is an unobservable vector of random errors of dimension N satisfying $E(\mathbf{e}_U) = \mathbf{0}_N$ and $Var(\mathbf{e}_U) = \sigma^2 \mathbf{V}_U$. Here $\mathbf{0}_N$ denotes a vector of zeros of dimension N, σ^2 is an unknown positive scale parameter and $\mathbf{V}_U$ is a known positive definite square matrix also of dimension N. Without loss of generality, we assume that the population units making up $\mathbf{y}_U$ are ordered so that the first n of them correspond to sample units, and the remaining $N - n$ correspond to the non-sample units. Using subscripts of s and r to denote sample and non-sample respectively, we then express $\mathbf{Z}_U$ and $\mathbf{V}_U$ in partitioned form as

$$\mathbf{Z}_U = \begin{bmatrix} \mathbf{Z}_s \\ \mathbf{Z}_r \end{bmatrix} \text{ and } \mathbf{V}_U = \begin{bmatrix} \mathbf{V}_{ss} & \mathbf{V}_{sr} \\ \mathbf{V}_{rs} & \mathbf{V}_{rr} \end{bmatrix}.$$

As stated above, (7.5) includes all the population models considered so far in this book. We will consider some of these as special cases in the next section.

We now consider optimal prediction of t_y under (7.5). In particular, we derive the BLUP for an arbitrary linear combination of the population Y-values in this case. To this end, let $\mathbf{a}_U$ denote a known vector of dimension N that is ordered identically to the vector $\mathbf{y}_U$ of population Y-values, and set

$$A_y = \mathbf{a}'_U \mathbf{y}_U = \sum_U a_i y_i. \tag{7.6}$$

We then have the following result.

Result 7.1 Consider a population for which (7.5) holds, with $\mathbf{Z}_U$ and $\mathbf{V}_U$ known, and suppose a sample s is drawn from this population using a sampling method that is non-informative given these quantities, with values of Y observed on the sample. Then the best linear unbiased predictor of the linear population parameter A_y defined by (7.6) is

$$\hat{A}_y^{BLUP} = \sum_s w_i y_i = \mathbf{w}'_s \mathbf{y}_s \tag{7.7a}$$

where the weight vector $\mathbf{w}_s$ is given by

$$\mathbf{w}_s = \mathbf{a}_s + \left\{ \mathbf{H}'_L \mathbf{Z}'_r + (\mathbf{I}_n - \mathbf{H}'_L \mathbf{Z}'_s) \mathbf{V}_{ss}^{-1} \mathbf{V}_{sr} \right\} \mathbf{a}_r. \tag{7.7b}$$

Here $\mathbf{a}_U = (\mathbf{a}'_s \ \mathbf{a}'_r)'$ denotes the sample/non-sample decomposition of the vector $\mathbf{a}_U$, $\mathbf{I}_n$ is the identity matrix of order n and $\mathbf{H}_L = \left(\mathbf{Z}'_s \mathbf{V}_{ss}^{-1} \mathbf{Z}_s \right)^{-1} \mathbf{Z}'_s \mathbf{V}_{ss}^{-1}$.

The Correlated General Linear Model

Proof of this general result requires matrix calculus and is given in Royall (1976) and Valliant et al. (2000, page 29). It is sketched out in Section 7.6 for the important special case where $\mathbf{a}_U = \mathbf{1}_N$.

From (7.7b) we see that we can write $\mathbf{w}_s = \mathbf{a}_s + \mathbf{U}_L \mathbf{a}_r$ where $\mathbf{U}_L = \mathbf{H}'_L \mathbf{Z}'_r + (\mathbf{I}_n - \mathbf{H}'_L \mathbf{Z}'_s) \mathbf{V}_{ss}^{-1} \mathbf{V}_{sr}$ is the $n \times (N-n)$ matrix that defines the prediction component of the weight vector $\mathbf{w}_s$. The prediction variance of $\hat{A}_y^{BLUP}$ under (7.5) is then easily seen to be

$$\mathrm{Var}(\hat{A}_y^{BLUP} - A_y) = \sigma^2 \mathbf{a}'_r \left(\mathbf{U}'_L \mathbf{V}_{ss} \mathbf{U}_L - \mathbf{U}'_L \mathbf{V}_{sr} - \mathbf{V}_{rs} \mathbf{U}_L + \mathbf{V}_{rr} \right) \mathbf{a}_r \quad (7.8)$$

An unbiased estimator of (7.8) is therefore

$$\hat{V}(\hat{A}_y^{BLUP}) = \hat{\sigma}^2 \mathbf{a}'_r \left(\mathbf{U}'_L \mathbf{V}_{ss} \mathbf{U}_L - \mathbf{U}'_L \mathbf{V}_{sr} - \mathbf{V}_{rs} \mathbf{U}_L + \mathbf{V}_{rr} \right) \mathbf{a}_r \quad (7.9)$$

where $\hat{\sigma}^2$ is an unbiased estimator of the scale parameter σ^2 in (7.5).

In order to define $\hat{\sigma}^2$ we first observe that $\hat{\beta} = \mathbf{H}_L \mathbf{y}_s$ is the BLUE (best linear unbiased estimator) of the regression parameter β in (7.5a). Hence $\hat{\mathbf{y}}_s = \mathbf{Z}_s \mathbf{H}_L \mathbf{y}_s = \mathbf{Z}_s \hat{\beta}$ is the BLUE of the expected value of $\mathbf{y}_s$ under (7.5). We therefore base our estimator $\hat{\sigma}^2$ on the scaled sum of squared residuals $(\mathbf{y}_s - \hat{\mathbf{y}}_s)' \mathbf{V}_{ss}^{-1} (\mathbf{y}_s - \hat{\mathbf{y}}_s) = \mathbf{y}'_s \mathbf{G}_s \mathbf{y}_s$, where

$$\mathbf{G}_s = (\mathbf{I}_n - \mathbf{H}'_L \mathbf{Z}'_s) \mathbf{V}_{ss}^{-1} (\mathbf{I}_n - \mathbf{Z}_s \mathbf{H}_L).$$

Now, under (7.5), $E(\mathbf{y}'_s \mathbf{G}_s \mathbf{y}_s) = \sigma^2 \sum_{i \in s} \sum_{j \in s} g_{ij} v_{ij}$, where $\mathbf{V}_{ss} = [v_{ij}]$ and $\mathbf{G}_s = [g_{ij}]$. Consequently, we put

$$\hat{\sigma}^2 = \left(\sum_{i \in s} \sum_{j \in s} g_{ij} v_{ij} \right)^{-1} \mathbf{y}'_s \mathbf{G}_s \mathbf{y}_s.$$

In practice the matrix $\mathbf{V}_U$ (and hence $\mathbf{V}_{ss}$) will depend on one or more unknown parameters. See for example its specification under the homogeneous strata population model (where it depends on the strata variances σ_h^2) and under the clustered population model (where it depends on the intra-cluster correlation ρ). Sample estimates of these parameters can be 'plugged into' the preceding formulae (7.7b) for the weights defining the BLUP (7.7a) and the estimator (7.9) of its prediction variance. This will mean that the weights depend on the sample data for Y, through the estimated variance components (such as ρ or σ_h^2), which will increase the prediction variance of $\hat{A}_y^{BLUP}$. Moreover, the variance estimator (7.8) does not reflect the extra variability caused by estimation of these variance components and so (7.9) will underestimate the true prediction variance of (7.7a). It can be complicated to correct (7.9) for this bias. We discuss some approaches to tackling this problem in Chapter 15.

7.3 Special Cases of the General Linear Population Model

In this section, we will state some special cases of the general linear model that are of interest, and derive the BLUP for t_y in each case. Note that in this case $\mathbf{a}_U = \mathbf{1}_N$, where $\mathbf{1}_N$ denotes a N-vector of ones. Consequently $\mathbf{a}_s = \mathbf{1}_n$ and $\mathbf{a}_r = \mathbf{1}_{N-n}$, with the obvious interpretations.

7.3.1 The Uncorrelated General Linear Population Model

This is model (7.1). In this case, $\mathbf{V}_U = \mathbf{I}_N$, $\mathbf{V}_{ss} = \mathbf{I}_n$ and $\mathbf{V}_{sr} = \mathbf{0}$, so

$$\mathbf{H}_L = \left(\mathbf{Z}_s' \mathbf{V}_{ss}^{-1} \mathbf{Z}_s\right)^{-1} \mathbf{Z}_s' \mathbf{V}_{ss}^{-1} = \left(\mathbf{Z}_s' \mathbf{Z}_s\right)^{-1} \mathbf{Z}_s'$$

and (7.7b) becomes

$$\begin{aligned} \mathbf{w}_s &= \mathbf{1}_n + \left\{ \mathbf{H}_L' \mathbf{Z}_r' + \left(\mathbf{I}_n - \mathbf{H}_L' \mathbf{Z}_s'\right) \mathbf{V}_{ss}^{-1} \mathbf{V}_{sr} \right\} \mathbf{1}_{N-n} \\ &= \mathbf{1}_n + \mathbf{H}_L' \mathbf{Z}_r' \mathbf{1}_{N-n} \\ &= \mathbf{1}_n + \mathbf{Z}_s \left(\mathbf{Z}_s' \mathbf{Z}_s\right)^{-1} \mathbf{Z}_r' \mathbf{1}_{N-n} \end{aligned}$$

which is equivalent to the BLUP weights obtained in Section 7.2. In the special case of the homogeneous model (3.1) from Chapter 3, there is a common mean so that $z_i = 1$ for every unit. Then $\mathbf{Z}_s = \mathbf{1}_n$ and $\mathbf{Z}_r = \mathbf{1}_{N-n}$, so the weights defining the BLUP become:

$$\begin{aligned} \mathbf{w}_s &= \mathbf{1}_n + \mathbf{Z}_s \left(\mathbf{Z}_s' \mathbf{Z}_s\right)^{-1} \mathbf{Z}_r' \mathbf{1}_{N-n} \\ &= \mathbf{1}_n + \mathbf{1}_n \left(\mathbf{1}_n' \mathbf{1}_n\right)^{-1} \mathbf{1}_{N-n}' \mathbf{1}_{N-n} \\ &= \mathbf{1}_n + \mathbf{1}_n n^{-1}(N-n) \\ &= \mathbf{1}_n \left(1 + \frac{N-n}{n}\right) \\ &= (N/n) \mathbf{1}_n. \end{aligned}$$

That is, these weights are all equal to N/n, as we saw in Chapter 3.

7.3.2 The Homogenous Strata Population Model

The homogeneous strata population model (4.1) is also a special case of (7.5). Let $I(i \in h)$ be the stratum h indicator function which takes the value 1 if population element i is in stratum h and is zero otherwise. The number of parameters p is equal to the number of strata, H. Column h of $\mathbf{Z}_U$, which we will denote $\mathbf{z}_{hU}$, is an $N \times 1$ vector containing a value of one for every element of stratum h and zeroes everywhere else. The variance-covariance matrix $\mathbf{V}_U$ is equal to a diagonal matrix with i^{th} diagonal component given by $\sum_h \sigma_h^2 I(i \in h)$.

Special Cases of the General Linear Population Model

For simplicity, we will consider the case where there are only $H = 2$ strata. (The generalisation to more strata is straightforward.) Assuming that units are ordered by strata, we can write $\mathbf{V}_U$ and $\mathbf{V}_{ss}$ in block matrix form as:

$$\mathbf{V}_U = \begin{bmatrix} \sigma_1^2 \mathbf{I}_{N_1} & \mathbf{0}_{N_1 \times N_2} \\ \mathbf{0}_{N_2 \times N_1} & \sigma_2^2 \mathbf{I}_{N_2} \end{bmatrix} \text{ and } \mathbf{V}_{ss} = \begin{bmatrix} \sigma_1^2 \mathbf{I}_{n_1} & \mathbf{0}_{n_1 \times n_2} \\ \mathbf{0}_{n_2 \times n_1} & \sigma_2^2 \mathbf{I}_{n_2} \end{bmatrix}$$

and similarly

$$\mathbf{Z}_U = \begin{bmatrix} \mathbf{1}_{N_1} & \mathbf{0}_{N_1} \\ \mathbf{0}_{N_2} & \mathbf{1}_{N_2} \end{bmatrix} \text{ and } \mathbf{Z}_s = \begin{bmatrix} \mathbf{1}_{n_1} & \mathbf{0}_{n_1} \\ \mathbf{0}_{n_2} & \mathbf{1}_{n_2} \end{bmatrix}.$$

Then

$$\begin{aligned}
\mathbf{H}_L &= \left(\mathbf{Z}_s' \mathbf{V}_{ss}^{-1} \mathbf{Z}_s\right)^{-1} \mathbf{Z}_s' \mathbf{V}_{ss}^{-1} \\
&= \left\{ \begin{bmatrix} \mathbf{1}_{n_1}' & \mathbf{0}_{n_2}' \\ \mathbf{0}_{n_1}' & \mathbf{1}_{n_2}' \end{bmatrix} \begin{bmatrix} \sigma_1^{-2} \mathbf{I}_{n_1} & \mathbf{0}_{n_1 \times n_2} \\ \mathbf{0}_{n_2 \times n_1} & \sigma_2^{-2} \mathbf{I}_{n_2} \end{bmatrix} \begin{bmatrix} \mathbf{1}_{n_1} & \mathbf{0}_{n_1} \\ \mathbf{0}_{n_2} & \mathbf{1}_{n_2} \end{bmatrix} \right\}^{-1} \\
&\quad \begin{bmatrix} \mathbf{1}_{n_1}' & \mathbf{0}_{n_2}' \\ \mathbf{0}_{n_1}' & \mathbf{1}_{n_2}' \end{bmatrix} \begin{bmatrix} \sigma_1^{-2} \mathbf{I}_{n_1} & \mathbf{0}_{n_1 \times n_2} \\ \mathbf{0}_{n_2 \times n_1} & \sigma_2^{-2} \mathbf{I}_{n_2} \end{bmatrix} \\
&= \begin{bmatrix} \sigma_1^{-2} \mathbf{1}_{n_1}' \mathbf{1}_{n_1} & 0 \\ 0 & \sigma_2^{-2} \mathbf{1}_{n_2}' \mathbf{1}_{n_2} \end{bmatrix}^{-1} \begin{bmatrix} \sigma_1^{-2} \mathbf{1}_{n_1}' & \mathbf{0}_{n_2}' \\ \mathbf{0}_{n_1}' & \sigma_2^{-2} \mathbf{1}_{n_2}' \end{bmatrix} \\
&= \begin{bmatrix} \sigma_1^2 n_1^{-1} & 0 \\ 0 & \sigma_2^2 n_2^{-1} \end{bmatrix} \begin{bmatrix} \sigma_1^{-2} \mathbf{1}_{n_1}' & \mathbf{0}_{n_2}' \\ \mathbf{0}_{n_1}' & \sigma_2^{-2} \mathbf{1}_{n_2}' \end{bmatrix} \\
&= \begin{bmatrix} n_1^{-1} \mathbf{1}_{n_1}' & \mathbf{0}_{n_2}' \\ \mathbf{0}_{n_1}' & n_2^{-1} \mathbf{1}_{n_2}' \end{bmatrix}.
\end{aligned}$$

All covariances are zero so $\mathbf{V}_{sr} = \mathbf{0}_{n \times (N-n)}$, and (7.7b) becomes

$$\begin{aligned}
\mathbf{w}_s &= \mathbf{1}_n + \left\{ \mathbf{H}_L' \mathbf{Z}_r' + (\mathbf{I}_n - \mathbf{H}_L' \mathbf{Z}_s') \mathbf{V}_{ss}^{-1} \mathbf{V}_{sr} \right\} \mathbf{1}_{N-n} \\
&= \mathbf{1}_n + \mathbf{H}_L' \mathbf{Z}_r' \mathbf{1}_{N-n} \\
&= \mathbf{1}_n + \begin{bmatrix} n_1^{-1} \mathbf{1}_{n_1} & \mathbf{0}_{n_1} \\ \mathbf{0}_{n_2} & n_2^{-1} \mathbf{1}_{n_2} \end{bmatrix} \begin{bmatrix} \mathbf{1}_{N_1-n_1}' & \mathbf{0}_{N_2-n_2}' \\ \mathbf{0}_{N_1-n_1}' & \mathbf{1}_{N_2-n_2}' \end{bmatrix} \begin{bmatrix} \mathbf{1}_{N_1-n_1} \\ \mathbf{1}_{N_2-n_2} \end{bmatrix} \\
&= \begin{bmatrix} \mathbf{1}_{n_1} \\ \mathbf{1}_{n_2} \end{bmatrix} + \begin{bmatrix} n_1^{-1} \mathbf{1}_{n_1} & \mathbf{0}_{n_1} \\ \mathbf{0}_{n_2} & n_2^{-1} \mathbf{1}_{n_2} \end{bmatrix} \begin{bmatrix} N_1 - n_1 \\ N_2 - n_2 \end{bmatrix} \\
&= \begin{bmatrix} \mathbf{1}_{n_1} \\ \mathbf{1}_{n_2} \end{bmatrix} + \begin{bmatrix} \mathbf{1}_{n_1} n_1^{-1} (N_1 - n_1) \\ \mathbf{1}_{n_2} n_2^{-1} (N_2 - n_2) \end{bmatrix} \\
&= \begin{bmatrix} \mathbf{1}_{n_1} (N_1/n_1) \\ \mathbf{1}_{n_2} (N_2/n_2) \end{bmatrix}.
\end{aligned}$$

That is, all sample units in stratum h have the same weight, N_h/n_h.

7.3.3 The Ratio Population Model

The ratio population model (5.2) corresponds to $p = 1$, with $\mathbf{Z}_U$ equal to the vector of population values of the auxiliary size variable and with $\mathbf{V}_U$ equal to the diagonal matrix with i^{th} diagonal component given by z_i. In this case

$$\begin{aligned}
\mathbf{H}_L &= \left(\mathbf{Z}'_s \mathbf{V}_{ss}^{-1} \mathbf{Z}_s\right)^{-1} \mathbf{Z}'_s \mathbf{V}_{ss}^{-1} \\
&= \left\{\begin{pmatrix} z_1 \\ \vdots \\ z_n \end{pmatrix}' diag\left(z_1^{-1}, \ldots, z_n^{-1}\right) \begin{pmatrix} z_1 \\ \vdots \\ z_n \end{pmatrix}\right\}^{-1} \begin{pmatrix} z_1 \\ \vdots \\ z_n \end{pmatrix}' diag\left(z_1^{-1}, \ldots, z_n^{-1}\right) \\
&= \left(\textstyle\sum_s z_i\right)^{-1} \mathbf{1}'_n
\end{aligned}$$

and (7.7b) becomes

$$\begin{aligned}
\mathbf{w}_s &= \mathbf{1}_n + \left\{\mathbf{H}'_L \mathbf{Z}'_r + \left(\mathbf{I}_n - \mathbf{H}'_L \mathbf{Z}'_s\right) \mathbf{V}_{ss}^{-1} \mathbf{V}_{sr}\right\} \mathbf{1}_{N-n} \\
&= \mathbf{1}_n + \mathbf{H}'_L \mathbf{Z}'_r \mathbf{1}_{N-n} \\
&= \mathbf{1}_n + \left(\textstyle\sum_s z_i\right)^{-1} \mathbf{1}_n \left(z_{n+1} \cdots z_N\right) \mathbf{1}_{N-n} \\
&= \left\{1 + \left(\textstyle\sum_s z_i\right)^{-1} \left(\textstyle\sum_r z_i\right)\right\} \mathbf{1}_n \\
&= \left(\frac{\sum_U z_i}{\sum_s z_i}\right) \mathbf{1}_n \\
&= \left(\frac{N}{n}\right)\left(\frac{\bar{z}_U}{\bar{z}_s}\right) \mathbf{1}_n.
\end{aligned}$$

The BLUP for t_y is then $\hat{t}_y^{BLUP} = \mathbf{w}'_s \mathbf{y}_s = \frac{\sum_s y_i}{\sum_s z_i} \sum_U z_i$, that is the ratio estimator.

7.3.4 Other Models

Other models that correspond to cases we have already seen include:

- The linear population model (5.5) corresponds to $p = 2$, with $\mathbf{z}_{1U} = \mathbf{1}_N$, $\mathbf{z}_{2U}$ equal to the vector of population values of the auxiliary size variable, and $\mathbf{V}_U = \mathbf{I}_N$.
- The separate ratio population model (5.12) has $p = H$ parameters, defined by the different slopes in each stratum. This corresponds to $\mathbf{z}_{hU}$ equalling a vector with i^{th} component given by $z_i I(i \in h)$. Similarly, $\mathbf{V}_U$ is a diagonal matrix with i^{th} diagonal component $\sum_h \sigma_h^2 z_i I(i \in h)$.
- The separate linear population model (5.14) has $p = 2H$ parameters, defined by the intercept and slope coefficients for the different strata. There are a number of ways that this model can be parameterised as a special case of (7.5). One of the most popular is the stratum by size interaction specification, where

$$\mathbf{Z}_U = [\mathbf{1}_N, \mathbf{d}_{1U}, \mathbf{d}_{2U}, \ldots, \mathbf{d}_{H-1,U}, \mathbf{z}_U, \mathbf{d}_{1U} \cdot \mathbf{z}_U, \mathbf{d}_{2U} \cdot \mathbf{z}_U, \ldots, \mathbf{d}_{H-1,U} \cdot \mathbf{z}_U].$$

Here $\mathbf{d}_{hU}$ denotes the stratum h indicator vector, that is the vector that contains a value of one for every population unit in stratum h and zero otherwise, and $\mathbf{d}_{hU} \bullet \mathbf{z}_U$ denotes the componentwise 'product' of this indicator with the auxiliary size vector $\mathbf{z}_U$. The matrix $\mathbf{V}_U$ in this case is the same as in the homogeneous strata population model.
- The clustered population model (6.1) has $p = 1$ with $\mathbf{z}_{1U} = \mathbf{1}_N$ and $\mathbf{V}_U$ is the block diagonal matrix $\mathbf{V}_U = diag(\mathbf{V}_{Ug} : g = 1, \ldots, Q)$ where $\mathbf{V}_{Ug}$ is the $M_g \times M_g$ matrix with diagonal elements equal to 1, and non-diagonal elements all equal to ρ. It is sometimes convenient to write this as $\mathbf{V}_{Ug} = (1 - \rho)\mathbf{I}_{M_g} + \rho \mathbf{1}_{M_g}\mathbf{1}'_{M_g}$. The sample component $\mathbf{V}_{ss}$ is defined similarly. Then (7.7) reduces to the BLUP derived in Chapter 6, that is (6.4) and (6.5).

7.4 Model Choice

The real strength of the general linear model and its associated BLUP is that we can create models for survey variables and derive associated efficient predictors that do not correspond to any of the special cases that we have seen so far.

One example of such a model is the *marginal post-stratification model*. Suppose that our auxiliary data consists of two factors (categorical variables) A and B, with J and K levels respectively. We could use a model where every category defined by the cross-classification of A and B is a post-stratum. This is a type of linear model and can be written using the usual regression shorthand as $Y = A^*B$, that is the saturated model with strata defined by the complete set of interactions between A and B. However, because these strata are imposed after the sample has been selected, some of the combinations making up A^*B might have little or no sample. In this case we might prefer to use a main effects only model, that is $Y = A + B$. This corresponds to setting

$$\mathbf{z}'_i = (I(A_i = 1) \cdots I(A_i = J) I(B_i = 2) \cdots I(B_i = K))$$

(The first level of B is dropped from the definition of $\mathbf{z}_i$ so that the model is of full rank.) More than two factors could be used if desired.

A mixture of continuous variables and factors could also be used. Powers or logarithms of continuous variables could be included to model curvature in the relationship between the values $\mathbf{y}_U$ and the auxiliary variables that define the columns of $\mathbf{Z}_U$. Some of the issues that should be considered in choosing an appropriate $\mathbf{Z}_U$ are:

- The linear model defined by $\mathbf{Z}_U$ should fit the sample data. The usual steps in fitting a linear model to sample data apply, including exploratory data analysis (e.g. scatterplot matrices), residuals plots, and use of significance tests or other model fitting diagnostics such as the Akaike Information Criterion

(AIC). The best model could depend on the particular Y-variable being considered. In practice, most surveys are multipurpose with many Y-variables, in which case the usual practice is to calculate a single set of weights for all Y-variables. This convenient and almost universal practice has the benefit of consistency for the different Y-variables; for example, if one Y-variable is by definition always less than another, this relationship will be preserved for all estimates of domain totals. It is in principle possible to use different weights for different variables and this would sometimes give improvements in efficiency. If a single set of weights is to be used, the $\mathbf{Z}_U$ chosen will be a compromise between the best set of explanatory variables for the different Y-variables.

- Auxiliary variables thought to be associated with the sampling process (including non-response) should be included in $\mathbf{Z}_U$ by default, since this can help to ensure the validity of the non-informative sampling assumption. For example, young men have lower response rates than other population groups in many household surveys, suggesting that one should include variables corresponding to age and gender in $\mathbf{Z}_U$.

- Not too many auxiliary variables should be included in $\mathbf{Z}_U$, since inclusion of unnecessary columns in $\mathbf{Z}_U$ can lead to instability in the BLUP weights (7.7b), with prediction variances potentially higher than if fewer auxiliary variables were used in this model matrix. There are no hard and fast rules on this, but one suggestion is that p/n should not be too large (say $p \leq n/10$) and $n - p$ should not be too small (say $n - p \geq 30$, to allow for variance estimation). We discuss this issue further in Chapter 13.

7.5 Optimal Sample Design

Finally, we comment on optimal sample design under the general linear population model. The prediction variance (7.8) depends on the particular values of $\mathbf{Z}_U$ and $\mathbf{V}_U$ that hold for the population. That is, it is specific to a particular Y-variable. Most surveys are multipurpose, with many variables of interest and many different target parameters, in which case the sample design that is optimal for one may be very different from the design that is optimal for another. Sometimes, it is possible to identify a few key variables and construct designs that are optimal for each, then 'average' these designs. If these variables all satisfy (7.5), that is the same $\mathbf{Z}_U$ and the same $\mathbf{V}_U$, with the only differences between them being different values of β and σ^2, then it is conceptually possible to search for an optimal sample s that minimises (7.8) for a given coefficient vector $\mathbf{a}_U$. Ideally, of course, one would want a design that worked for many different values of $\mathbf{a}_U$, corresponding to different target parameters, in which case the optimal design problem becomes one of choosing s in order to make the $(N-n) \times (N-n)$ matrix within the parentheses on the right hand side of (7.8) as 'small' as possible. This type of problem has been explored in the experimental design literature, where 'small' has usually been interpreted as corresponding to a small trace or a small determinant, but not with the type of $\mathbf{Z}_U$ and $\mathbf{V}_U$ matrices observed in survey practice, and almost always assuming uncorrelated population elements. However, so far there has been little similar development in the survey sampling literature, to a large extent because 'big p' $\mathbf{Z}_U$ matrices are still relatively rare in survey applications. Most survey design studies tend to focus on choice of stratification variables (i.e. specification of $\mathbf{Z}_U$). For each key variable, the sample is allocated between the different strata using the optimal allocation rules discussed in Chapter 4. These initial allocations are suitably averaged to provide the final design.

7.6 Derivation of BLUP Weights

We obtain (7.7b) for the important special case $\mathbf{a}_U = \mathbf{1}_N$. First, we note that in this case (7.7b) can be equivalently written

$$\begin{aligned}\mathbf{w}_s &= \mathbf{1}_n + \mathbf{V}_{ss}^{-1}\mathbf{Z}_s\left(\mathbf{Z}_s'\mathbf{V}_{ss}^{-1}\mathbf{Z}_s\right)^{-1}\mathbf{Z}_r'\mathbf{1}_{N-n} \\ &+ \left\{\mathbf{I}_n - \mathbf{V}_{ss}^{-1}\mathbf{Z}_s\left(\mathbf{Z}_s'\mathbf{V}_{ss}^{-1}\mathbf{Z}_s\right)^{-1}\mathbf{Z}_s'\right\}\mathbf{V}_{ss}^{-1}\mathbf{V}_{sr}\mathbf{1}_{N-n} \\ &= \mathbf{1}_n + \mathbf{V}_{ss}^{-1}\left\{\mathbf{V}_{sr}\mathbf{1}_{N-n} - \mathbf{Z}_s\left(\mathbf{Z}_s'\mathbf{V}_{ss}^{-1}\mathbf{Z}_s\right)^{-1}\left(\mathbf{Z}_s'\mathbf{V}_{ss}^{-1}\mathbf{V}_{sr}\mathbf{1}_{N-n} - \mathbf{Z}_r'\mathbf{1}_{N-n}\right)\right\}.\end{aligned}$$

Let $\mathbf{w}_s = \mathbf{1}_n + \mathbf{u}_s$. By definition, the BLUP weights must minimise $E\left(\mathbf{w}_s'\mathbf{y}_s - \mathbf{1}_N'\mathbf{y}_U\right)^2$ subject to $E\left(\mathbf{w}_s'\mathbf{y}_s - \mathbf{1}_N'\mathbf{y}_U\right) = 0$, or equivalently, they minimise $E\left(\mathbf{u}_s'\mathbf{y}_s - \mathbf{1}_{N-n}'\mathbf{y}_r\right)^2$ subject to $\mathbf{Z}_s'\mathbf{u}_s = \mathbf{Z}_r'\mathbf{1}_{N-n}$. Now

$$\begin{aligned}E\left(\mathbf{u}_s'\mathbf{y}_s - \mathbf{1}_{N-n}'\mathbf{y}_r\right)^2 &= E\left(\mathbf{u}_s'\mathbf{y}_s - \mathbf{1}_{N-n}'\mathbf{y}_r\right)\left(\mathbf{y}_s'\mathbf{u}_s - \mathbf{y}_r'\mathbf{1}_{N-n}\right) \\ &= E\left(\mathbf{u}_s'\mathbf{y}_s\mathbf{y}_s'\mathbf{u}_s - 2\mathbf{u}_s'\mathbf{y}_s\mathbf{y}_r'\mathbf{1}_{N-n} + \mathbf{1}_{N-n}'\mathbf{y}_r\mathbf{y}_r'\mathbf{1}_{N-n}\right) \\ &= \mathbf{u}_s'E\left(\mathbf{y}_s\mathbf{y}_s'\right)\mathbf{u}_s - 2\mathbf{u}_s'E\left(\mathbf{y}_s\mathbf{y}_r'\right)\mathbf{1}_{N-n} + \mathbf{1}_{N-n}'E\left(\mathbf{y}_r\mathbf{y}_r'\right)\mathbf{1}_{N-n} \\ &= \mathbf{u}_s'\left(\mathbf{V}_{ss} + \mathbf{Z}_s\beta\beta'\mathbf{Z}_s'\right)\mathbf{u}_s - 2\mathbf{u}_s'\left(\mathbf{V}_{sr} + \mathbf{Z}_s\beta\beta'\mathbf{Z}_r'\right)\mathbf{1}_{N-n} \\ &\quad + \mathbf{1}_{N-n}'\left(\mathbf{V}_{rr} + \mathbf{Z}_r\beta\beta'\mathbf{Z}_r'\right)\mathbf{1}_{N-n}.\end{aligned}$$

The terms involving β vanish because of the condition $\mathbf{Z}_s'\mathbf{u}_s = \mathbf{Z}_r'\mathbf{1}_{N-n}$, so that

$$E\left(\mathbf{u}_s'\mathbf{y}_s - \mathbf{1}_{N-n}'\mathbf{y}_r\right)^2 = \mathbf{u}_s'\mathbf{V}_{ss}\mathbf{u}_s - 2\mathbf{u}_s'\mathbf{V}_{sr}\mathbf{1}_{N-n} + \mathbf{1}_{N-n}'\mathbf{V}_{rr}\mathbf{1}_{N-n}.$$

We wish to minimise this prediction variance subject to $\mathbf{Z}_s'\mathbf{u}_s = \mathbf{Z}_r'\mathbf{1}_{N-n}$. The Lagrangian function for this is

$$L = \mathbf{u}_s'\mathbf{V}_{ss}\mathbf{u}_s - 2\mathbf{u}_s'\mathbf{V}_{sr}\mathbf{1}_{N-n} + 2\lambda'\left(\mathbf{Z}_s'\mathbf{u}_s - \mathbf{Z}_r'\mathbf{1}_{N-n}\right)$$

so differentiating we obtain $\frac{dL}{d\mathbf{u}_s} = 2\mathbf{V}_{ss}\mathbf{u}_s - 2\mathbf{V}_{sr}\mathbf{1}_r + 2\mathbf{Z}_s\lambda$. Setting this expression to zero and solving for $\mathbf{u}_s$, we obtain $\mathbf{u}_s = \mathbf{V}_{ss}^{-1}(\mathbf{V}_{sr}\mathbf{1}_{N-n} - \mathbf{Z}_s\lambda)$.

Pre-multiplying both sides by $\mathbf{Z}'_s$ and making use of the unbiasedness constraint yields $\mathbf{Z}'_s \mathbf{u}_s = \mathbf{Z}'_s \mathbf{V}_{ss}^{-1}(\mathbf{V}_{sr}\mathbf{1}_{N-n} - \mathbf{Z}_s \lambda) = \mathbf{Z}'_r \mathbf{1}_{N-n}$, so that

$$\lambda = \left(\mathbf{Z}'_s \mathbf{V}_{ss}^{-1} \mathbf{Z}_s\right)^{-1} \left(\mathbf{Z}'_s \mathbf{V}_{ss}^{-1} \mathbf{V}_{sr} \mathbf{1}_{N-n} - \mathbf{Z}'_r \mathbf{1}_{N-n}\right).$$

Replacing λ in the optimal value for $\mathbf{w}_s$ gives the required result

$$\mathbf{u}_s = \mathbf{V}_{ss}^{-1} \left\{ \mathbf{V}_{sr} \mathbf{1}_{N-n} - \mathbf{Z}_s \left(\mathbf{Z}'_s \mathbf{V}_{ss}^{-1} \mathbf{Z}_s\right)^{-1} \left(\mathbf{Z}'_s \mathbf{V}_{ss}^{-1} \mathbf{V}_{sr} \mathbf{1}_{N-n} - \mathbf{Z}'_r \mathbf{1}_{N-n}\right) \right\}.$$

PART II
Robust Model-Based Survey Methods

In the real world, models only approximate reality. They are never literally true. However, the optimal results developed in the previous chapters depended on the assumed model actually describing the population of interest. When we take a model-based approach to survey inference it is therefore crucial that we first check to see whether our inference is sensitive to misspecification of this model. In particular, if this inference remains valid, or approximately valid, for reasonable alternative model specifications for the population and sample data, then we say that it is *robust*. If not, we must be cautious, since sensitivity to model misspecification could make our inference seriously misleading.

When we assume a particular model for a population in order to make some inference about it (e.g. an inference about the population total t_y of a variable Y defined on that population), we are essentially 'working' with that model. Consequently we will refer to an assumed model as a *working* model below in order to emphasise that such a model is really a tool we use to guide our inference.

There are two common ways in which misspecification of a working model can occur. The first is what we may think of as global misspecification. That is, suppose we assume that the population follows a particular model (which we denote by ξ), but in reality another model (which we denote by η) describes it better. Our inference is therefore based on assumptions that do not 'fit' reality. In effect, our working model ξ is incorrect. Questions that we could ask ourselves in this case are:

- Is our (ξ-optimal) estimator still unbiased under η?
- Is it still efficient under η?
- Are our (ξ-based) prediction intervals still valid?

The other type of misspecification occurs when a few sample data values clearly do not follow the working model. We can think of this as local misspecification, in that it seems reasonable to assume that the majority of the population follow the working model, but this model does not properly describe the entire population. The evidence for this is the so-called *outliers* (relative to the working model) that we see in the sample data. In many cases these outliers are so extreme that ignoring them and proceeding on the basis that the working model ξ applies to the entire population can lead to seriously biased inference. Unfortunately, however, there are too few of these values in our sample to allow us to modify ξ in order to accommodate them. In this case the question we should ask ourselves is:

- How should we modify our (ξ-optimal) inference in order to make it robust to these outlying values?

In the following three chapters we explore methods for protecting inferences based on standard population models against both these types of misspecification. In Chapters 8 and 9 we focus on robust prediction and robust estimation of prediction variance under global misspecification, while in Chapter 10 we consider outlier robustness issues. Note that because we will be considering statistical properties (bias, variance) with respect to both the assumed working model as well as the 'true' model, we will often use a subscript to identify the model underpinning the property. Thus, E_ξ and Var_ξ will denote expectation and variance assuming the working model is true, while E_η and Var_η will denote expectation and variance with respect to an alternative (and better) model for the population. The extension of this notation to other statistical measures that depend on a model for the population is straightforward.

8 Robust Prediction Under Model Misspecification

8.1 Robustness and the Homogeneous Population Model

Suppose our working model ξ is the second order homogeneous population model defined by (3.1), but an alternative model actually characterises the population values. Is the BLUP $\hat{t}_y^E$ under (3.1), that is the expansion estimator given by (3.2), still unbiased and optimal under this alternative model? To answer this question we first need to specify the alternative model (or models). Conceivably the simplest alternative to ξ is a model where the population units have different variances, but the assumption of a common mean is retained. Denoting this alternative model by η, we have

$$E_\eta(y_i) = \mu \qquad (8.1a)$$

$$Var_\eta(y_i) = \sigma_i^2 \qquad (8.1b)$$

$$Cov_\eta(y_i, y_j) = 0, \text{ for } i \neq j. \qquad (8.1c)$$

Since the population Y-values still have a common mean (μ) under η, it is easy to see that $\hat{t}_y^E = N/n \sum_s y_i$ is unbiased for t_y under η. However, it is a different story as far as variance estimation is concerned. To start, we note that the actual prediction variance of $\hat{t}_y^E$ under η is

$$Var_\eta\left(\hat{t}_y^E - t_y\right) = (N-n)^2 \left\{ n^{-2} \sum_s \sigma_i^2 + (N-n)^{-2} \sum_r \sigma_i^2 \right\}. \qquad (8.2)$$

We can rewrite this as

$$Var_\eta\left(\hat{t}_y^E - t_y\right) = N^2 n^{-1}(1 - n/N)\left\{\bar{\sigma}_s^2 + \frac{n}{N}\left(\bar{\sigma}_r^2 - \bar{\sigma}_s^2\right)\right\} \qquad (8.3)$$

where $\bar{\sigma}_s^2 = n^{-1}\sum_s \sigma_i^2$ is the mean of the σ_i^2 for the sampled units, and $\bar{\sigma}_r^2 = (N-n)^{-1}\sum_r \sigma_i^2$ is the corresponding mean for the non-sampled units. An unbiased estimator of the prediction variance of $\hat{t}_y^E$ under the working model ξ is given by (3.6):

$$\hat{V}\left(\hat{t}_y^E\right) = \frac{N^2}{n}\left(1 - \frac{n}{N}\right)s_y^2$$

where $s_y^2 = (n-1)^{-1}\sum_s (y_i - \bar{y}_s)^2$ is the sample variance of Y. Is this estimator also unbiased for the prediction variance of $\hat{t}_y^E$ under η? Unfortunately the answer is no. It can be shown that the expected value of $\hat{V}\left(\hat{t}_y^E\right)$ under η is

$$E_\eta\left(\hat{V}\left(\hat{t}_y^E\right)\right) = (N^2/n)(1-n/N)n^{-1}\sum_s \sigma_i^2 = (N^2/n)(1-n/N)\bar{\sigma}_s^2, \quad (8.4)$$

which is not the same as (8.3), although the difference is small if the sampling fraction n/N is small, as is often the case.

Why is this important? Suppose that we wish to construct a prediction interval for t_y using $\hat{t}_y^E$ and $\hat{V}\left(\hat{t}_y^E\right)$. The validity of this prediction interval requires that the large sample distribution of the statistic

$$\left(\hat{t}_y^E - t_y\right)/\sqrt{\hat{V}\left(\hat{t}_y^E\right)}$$

is at least approximately normal with mean zero and unit standard deviation. However, the variance of the numerator of this statistic should match the expected value of its squared denominator if it is to have this large sample distribution. See, for example, Cressie (1982). What we have shown above is that this is not true when η holds.

There are two basic approaches we can take to resolve this situation. The first is to not use the usual unbiased variance estimator under the working model ξ, but instead develop an alternative estimator of the prediction variance of $\hat{t}_y^E$ that is at least approximately unbiased under η. Since ξ is a special case of η, such an estimator can be constructed to be exactly unbiased if ξ is in fact true. We consider this approach in Chapter 9. The second approach is to select a sample in such a way that the bias in $\hat{V}\left(\hat{t}_y^E\right)$ is zero (or approximately zero) under η and use the prediction interval based on $\hat{t}_y^E$ and $\hat{V}\left(\hat{t}_y^E\right)$ with this specially selected sample. This is the approach we now consider.

8.1.1 Safe Sampling for the Homogeneous Population Model

Comparing (8.3) to (8.4), it is clear that the bias is approximately zero if either the sampling fraction is very small, or if the sample is selected in such a way that $\bar{\sigma}_s^2 \cong \bar{\sigma}_r^2$. In this case, we can say that our inference is 'safe'.

Clearly we cannot choose the sample to exactly satisfy $\bar{\sigma}_s^2 = \bar{\sigma}_r^2$ if we do not know the σ_i. However, intuitively we expect that a large sample selected via *simple random sampling* (SRS) will be just as likely to have $\bar{\sigma}_s^2$ less than $\bar{\sigma}_r^2$ as the other way around. That is, 'on average' $\bar{\sigma}_s^2 = \bar{\sigma}_r^2$ will be at least approximately true. Consequently, provided the sample size is large, SRS represents a safe sampling strategy for prediction interval estimation under the model η. Even if we do not use SRS, if the sampling fraction n/N is small, our inferences will be safe anyway, as we can see by comparing (8.3) and (8.4).

Note that this does not mean that SRS is the only safe sampling strategy in this situation. Again, depending upon the information we have regarding the possible spread of values of the σ_i in the population, we may be able to specify other sampling schemes that also make for safe inference. For example, if there is an auxiliary variable Z which we believe might be related to σ_i, then an alternative approach is to order the population units according to these Z-values and then sample by selecting every k^{th} population unit on this ordered list, where k is the integer part of N/n. Such an *ordered systematic sample* also stands a good chance of satisfying (8.3).

8.1.2 Robustness of Optimality of the Expansion Estimator

What about the optimality of the expansion estimator $\hat{t}_y^E$ under η? In Chapter 3 we proved that $\hat{t}_y^E$ is the BLUP under the working model ξ. However, although $\hat{t}_y^E$ is unbiased under η, it is not true that it is generally also the BLUP under η. We denote this different η-BLUP by $\hat{t}_{\eta y}$. What is the efficiency loss from using $\hat{t}_y^E$ instead of $\hat{t}_{\eta y}$ when η is true?

In order to answer this question we need to compare $Var_\eta(\hat{t}_{\eta y} - t_y)$ with $Var_\eta(\hat{t}_y^E - t_y)$. Firstly, notice that

$$\hat{t}_{\eta y} = \sum_s y_i + (N-n)\left(\sum_s y_i \sigma_i^{-2}\right)\left(\sum_s \sigma_i^{-2}\right)^{-1}$$

with

$$Var_\eta(\hat{t}_{\eta y} - t_y) = (N-n)^2 \left(\sum_s \sigma_i^{-2}\right)^{-1} + \sum_r \sigma_i^2.$$

In contrast, from (8.2)

$$Var_\eta(\hat{t}_y^E - t_y) = ((N-n)^2/n^2)\sum_s \sigma_i^2 + \sum_r \sigma_i^2.$$

The gain in precision from using the η-BLUP $\hat{t}_{\eta y}$ instead of the ξ-BLUP $\hat{t}_y^E$ is the difference between $Var_\eta(\hat{t}_{\eta y} - t_y)$ and $Var_\eta(\hat{t}_y^E - t_y)$, which is

$$((N-n)^2/n)\left[n^{-1}\sum_s \sigma_i^2 - \left(n^{-1}\sum_s \sigma_i^{-2}\right)^{-1}\right].$$

The term in the square brackets is the difference between the arithmetic mean and the harmonic mean of the values of σ_i^2 for the sample units. This difference is always greater than or equal to zero, with equality holding only when σ_i^2 are constant.

Another question is: what is the optimal sample design for $\hat{t}_{\eta y}$ under model η? If N is much larger than n, then

$$Var_\eta(\hat{t}_{\eta y} - t_y) = (N-n)^2 \left(\sum_s \sigma_i^{-2}\right)^{-1} + \sum_r \sigma_i^2 \approx (N-n)^2 \left(\sum_s \sigma_i^{-2}\right)^{-1}$$

which is minimised when s contains those units with the smallest values of σ_i. We would not be able to select this sample in practice, as we would not know σ_i, but it is still of interest to see how much we lose by not being able to use this design. This extreme sample will usually be quite different from the 'σ-balanced' sample defined by (8.3) that guarantees the validity of prediction intervals based on $\hat{t}_y^E$ and $\hat{V}\left(\hat{t}_y^E\right)$. Consequently, using such a safe sample and estimating t_y by $\hat{t}_y^E$ does not lead to robustness of optimality for this estimator. This is an example of the *insurance premium* one typically has to pay for using a robust approach. Such premiums also occur with other population models and estimators, as we shall see in the following sections.

8.2 Robustness and the Ratio Population Model

8.2.1 Robustness to Non-zero Intercept

We now turn to misspecification issues that arise in populations where a regression model is assumed. In particular, suppose that we have an auxiliary variable Z and the ratio population model (5.2). As shown in Section 5.3, the optimal sampling strategy for minimising the prediction variance of the BLUP (the ratio estimator) in this case is to select the extreme sample consisting of the n population units with largest values of Z.

In practice, however, such extreme samples are hardly ever chosen. The reason for this is easy to understand – what if the assumed model is wrong? As it stands, however, this is not really an adequate reason for not adopting this optimal approach. Models are always wrong, since they only approximate reality. The real issue is the sensitivity of the above optimal strategy to misspecification of the model.

In order to demonstrate this sensitivity, suppose that our working model, ξ, is the ratio model, (5.2), while the true model, η, is similar but also includes an intercept:

$$E_\eta(y_i|z_i) = \alpha + \beta z_i \tag{8.5a}$$

$$Var_\eta(y_i|z_i) = \sigma^2 z_i \tag{8.5b}$$

$$Cov_\eta(y_i, y_j|z_i, z_j) = 0 \text{ for all } i \neq j \tag{8.5c}$$

The ξ–BLUP is the ratio estimator, defined in (5.4): $\hat{t}_y^R = (\bar{y}_s/\bar{z}_s)t_z = (N\bar{z}_U/n\bar{z}_s)\sum_s y_i$. What are the properties of this estimator when η is true? The expected value of the estimator under η is:

$$\begin{aligned}
E_\eta\left(\hat{t}_y^R - t_y\right) &= E_\eta\left(\frac{N\bar{z}_U}{n\bar{z}_s}\sum_s y_i - \sum_U y_i\right) \\
&= \frac{N\bar{z}_U}{n\bar{z}_s}\sum_s(\alpha + \beta z_i) - \sum_U(\alpha + \beta z_i) \\
&= \frac{N\bar{z}_U}{n\bar{z}_s}(n\alpha + \beta n\bar{z}_s) - (N\alpha + \beta N\bar{z}_U) \\
&= \alpha(N - n)\left(\frac{\bar{z}_r}{\bar{z}_s} - 1\right).
\end{aligned}$$

For $\alpha > 0$ this bias will be negative (and large in absolute value) when s is the extreme sample described above. Hence adopting the optimal strategy under the ratio population model leaves us vulnerable to a possible large bias if in fact the population regression function contains an intercept.

It is easy to see that the preceding expression for the bias in the ratio estimator induced by the presence of an intercept in the population regression function is zero when the sample s satisfies $\bar{z}_s = \bar{z}_U$, since then $\bar{z}_s = \bar{z}_r$. That is, selecting a sample that is first order balanced on the auxiliary variable Z represents a robust sampling strategy for the ratio estimator if we have reason to believe that the regression of Y on Z may not go through the origin.

An interesting feature of first order balanced sampling is that the ratio estimator simplifies to

$$\hat{t}_y^R = (\bar{y}_s/\bar{z}_s)\, t_z = (\bar{y}_s/\bar{z}_s) N\, \bar{z}_U = N \bar{y}_s$$

which is the expansion estimator. In other words, under first order balancing, not only is the ratio estimator robust to a non-zero intercept, but even the simple expansion estimator is equivalent to the ratio estimator and so also unbiased under η. Moreover, the η-BLUP can also be shown (Royall and Herson, 1973a) to be equal to the expansion estimator in this case. So, provided we select such a balanced sample, the homogenous population model (3.1), the ratio population model (5.2) and model (8.5) all have the same BLUP.

8.2.2 Robustness to Polynomial Terms

Suppose that our working model is still the ratio model, but that the true model η involves a polynomial of order K:

$$E_\eta(y_i|z_i) = \beta_0 + \beta_1 z_i + \beta_2 z_i^2 + \ldots + \beta_K z_i^K \tag{8.6a}$$

$$Var_\eta(y_i|z_i) = \sigma^2 z_i \tag{8.6b}$$

$$Cov_\eta(y_i, y_j|z_i, z_j) = 0 \text{ for all } i \neq j \tag{8.6c}$$

How does the ratio estimator perform under this model? We will use the notation

$$\overline{(z^k)}_s = n^{-1} \sum_s z_i^k$$
$$\overline{(z^k)}_U = N^{-1} \sum_U z_i^k$$

for the kth sample and population moment of Z, respectively. The expected value of the ratio estimator under model η is:

$$\begin{aligned}
E_\eta\left(\hat{t}_y^R - t_y\right) &= E_\eta\left[\frac{N\bar{z}_U}{n\bar{z}_s} \sum_s y_i - \sum_U y_i\right] \\
&= \frac{N\bar{z}_U}{n\bar{z}_s} \sum_s \left(\beta_0 + \beta_1 z_i + \ldots + \beta_K z_i^K\right) \\
&\quad - \sum_U \left(\beta_0 + \beta_1 z_i + \ldots + \beta_K z_i^K\right) \\
&= \frac{N\bar{z}_U}{\bar{z}_s} \left\{\beta_0 + \beta_1 \bar{z}_s + \beta_2 \overline{(z^2)}_s \ldots + \beta_K \overline{(z^K)}_s\right\} \\
&\quad - N \left\{\beta_0 + \beta_1 \bar{z}_U + \beta_2 \overline{(z^2)}_U \ldots + \beta_K \overline{(z^K)}_U\right\} \\
&= N\left[\beta_0 \left(\frac{\bar{z}_U}{\bar{z}_s} - 1\right) + \beta_2 \left\{\frac{\bar{z}_U}{\bar{z}_s}\overline{(z^2)}_s - \overline{(z^2)}_U\right\}\right. \\
&\quad \left. + \ldots + \beta_K \left\{\frac{\bar{z}_U}{\bar{z}_s}\overline{(z^K)}_s - \overline{(z^K)}_U\right\}\right].
\end{aligned}$$

It is clear that a sampling scheme such that $\overline{(z^k)}_s = \overline{(z^k)}_U$ for $k = 1, \ldots, K$ will ensure that the bias is zero. We call any sampling method with this property *balanced sampling of order K*. The sample is chosen such that it is consistent with the population in terms of the first K moments of Z. If we are using the ratio model, but are concerned that the true model might be a polynomial of order up to K, then we can balance the sample to K^{th} order, and be confident that the ratio estimator will be unbiased.

It can be shown (Royall and Herson, 1973a) that if balanced sampling of order K is used, the simple expansion estimator, the ratio estimator and the η-BLUP for model (8.6) will all be equal to $\hat{t}_y^E = N\bar{y}_s$.

8.2.3 The Downside of Balanced Sampling

Balanced sampling is not cost free. When the sample is first or higher order balanced, $\bar{z}_s = \bar{z}_U$, and the prediction variance (5.7) of the ratio estimator becomes

$$Var_\xi \left(\hat{t}_y^R - t_y \,|\, \text{balance} \right) = \sigma^2 (N-n)(N/n)\bar{z}_U$$

when the working ratio population model ξ is true. In contrast, under the extreme sample consisting of the n population elements with largest values of Z and this model,

$$Var_\xi \left(\hat{t}_y^R - t_y \,|\, \text{extreme} \right) = \min_s \left\{ \sigma^2 (N-n)(N/n)\bar{z}_U (\bar{z}_r / \bar{z}_s) \right\}.$$

The relative efficiency of using a balanced sample instead of the extreme sample when the working model ξ is correct is therefore

$$\frac{Var_\xi \left(\hat{t}_y^R - t_y \,|\, \text{extreme} \right)}{Var_\xi \left(\hat{t}_y^R - t_y \,|\, \text{balance} \right)} = \min_s (\bar{z}_r / \bar{z}_s) = (N-n)^{-1} \left\{ N \frac{\bar{z}_U}{\max_s (\bar{z}_s)} - n \right\}.$$

If the population distribution of Z is highly skewed, $\bar{z}_U \ll \max_s \bar{z}_s$, and this ratio can be very small. Adopting a balanced sampling strategy can thus lead to a large loss of efficiency when the working model ξ is in fact correct. That is, as we found out in Section 8.1, robustness to model misspecification can have a large insurance premium in terms of efficiency loss.

8.2.4 Improving on Balanced Sampling using Stratification

Balanced sampling is inefficient if the working model (the ratio population model) actually holds. Given this, can we find alternative sampling and estimation strategies that have similar robustness to model misspecification as balanced sampling and ratio estimation, but which are more efficient under a ratio population model? Two approaches one can take are:

1. Select a *stratified balanced sample* and use the separate ratio estimator (5.13), which in this case corresponds to the stratified expansion estimator. Provided

the strata are chosen appropriately and optimal allocation is used, Royall and Herson (1973b) show that this strategy has a smaller prediction variance than unstratified balanced sampling and ratio estimation under the ratio population model, as well as superior robustness.

2. Select what is called an *overbalanced sample* and carry out prediction using the BLUP under a working model where $E_\xi(y_i|z_i) = \beta z_i$ and $Var_\xi(y_i|z_i) \propto z_i^2$ rather than $Var_\xi(y_i|z_i) \propto z_i$. This strategy has been shown by Scott et al. (1978) to also lead to a smaller prediction variance than balanced sampling and ratio estimation under the ratio population model.

We will develop the first strategy in this subsection, and the second in Subsection 8.2.5. All results are given without proof, and the working model ξ is the ratio population model. Our development is based on Royall and Herson (1973b). A third approach suggested by Valliant et al. (2000, Section 3.3) is based on a model with $E_\xi(y_i|z_i) = \beta_1 z_i^{1/2} + \beta_2 z_i$ and $Var_\xi(y_i|z_i) \propto z_i$, with the sample balanced on $\sqrt{z}$.

The population is divided into strata defined by ranges of Z, and a balanced sample is selected within each stratum, so that $\bar{z}_{sh} = \bar{z}_h$. The separate ratio estimator, defined by (5.13), is used. Its prediction variance under the ratio population model ξ is given by:

$$Var_\xi \left(\hat{t}_y^{SR} - t_y \right) = \sigma^2 \sum_h \left(N_h^2 / n_h \right) \{1 - (n_h/N_h)\} \bar{z}_{sh}.$$

This is minimised subject to an overall sample size of n when the sample stratum allocation is proportional to $N_h \sqrt{\bar{z}_{sh}}$, in which case

$$Var_\xi \left(\hat{t}_y^{SR} - t_y \mid n_h \propto N_h \sqrt{\bar{z}_{sh}} \right) \sigma^2 \left\{ n^{-1} \left(\sum_h N_h \sqrt{\bar{z}_{sh}} \right)^2 - N \bar{z}_U \right\}.$$

This is always less than or equal to the prediction variance of the ratio estimator under the ratio population model and simple balanced sampling.

The strategy of stratified balanced sampling and the separate ratio estimator is *qualitatively* more robust than the strategy consisting of simple balanced sampling and the ordinary ratio estimator, since the former can accommodate alternative models for the population data where regression coefficients vary between the strata. The real world is highly unlikely to be exactly linear or even polynomial, and the separate ratio estimator can more closely approximate underlying non-linear behaviour in the population.

The prediction variance of the separate ratio estimator under the ratio population model ξ can always be made smaller by increasing the number of strata. However, selection of a balanced sample within each stratum requires the stratum sample sizes to be large enough to enable such a sample to be selected in the first place. Reconciling these two aims points to the desirability of equal allocation across the strata. This is achieved by stratifying such that t_{zh} are equal

8.2.5 Improving on Balanced Sampling using Overbalancing

Another robust strategy that improves on simple balanced sampling and ratio estimation is the overbalanced strategy developed by Scott et al. (1978). Let

$$\hat{t}_y^R(\gamma) = \sum_s y_i + \frac{\sum_s y_i z_i^{1-2\gamma}}{\sum_s z_i^{2-2\gamma}} \sum_r z_i \qquad (8.7)$$

denote the BLUP for t_y under a working model ξ corresponding to the gamma population model (5.1), where $E_\xi(y_i|z_i) = \beta z_i$ and $Var_\xi(y_i|z_i) \propto z_i^{2\gamma}$. See exercise E.14. It can then be shown that this estimator remains unbiased for t_y under the alternative model η where $E_\eta(y_i|z_i) = \sum_{k=0}^{K} \beta_k z_i^k$ when the sample is γ-balanced up to order K. That is, for $k = 0, 1, \ldots, K$,

$$\left(\overline{z^k z^{1-2\gamma}}\right)_s \Big/ \left(\overline{z^{2-2\gamma}}\right)_s = \left(\overline{z^k}\right)_r \Big/ \bar{z}_r. \qquad (8.8)$$

Note that $\gamma = 0.5$ corresponds to the simple ratio estimator and (simple) balancing up to order K. Also, under γ-balancing with $K \geq 0$, the ξ-BLUP reduces to

$$\hat{t}_y^R(\gamma) = \sum_s y_i + (N-n) \sum_s w_i(\gamma) y_i \Big/ \sum_s w_i(\gamma)$$

where $w_i(\gamma) = z_i^{1-2\gamma}$.

The case of $\gamma = 1$ is of particular interest, and Scott et al. (1978) termed a γ-balanced sample with $\gamma = 1$ an *overbalanced* sample. Under overbalanced sampling the BLUP (8.7) becomes:

$$\hat{t}_y^R(\gamma) = \sum_s y_i + n^{-1} \left(\sum_s y_i/z_i\right) \left(\sum_r z_i\right). \qquad (8.9)$$

and condition (8.8) becomes

$$\left(\overline{z^{k-1}}\right)_s = \left(\overline{z^k}\right)_r \Big/ \bar{z}_r \qquad (8.10)$$

for $k = 0, 1, \ldots, K$. A sampling method that selects population units with probability proportional to the auxiliary size variable Z will tend to result in samples that are approximately overbalanced. The overbalancing strategy defined by (8.9) and (8.10) has the remarkable property that it gives lower prediction variance than using the ratio estimator and balanced sampling, *even when the ratio population model applies*, that is when $\gamma = 0.5$, although the improvement is usually small in this case. Overbalancing gives more substantial gains when the variance is given by

$$Var_\eta(y_i|z_i) \propto az_i + bz_i^2. \qquad (8.11)$$

with both a and b non-zero.

In summary, the overbalancing strategy (8.9) and (8.10) is an alternative way of achieving robustness to possible polynomial terms in the model. It is slightly more efficient than ratio estimation with balanced sampling, even when the ratio population model is true, and is much more efficient when the variances are given by (8.11). A word of warning about using an overbalanced sampling strategy, however. The theoretical results on efficiency gains above are specific to situations where the underlying regression relationship between Y and Z can be expressed as a polynomial of order K. There is no guarantee that these gains persist if the relationship is not polynomial. In comparison, the stratification-based strategy described in Subsection 8.2.4 can accommodate many more types of non-linearity in the population of interest.

8.2.6 Balance and Optimality Results for General Linear Predictors

The discussion so far has been almost entirely focused on ratio-type predictors. However, the idea of balance extends quite naturally to arbitrary linear predictors and multiple linear regression models (not just polynomial models). In what follows therefore we use the same vector notation as in Chapter 7.

Let $\mathbf{W}$ be an arbitrary $(N-n) \times n$ matrix. Any linear predictor of t_y is then a special case of the $\mathbf{W}$-weighted linear predictor $\hat{t}_y(\mathbf{W}) = \sum_s y_i + \mathbf{1}'_{N-n}\mathbf{W}\mathbf{y}_s$. It is easy to see that the prediction bias of this estimator under the general linear model (7.1) is $\mathbf{1}'_{N-n}(\mathbf{W}\mathbf{Z}_s - \mathbf{Z}_r)\beta$. This bias is zero if the sample is $\mathbf{W}$-balanced on $\mathbf{Z}_U$, that is $\mathbf{1}'_{N-n}\mathbf{W}\mathbf{Z}_s = \mathbf{1}'_{N-n}\mathbf{Z}_r$.

A general theorem (Chambers, 1982; see also Tallis, 1978; Tam, 1986) can then be stated that identifies the conditions for $\hat{t}_y(\mathbf{W})$ to be the BLUP of t_y under (7.1). Suppose a $\mathbf{W}$-balanced sample is selected. Then $\hat{t}_y(\mathbf{W})$ is the BLUP of t_y under (7.1) if and only if $\mathbf{V}_{ss}^{1/2}\left(\mathbf{W} - \mathbf{V}_{rs}\mathbf{V}_{ss}^{-1}\right)' \mathbf{1}_{N-n}$ is in the vector space spanned by the columns of $\mathbf{V}_{ss}^{-1/2}\mathbf{Z}_s$. Here $\mathbf{V}_{ss}$ and $\mathbf{V}_{rs}$ are defined by the usual sample/non-sample decomposition of the variance-covariance matrix $\mathbf{V}$ of the linear model errors $\mathbf{y}_U - \mathbf{Z}_U\beta$.

All the results on balancing developed so far can be obtained as special cases of this theorem.

8.3 Robustness and the Clustered Population Model

Suppose that our working model ξ is the clustered population model (6.1). There are several ways in which this model might be misspecified. The mean, variance or intra-cluster correlation might vary by cluster, for example depending on the cluster size or other cluster characteristics. The most serious model failure would be if the expected values were not constant, since this could result in the BLUP, defined by the working model (6.4) and (6.5), being biased.

One way of improving on this working model BLUP would be to use a more complex model, for example by using auxiliary variables summarising cluster and individual characteristics. The general linear model, (7.1), with an appropriate covariance specification, can be used for this purpose. Another approach, which we will develop in this section, is to continue to use (6.1) as the working model, but to use a sample design such that the BLUP under this working model is approximately unbiased even if (6.1) is incorrect. We therefore suppose that the true model η is given by:

$$E_\eta(y_i | i \in g) = \mu_i \tag{8.12a}$$

$$Var_\eta(y_i | i \in g) = \sigma^2 \tag{8.12b}$$

$$Cov_\eta(y_i, y_j | i \neq j, i \in g, j \in f) = \rho\sigma^2 I(g = f). \tag{8.12c}$$

We suppose that (6.4) and (6.5) are being used to estimate t_y, and that the optimal design for fixed cost ignoring listing costs, described in Section 6.4, is being used. In this design, $m_g = m = \sqrt{c_1 c_2^{-1}(1-\rho)\rho^{-1}}$, with q determined by the cost constraint $C = c_1 q + c_2 q m$. As discussed in Section 6.4, any set of q clusters can be selected and the design is still optimal under (6.1) and the cost model (6.10) with $c_0 = c_3 = 0$. Equation (6.7) gives an approximation for the (constant within cluster) BLUP weights under the working model, which become

$$w_g = 1 + (N-n) \frac{(1-\rho+m\rho)^{-1}}{\sum_{g \in s} m(1-\rho+m\rho)^{-1}} = \frac{N}{n}$$

when $m_g = m$. Hence the bias of the working model BLUP under η is approximately equal to

$$\begin{aligned} E_\eta\left(\hat{t}_y - t_y\right) &= E_\eta\left(\sum_s w_g \sum_{s_g} y_i - \sum_U \sum_{U_g} y_i\right) \\ &= \sum_s w_g \sum_{s_g} \mu_i - \sum_U \sum_{U_g} \mu_i \\ &\cong \frac{N}{n} m \sum_s \bar{\mu}_{gs} - \sum_U M_g \bar{\mu}_g \end{aligned}$$

where $\bar{\mu}_g = M_g^{-1} \sum_{U_g} \mu_i$ and $\bar{\mu}_{gs} = m_g^{-1} \sum_{s_g} \mu_i$.

It is easy to see that this bias is equal to zero if

$$\bar{\mu}_{gs} = \bar{\mu}_g \tag{8.13a}$$

and

$$q^{-1} \sum_s \bar{\mu}_g = N^{-1} \sum_U M_g \bar{\mu}_g. \tag{8.13b}$$

The first condition implies first order balanced sampling of elements with respect to μ_i within each selected cluster. This can be approximately achieved by taking a simple random sampling without replacement from each selected cluster. The second condition consists of *weighted balanced sampling* of clusters, with respect to $\bar{\mu}_g$. The unweighted mean of $\bar{\mu}_g$ for sampled clusters g, is required to equal to the weighted mean of $\bar{\mu}_g$ for all clusters in the population, weighting by M_g. This is similar to the overbalancing condition (8.10) discussed in Section 8.2, with k equal to 1, and the size variable equal to M_g. It can be approximately achieved by selecting clusters using a probability sampling scheme where each cluster's probability of selection is proportional to its size, M_g.

It is clear that the relationship between $\bar{\mu}_g$ and M_g is central here. If there were no relationship, then the sample mean of $\bar{\mu}_g$ would be approximately equal to the weighted sample mean of $\bar{\mu}_g$, and this would in turn be approximately equal to the weighted population mean, so that (8.13b) would be approximately true, provided that sampling is non-informative. However, a relationship between $\bar{\mu}_g$ and M_g could well occur in practice, for example:

- For operational convenience in household surveys, clusters in rural areas are usually smaller in population but larger in area than clusters in urban areas. Otherwise travel within a cluster would be very expensive in rural areas. Incomes are often higher in urban than in rural areas; health characteristics and the mix of ethnicities will also differ between urban and rural areas. This could lead to an apparent relationship between cluster size and cluster mean for some variables of interest.
- A survey of employees may use the employing businesses as clusters. Business size could be related to a range of employee characteristics, such as income, occupation and level of education.

To guard against this, weighted balanced sampling is a sensible approach. There is little or no 'insurance penalty' for robustness in this case, since our approximation (6.8) to the prediction variance does not depend on which clusters are selected in s.

The sample design where clusters are selected by probability proportional to their size, with equal-sized simple random samples from each selected cluster, is an example of an equal probability selection method or *epsem* design: every element in the population has the same chance of being selected in the sample. The design is widely used in practice; see for example Foreman (1991, Section 7.5).

8.4 Non-parametric Prediction

If we are unsure about the specification of a parametric working model for inference about t_y, an obvious strategy is to use a non-parametric one. Although inferences based on non-parametric models are typically less efficient than those based on (properly specified) parametric models, they have the advantage of being valid under a much wider range of population structures. As a consequence, they have an inbuilt robustness to model misspecification.

We assume the existence of a variable Y defined on a population of interest for which sample data values are available, and an auxiliary variable Z also defined on this population whose population values are all known. As usual, sampling is assumed to be non-informative given Z (see Section 1.4). The relationship between Y and Z in the population is then characterised by

$$y_i = f(z_i) + e_i \qquad (8.14)$$

where f is an unknown (but reasonably smooth) regression function and the errors e_i have zero mean and constant variance, σ^2.

Prediction of the population total t_y under (8.14) requires an estimate $\hat{f}$ of f. If we have such an estimate we can then easily write down a predictor of t_y of the form

$$\hat{t}_y^{NP} = \sum_s y_i + \sum_r \hat{f}(z_i). \qquad (8.15)$$

Note that (8.15) requires access to individual non-sample Z-values. This can be a practical problem in some situations, particularly for so-called 'secondary' survey data analysts who do not have access to the sampling frame. In contrast it is easy to see that linear regression predictors (i.e. predictors based on a parametric linear model for the regression of Y on Z) only require that we know sample values and the population total of Z.

Non-parametric smoothing of the sample Y-values against the sample Z-values is a standard way of calculating $\hat{f}$. There is an extensive literature on non-parametric smoothing methods (e.g. Härdle, 1990; Wand and Jones, 1995). We restrict ourselves here to *kernel* smoothers since these are widely used and easy to interpret. Kernel smoothers are local smoothers, that is they are defined by a regression fit at any particular value z_0 of Z that only depends on those sample units whose Z values are close to z_0. In this context, 'close' is defined in terms of a neighbourhood $nbd(z_0)$ of z_0 whose 'shape' is determined by a unimodal symmetric density function (the kernel function) located at z_0 and whose 'size' is determined by a *bandwidth* parameter that specifies the support of this density. It is well known that the efficiency of a kernel smoother depends mainly on the choice of the bandwidth parameter.

There are then two standard ways of carrying out kernel smoothing of Y with respect to Z. Both of these are usually implemented by calculating a set of weights $\{u_i(z_0); i \in s\}$ that reflect how close the Z-value of a sample unit is to z_0. Typically this means that the weights $u_i(z_0)$ are zero outside $nbd(z_0)$, and get smaller the further z_i is from z_0. These weights are usually of the form

$$u_i(z_0) = \frac{K\left(b^{-1}(z_i - z_0)\right)}{\sum_s K\left(b^{-1}(z_j - z_0)\right)} \qquad (8.16)$$

where K denotes the kernel function and b denotes the bandwidth. Under *local mean smoothing* the non-parametric fit at z_0 is defined by the weighted average of the sample Y-values. That is

Non-parametric Prediction

$$\hat{f}(z_0) = \sum_s u_i(z_0) y_i. \qquad (8.17)$$

It follows that the predictor of t_y defined by (8.15) is then equal to

$$\hat{t}_y^{NP} = \sum_s y_i + \sum_{j \in r} \hat{f}(z_i) = \sum_s y_i + \sum_s m_i y_i = \sum_s w_i^{NP} y_i \qquad (8.18a)$$

where

$$m_i = \sum_{j \in r} u_i(z_j) = \sum_{j \in r} \frac{K\left(b^{-1}(z_i - z_j)\right)}{\sum_{k \in s} K\left(b^{-1}(z_k - z_j)\right)}; \text{ and} \qquad (8.18b)$$

$$w_i^{NP} = 1 + m_i. \qquad (8.18c)$$

Under *local linear smoothing*, we estimate $f(z_0)$ using the weighted linear regression of Y on the sample values of Z using the weights $u_i(z_0)$. The non-parametric fit at z_0 is

$$\hat{f}(z_0) = a(z_0) + b(z_0) z_0 \qquad (8.19)$$

where $a(z_0)$ and $b(z_0)$ are the intercept and slope parameter estimates derived from a weighted least squares fit of Y on Z that uses the weights $u_i(z_0)$ defined in (8.16) above. Since both $a(z_0)$ and $b(z_0)$ are linear functions of the sample Y-values, we can write the fitted value of the local linear smooth at z_0 as a weighted sum

$$\hat{f}(z_0) = \sum_s v_i(z_0) y_i \qquad (8.20a)$$

where

$$v_i(z_0) = u_i(z_0) \left\{ 1 + \frac{(z_i - \bar{z}_{u0})(z_0 - \bar{z}_{u0})}{\sum_s u_j(z_0) z_j (z_j - \bar{z}_{u0})} \right\} \qquad (8.20b)$$

with $\bar{z}_{u0} = \sum_s u_j(z_0) z_j$. When we use (8.20) to estimate $f(z_0)$, then $\hat{t}_y^{NP}$ defined by (8.15) becomes

$$\hat{t}_y^{NP} = \sum_s y_i + \sum_{j \in r} \sum_{i \in s} v_i(z_j) y_i = \sum_s y_i + \sum_s m_i y_i = \sum_s w_i^{NP} y_i \qquad (8.21a)$$

where

$$m_i = \sum_{j \in r} v_i(z_j); \text{ and} \qquad (8.21b)$$

$$w_i^{NP} = 1 + m_i. \qquad (8.21c)$$

In (8.18) and (8.21), w_i^{NP} and m_i are heavily dependent on the bandwidth, b.

It is easy to see that this approach to smoothing can be easily extended to local polynomial smoothing. In practice, however, it is rare to see anything more complex than local linear smoothing being used.

In both local mean and local linear smoothing, the non-parametric weights w_i^{NP} and m_i depend only on the population values of Z. The prediction variance of the non-parametric predictor under the model (8.14) follows directly:

$$\text{Var}\left(\hat{t}_y^{NP} - t_y\right) = \text{Var}\left(\sum_s m_i y_i - \sum_r y_i\right) = \sigma^2 \left\{\sum_s m_i^2 + (N-n)\right\}.$$

Provided that we can estimate σ^2, we can estimate this prediction variance. However, one must be careful here. Our focus on estimation of the prediction variance is based on an implicit assumption of unbiased prediction. But non-parametric predictors are not unbiased predictors, since the non-parametric regression fit $\hat{f}(z)$ at an arbitrary value z of Z is generally not an unbiased estimator of $f(z)$. In particular, choice of the bandwidth b in (8.16) usually involves a trade-off between an unbiased, but volatile, fit (small b) and a biased, but stable, fit (large b). Consequently we need an estimator of the prediction MSE of the non-parametric predictor, which can be very complicated to derive. It is for this reason that modern approaches to estimation of this MSE tend to use numerically intensive re-sampling methods like the bootstrap. See Chapter 12.

Although non-parametric smoothing as an alternative to parametric regression has a long history in statistics, the application of these ideas to survey inference, and particularly to prediction, is more recent. See Kuo (1988), Dorfman and Hall (1992), Chambers et al. (1993), Kuk (1993) and Dorfman (1994). The exception is stratification on the auxiliary variable Z, combined with separate prediction within each stratum, which is a standard sampling strategy. This strategy can be interpreted as a crude form of non-parametric regression based estimation, with a fitted regression function given by the non-linear step function

$$\hat{f}(z_i) = \sum_h I(i \in h)\bar{y}_{sh}.$$

In particular, it is this non-parametric regression interpretation of stratified estimation that allowed us to claim earlier that this method of prediction is qualitatively robust.

Before closing this discussion it should also be noted that any use of kernel-based methods of non-parametric prediction requires one to specify the kernel function and the bandwidth. The former choice is not critical. There is very little variation in efficiency between different kernel function specifications. The more important choice is that of the bandwidth. Here we note that standard asymptotic results regarding appropriate choice of bandwidth in non-parametric kernel-based smoothing for regression function estimation do not necessarily apply when the method is used for prediction. For example, it is known that the optimal bandwidth in the former case is proportional to the inverse of the fifth root of the sample size. See, for example, Härdle (1990, Sections 5.1–5.2). In the context of predicting a finite population total using local mean smoothing, however, Dorfman (1994) shows that, for local mean smoothing (often referred to as *Nadaraya–Watson smoothing*, after its initial developers), the optimal bandwidth is smaller, typically less than $Cn^{-1/4}$, where C is a constant that depends on the characteristics of the target population. This is an area of ongoing research.

The interested reader is referred to Breidt and Opsomer (2009) for a review of recent results on the use of non-parametric methods in surveys.

The following example, extracted from Dorfman (1994), illustrates the gains from adopting a non-parametric approach to prediction. The population here consists of $N = 400$ businesses that participated in the 1991 Occupational Compensation Survey carried out by the United States Bureau of Labor Statistics. The variable of interest Y is the total wages paid to workers in a selected group of occupations and Z is the total workforce of the business, including those in the selected group. A simulation study was carried out, consisting of taking 100 samples, using stratified random sampling based on 3 strata, of sizes 202, 114 and 84 determined by $Z < 250$, $250 \leq Z < 1000$ and $Z \geq 1000$ respectively. Samples of size 20 were taken from each of these three strata, and a number of different predictors of the population total of Y were considered. In Table 8.1 we show the average relative prediction error (RBIAS) and the square root of the average squared prediction error (RMSE) over the simulations for seven of these predictors.

Table 8.1 Simulation results from Dorfman (1994). All non-parametric predictors use the kernel function $K(u) = \exp(-u^2/2)$, with bandwidths b chosen on the basis that they gave a reasonable fit (judged on the basis of a visual inspection) to data generated from a single, randomly chosen sample.

Predictor	RBIAS ($\times 10^{-6}$)	RMSE ($\times 10^{-6}$)
Stratified expansion estimator (4.2)	0.035	6.34
Ratio estimator (5.4) based on Z, ignoring stratification	0.067	6.94
Generalised ratio estimator (5.3) based on Z, with $v(z) = z^2$, ignoring stratification	-0.063	4.56
Separate ratio estimator (5.13), based on Z	0.042	6.32
Local mean-based non-parametric predictor (8.18) with $u_i(z_j) \propto K\left(b^{-1}(z_i - z_j)\right)$, $b = 0.25$	0.040	6.50
Local mean-based non-parametric predictor (8.18) with $u_i(z_j) \propto K(b^{-1}(z_i - z_j))$, $b = 0.5$	0.013	5.67
Local mean-based non-parametric predictor (8.18) with $u_i(z_j) \propto K\left(b^{-1}(z_i - z_j)\right)$, $b = 0.75$	0.001	5.40

Inspection of the results in Table 8.1 shows clearly that the non-parametric approach can result in substantial gains. With the exception of the Generalised Ratio Estimator, the 'longer bandwidth' non-parametric predictors ($b = 0.5,\ 0.75$) outperformed all of the parametric predictors with respect to RMSE. If one takes low values of RBIAS into account as well, it seems clear that these two non-parametric predictors performed best in the simulation study. However, the poor performance of the shortest bandwidth ($b = 0.25$) predictor indicates the need for a robust algorithm for specifying a suitable bandwidth to use with the non-parametric approach. Choice of bandwidth in the context of local linear regression is discussed by Opsomer and Miller (2003) and Dorfman (2009b).

We return to use of non-parametric methods for model-based inference in Chapter 10, where we consider their use in robust bias correction.

9 Robust Estimation of the Prediction Variance

The main emphasis in the previous chapter was protection against misspecification of the population regression function $E(Y|Z)$. However, in Section 8.1 we saw that even if this function is correctly specified, misspecification of the conditional variance function $Var(Y|Z)$ can lead to invalid inference, in the sense that prediction variances are then estimated with bias, and inference (e.g. prediction intervals) becomes invalid. In this section we explore methods for estimating prediction variances that make minimal assumptions about $Var(Y|Z)$. To start, we motivate these methods by examining the case of the ratio estimator in some detail. As in the previous chapter, we use ξ to denote a working (i.e. assumed) model and η to denote the 'true' model underpinning the population values of Y. Moments under the working model are subscripted by ξ and moments under the true model are subscripted by η. We use expressions like η-expectation and η-bias to denote expectation and bias under η, with ξ-expectation, ξ-bias and so on similarly defined.

9.1 Robust Variance Estimation for the Ratio Estimator

Suppose the working model ξ is the ratio population model (5.2), and so we use the ratio estimator to estimate the population total of Y. Under this model an unbiased estimator of the prediction variance of the ratio estimator is given by (5.8). This estimator can be equivalently written

$$\hat{V}\left(\hat{t}_y^R\right) = \hat{\sigma}^2(N^2/n)(1-n/N)\left(\bar{z}_r\bar{z}_U\right)/\bar{z}_s \qquad (9.1)$$

where $\hat{\sigma}^2$ is defined immediately above (5.8) and is an unbiased estimator of the parameter σ^2 in (5.2b). Unfortunately, this standard 'plug-in' approach to variance estimation is non-robust to misspecification of $Var(Y|Z)$.

To see this, suppose the true model η for the population has $E_\eta(y_i|z_i) = \beta z_i$ and $Var_\eta(y_i|z_i) = \sigma^2 v_i$, where v_i is an unknown positive value. Since the mean function under η is unchanged from that assumed under ξ, the ratio estimator remains η-unbiased, but now its prediction variance is

$$Var_\eta\left(\hat{t}_y^R - t_y\right) = \sigma^2(N^2/n)(1-n/N)\left\{(1-n/N)\bar{v}_s(\bar{z}_r/\bar{z}_s)^2 + (n/N)\bar{v}_r\right\}. \qquad (9.2)$$

Furthermore, under η the estimator $\hat{\sigma}^2$ is no longer unbiased, since

$$E_\eta(\hat{\sigma}^2) = \sigma^2 n^{-1} \sum_s \frac{v_i}{x_i} \left\{ 1 - \frac{x_i}{n\bar{x}_s}\left(2 - \frac{x_i}{\bar{x}_s}\frac{\bar{v}_s}{v_i}\right)\right\}\left\{1 - \frac{x_i}{n\bar{x}_s}\right\}^{-1}$$

$$= \sigma^2 \overline{(v/z)}_s + O(n^{-1})$$

where $\overline{(v/z)}_s$ denotes the sample mean of the ratios v_i/z_i.

Does balancing help in this situation? Unfortunately, the answer is no. To see this, suppose our sample is first order balanced on Z, that is $\bar{z}_s = \bar{z}_r = \bar{z}_U$. The η-expectation of the prediction variance estimator (9.1) then satisfies

$$E_\eta\left\{\hat{V}\left(\hat{t}_y^R\right)\right\} \cong \sigma^2(N^2/n)\overline{(v/z)}_s \bar{z}_s$$

where '$\cong$' indicates that the expression on the right hand side of the equation is the leading $O(N^2 n^{-1})$ term of the expression on the left, which in this case is the η-expectation of (9.1). In contrast, the actual prediction variance (9.2) of the ratio estimator under η and balanced sampling satisfies

$$Var_\eta\left(\hat{t}_y^R - t_y\right) \approx \sigma^2(N^2/n)\bar{v}_s$$

so the ξ-based variance estimator (9.1) will tend to be biased high when $v(z) < z$ (i.e. when $v(z)$ is proportional to a power of z less than 1) since in this case $\bar{v}_s/\bar{z}_s < \overline{(v/z)}_s$. The variance estimator will be biased low when $v(z) > z$.

If the sampling fraction is small, then (9.2) is approximately equal to

$$Var_\eta\left(\hat{t}_y^R - t_y\right) \approx \sigma^2(N^2/n)(\bar{z}_r/\bar{z}_s)^2 \bar{v}_s = \sigma^2(N^2/n)(\bar{z}_r/\bar{z}_s)^2 n^{-1}\sum_s v_i. \quad (9.3)$$

Let $\sigma_i^2 = \sigma^2 v_i$. To obtain an approximately unbiased estimator of (9.3), note that we can estimate σ_i^2 using $\hat{\sigma}_i^2 = (y_i - bz_i)^2$ where $b = \bar{y}_s/\bar{z}_s$. This is approximately unbiased, as

$$E_\eta\left(\hat{\sigma}_i^2\right) = Var_\eta\left(y_i - bz_i\right)$$
$$= Var_\eta(y_i) + z_i^2\, Var_\eta(b) - 2z_i\, Cov_\eta(y_i, b)$$
$$= Var_\eta(y_i) + O\left(n^{-1}\right)$$
$$= \sigma_i^2 + O\left(n^{-1}\right).$$

This suggests the following estimator of the prediction variance of the ratio estimator

$$\hat{V}_{rob1}\left(\hat{t}_y^R\right) = (N^2/n)n^{-1}(\bar{z}_r/\bar{z}_s)^2 \sum_s (y_i - bz_i)^2. \quad (9.4)$$

It is clear that $\hat{V}_{rob1}\left(\hat{t}_y^R\right)$ is approximately unbiased under η.

We can improve on (9.4). Ideally, we would like our robust estimator to be approximately unbiased under η, and to be exactly unbiased under the working ratio population model ξ. Our first attempt $\hat{V}_{rob1}(\hat{t}_y^R)$ has the former property but not the latter:

$$E_\xi\left(\hat{V}_{rob1}(\hat{t}_y^R)\right) = (N^2/n)n^{-1}(\bar{z}_r/\bar{z}_s)^2 E_\xi\left\{\sum_s \left(y_i - \frac{\bar{y}_s}{\bar{z}_s}z_i\right)^2\right\}$$

$$= (N^2/n)(\bar{z}_r^2/\bar{z}_s)\sigma^2(1-n^{-1})\left(1 - n^{-1}s_z^2\bar{z}_s^{-2}\right).$$

Comparing this expression with the prediction variance (5.7b), we can see that $E_\xi\left(\hat{V}_{rob1}(\hat{t}_y^R)\right)$ is equal to $Var_\xi(\hat{t}_y^R - t_y)$ multiplied by a factor

$$(\bar{z}_r/\bar{z}_U)(1-n^{-1})\left(1 - n^{-1}s_z^2\bar{z}_s^{-2}\right). \tag{9.5}$$

We can create an unbiased variance estimator under ξ by dividing $\hat{V}_{rob1}(\hat{t}_y^R)$ by this factor:

$$\hat{V}_{rob2}(\hat{t}_y^R) = (N^2/n)\bar{z}_r\bar{z}_U\bar{z}_s^{-2}\left(1 - n^{-1}s_z^2\bar{z}_s^{-2}\right)^{-1}(n-1)^{-1}\sum_s(y_i - bz_i)^2. \tag{9.6}$$

This estimator will still be approximately unbiased under η because the factor in (9.5) is approximately equal to 1 when the sampling fraction is small. $\hat{V}_{rob2}(\hat{t}_y^R)$ was first described by Royall and Eberhardt (1975).

Another estimator of $V(\hat{t}_y^R - t_y)$ that has been proposed is

$$\hat{V}_{SRS}(\hat{t}_y^R) = (N^2/n)(1-n/N)(n-1)^{-1}\sum_s(y_i - bz_i)^2. \tag{9.7}$$

This is the traditional estimator of the variance of the ratio estimator under simple random sampling (see Cochran, 1977, formula 6.9, page 155). In general, (9.7) will be biased, even if the sampling fraction is small. If balanced sampling is used, so that $\bar{z}_r = \bar{z}_s$, it is equivalent to $\hat{V}_{rob1}(\hat{t}_y^R)$.

Royall and Cumberland (1981) explored the properties of $\hat{V}_{rob2}(\hat{t}_y^R)$, $\hat{V}_{SRS}(\hat{t}_y^R)$, and a number of other estimators in a wide-ranging simulation study. They concluded that $\hat{V}_{rob2}(\hat{t}_y^R)$ works well in a variety of situations where the ratio population model seems reasonable.

9.2 Robust Variance Estimation for General Linear Estimators

Suppose that our working model is the general linear model, (7.1). The BLUP, stated in (7.3), can be written in weighted form:

$$\hat{t}_y^w = \sum_s w_i y_i$$

where the sample weight w_i can depend on the values of one or more auxiliary Z-variables and can also be sample dependent, in the sense that it can also depend on the Z-values of all the sample units. However, w_i is not a function of the sample Y-values.

Our aim is to develop a robust estimator for the prediction variance of $\hat{t}_y^w$, using the ideas described in the previous section. To start, we set up an alternative model, η, which is the same as (7.1) except that the variances may be different for every unit:

$$E_\eta(y_i|\mathbf{z}_i) = \mathbf{z}_i'\beta \tag{9.8a}$$

$$Var(y_i|\mathbf{z}_i) = \sigma_i^2 \tag{9.8b}$$

$$Cov(y_i, y_j|\mathbf{z}_i, \mathbf{z}_j) = 0 \text{ for all } i \neq j. \tag{9.8c}$$

The prediction variance of $\hat{t}_y^w$ under (9.8) is

$$Var_\eta\left(\hat{t}_y^w - t_y\right) = \sum_s (w_i - 1)^2 \sigma_i^2 + \sum_r \sigma_i^2. \tag{9.9}$$

To estimate this variance, we need to plug in estimates of σ_i^2. One option is to assume that $\sigma_i^2 = \sigma^2$, as in model (7.1), and then to estimate using

$$\hat{\sigma}^2 = (n-p)^{-1} \sum_s \left(y_i - \mathbf{z}_i'\hat{\beta}_{ols}\right)^2 \tag{9.10}$$

where $\hat{\beta}_{ols}$ denotes the OLS estimate of β in (9.8a). This gives us

$$\hat{V}\left(\hat{t}_y^w\right) = \sum_s (w_i - 1)^2 \hat{\sigma}^2 + \sum_r \hat{\sigma}^2$$

$$= \hat{\sigma}^2 \left\{\sum_s (w_i - 1)^2 + N - n\right\}$$

$$= \hat{\sigma}^2 \left[\sum_s \left\{\mathbf{t}'_{zr}\left(\sum_s \mathbf{z}_j\mathbf{z}_j'\right)^{-1}\mathbf{z}_i\right\}^2 + N - n\right]$$

$$= \hat{\sigma}^2 \left\{\mathbf{t}'_{zr}\left(\mathbf{Z}'_s\mathbf{Z}_s\right)^{-1}\mathbf{t}_{zr} + N - n\right\}$$

which is the same as we obtained in (7.4).

How biased will $\hat{V}\left(\hat{t}_y^w\right)$ be if the true variances σ_i^2 are not all equal? Under model (9.8), it is straightforward to show that $E\left(\hat{\sigma}^2\right) \approx n^{-1}\sum_s \sigma_i^2 = \overline{(\sigma^2)}_s$. Moreover, the terms involving $(N-n)$ are negligible relative to the other terms if $n \ll N$, so that

$$E_\eta\left(\hat{V}\left(\hat{t}_y^w\right)\right) - Var_\eta\left(\hat{t}_y^w - t_y\right) \cong \overline{(\sigma_i^2)}_s \mathbf{t}'_{zr}\left(\mathbf{Z}'_s\mathbf{Z}_s\right)^{-1}\mathbf{t}_{zr} - \sum_s (w_i - 1)^2 \sigma_i^2$$

$$= \overline{(\sigma_i^2)}_s \sum_s (w_i - 1)^2 - \sum_s (w_i - 1)^2 \sigma_i^2$$

$$= -\sum_s (w_i - 1)^2 \left(\sigma_i^2 - \overline{(\sigma^2)}_s\right).$$

If the true variances are all equal, then $\sigma_i^2 = \overline{(\sigma^2)}_s = \sigma^2$, so that $\hat{V}\left(\hat{t}_y^w\right)$ will be unbiased. If the σ_i^2 are not all equal, this variance estimator will be biased in general. The extent of the bias will depend on the extent to which the values of σ_i^2 correlate with the values of $(w_i - 1)^2$.

We can construct a robust prediction variance estimator for $\hat{t}_y^w$ by estimating σ_i^2 for each unit i in sample, rather than assuming $\sigma_i^2 = \sigma^2$. An obvious candidate for this estimator is the squared residual, $\left(y_i - z_i'\hat{\beta}_{ols}\right)^2$. We make a slight adjustment to the squared residual to reduce its bias, and use

$$\hat{\sigma}_i^2 = \frac{n}{n-p}\left(y_i - z_i'\hat{\beta}_{ols}\right)^2.$$

We substitute these $\hat{\sigma}_i^2$ for $i \in s$ in (9.9) to give us a robust estimator. We cannot directly estimate σ_i^2 for $i \in r$, so we are forced to use $\hat{\sigma}^2$ from (9.10) for these elements. This gives us the following robust estimator of $Var_\eta\left(\hat{t}_y^w - t_y\right)$:

$$\hat{V}_{rob}\left(\hat{t}_y^w\right) = \sum_s (w_i - 1)^2 \hat{\sigma}_i^2 + \sum_r \hat{\sigma}^2$$

$$= \sum_s (w_i - 1)^2 \frac{n}{n-p}\left(y_i - z_i'\hat{\beta}_{ols}\right)^2 + \frac{N-n}{n-p}\sum_s \left(y_i - z_i'\hat{\beta}_{ols}\right)^2$$

$$= (n-p)^{-1}\sum_s \left\{n(w_i - 1)^2 + N - n\right\}\left(y_i - z_i'\hat{\beta}_{ols}\right)^2.$$

Application of this robust approach to prediction variance estimation for general regression-based predictors is discussed further in Royall and Cumberland (1978, 1981).

9.3 The Ultimate Cluster Variance Estimator

Recollect the two level hierarchical population model defined by (6.1). As we have already seen, this model is very useful for social populations where natural groupings (clusters) occur and where it is reasonable to assume that all population units have the same mean and variance, but where we also expect different units from the same cluster to be correlated. In particular, (6.1) assumes the population is made up of homogeneous subpopulations, defined by the clusters, with all clusters sharing the same mean (μ), variance (σ^2) and intracluster correlation (ρ). Predictors of the population total t_y for such populations will then have prediction variances that depend on ρ. See, for example, (6.6). Although estimates of these prediction variances can be defined using estimates of σ^2 and ρ obtained from multilevel modelling software, these inputs are not always easily available, and so there is interest in developing methods of variance estimation under (6.1) that do not depend on knowing how to estimate σ^2 and ρ, and which are robust to misspecification of the variances and the covariances within clusters. We now describe one widely used method that has this property.

In what follows we use ξ to denote our working model (6.1). We also use the notation introduced in Chapter 6 without further comment. We motivate our approach by noting that homogeneity of clusters implies that sample weights will be the same for all elements in a cluster (which is certainly true of the BLUP weights (6.4) and (6.5) developed in Section 6.2). Consequently a general linear estimator for t_y in this situation will take the form

$$\hat{t}_y^w = \sum_{g \in s} w_g \sum_{i \in s_g} y_i. \tag{9.11}$$

Suppose that instead of (6.1), the following model η applies:

$$E(y_i | i \in g) = \mu \tag{9.12a}$$

$$Var(y_i | i \in g) = \sigma_i^2 \tag{9.12b}$$

$$Cov(y_i, y_j | i \neq j, i \in g, j \in f) = \sigma_{ij} I(g = f). \tag{9.12c}$$

If the sampling fraction of clusters, q/Q, is small, then the prediction variance of $\hat{t}_y^w$ will be approximately equal to

$$Var_\eta \left(\hat{t}_y^w - t_y \right) = Var_\eta$$

$$\left\{ \sum_{g \in s} (w_g - 1) \sum_{i \in s_g} y_i - \sum_{g \in s} \sum_{i \in r_g} y_i - \sum_{g \in r} \sum_{i \in U_g} y_i \right\}$$

$$\cong Var_\eta \left\{ \sum_{g \in s} (w_g - 1) \sum_{i \in s_g} y_i \right\}$$

$$= \sum_{g \in s} (w_g - 1)^2 Var_\eta \left(\sum_{i \in s_g} y_i \right).$$

We could proceed by replacing $Var_\eta \left(\sum_{i \in s_g} y_i \right)$ by its value under the working model ξ:

$$Var_\xi \left(\sum_{i \in s_g} y_i \right) = m_g \left\{ 1 + (m_g - 1) \rho \right\} \sigma^2$$

and plugging in estimators of ρ and σ^2. However, this would require us to explicitly estimate ρ and σ^2, which we would rather avoid, and the result would be biased under the more general model η. Alternatively, we could use the expression for this variance under the actual model η:

$$Var_\eta \left(\sum_{i \in s_g} y_i \right) = \sum_{i \in s_g} \sigma_i^2 + \sum_{i \neq j \in s_g} \sigma_{ij}$$

and plug in estimators of σ_i^2 and σ_{ij}. But this would require us to correctly specify these quantities, which could be problematic.

Instead, we will estimate $Var_\eta\left(\sum_{i\in s_g} y_i\right)$ directly. First, note that $E_\eta\left(\sum_{i\in s_g} y_i\right) = m_g\mu$. We can estimate μ using the overall sample mean $\hat{\mu} = \bar{y}_s = n^{-1}\sum_{g\in s}\sum_{i\in s_g} y_i$, and then estimate $Var_\eta\left(\sum_{i\in s_g} y_i\right)$ by the squared difference between $\sum_{i\in s_g} y_i$ and its expectation, that is using

$$\hat{V}\left(\sum_{i\in s_g} y_i\right) = \left(\sum_{i\in s_g} y_i - m_g\bar{y}_s\right)^2.$$

This will be approximately unbiased under model (9.12). We then estimate the prediction variance of $\hat{t}_y^w$ by

$$\hat{V}_{UC}\left(\hat{t}_y^w\right) = \sum_{g\in s}(w_g-1)^2 \hat{V}\left(\sum_{i\in s_g} y_i\right)$$
$$= \sum_{g\in s}(w_g-1)^2\left(\sum_{i\in s_g} y_i - m_g\bar{y}_s\right)^2. \qquad (9.13)$$

This estimator is sometimes referred to as the *ultimate cluster variance estimator*. It is straightforward to calculate and does not require estimation of either the variance σ^2 or the intracluster correlation ρ.

10 Outlier Robust Prediction

The previous two chapters have focused on robustness to model misspecification. However, in practice it is often the case that while most of our sample data do appear to follow the working model, there are isolated data values in our sample that clearly do not follow this model. Recollect the Cities' population shown in Fig. 5.2. These outliers (or 'wild' data values) are a common feature of many sample surveys, particularly those of highly skewed economic populations. Ignoring these outlying values and calculating an estimate for t_y on the basis that the working model applies to the entire population can then lead to a completely unrealistic value for this estimate.

10.1 Strategies for Outlier Robust Prediction

What can we do if we observe outliers in our sample data? To start, we discuss this problem in the context of weighted linear prediction. That is, we focus on the common situation where our predictor of t_y can be written in the form $\hat{t}_y^w = \sum_s w_i y_i$ where the sample weight w_i is allowed to depend on all the sample values of the auxiliary variable Z but is not a function of the sample Y-values. An outlier may then be a value y_i completely unlike any other sample Y-value. However, it can also be a value y_i that is associated with a large weight w_i so the product $w_i y_i$ is large.

The interplay between the sample weight and the sample value when identifying an outlier leads to two basic approaches to dealing with these cases.

- Modify the sample weights of the outliers by reducing them relative to those of sample units that are not outliers. However, leave their Y-values unchanged.
- Modify the Y-values of the outliers so as to make them more 'acceptable', but leave their sample weights unchanged.

Both the above approaches are reflected in the following practical strategies for dealing with outliers that are in common use:

(a) Post-stratify sample outliers by placing them in a special stratum with smaller (typically unit) weights.
(b) Replace an outlier value in the current survey by a more acceptable value, either subjectively chosen or based on historical information about the sample unit generating the outlier.

(c) Force all sample weights to lie in some acceptable range (e.g. $\pm k\%$ of a target value).

(d) Replace extreme values above an upper cutoff or below a lower cutoff by the cutoff value, a procedure known as *winsorizing* (Chambers and Kokic, 1993).

Strategies (a) and (b) above presuppose that sample outliers are identifiable *a priori*. This is a strong assumption. In some cases the complexity of the survey data structure and the time limitations on survey processing mean that an outlier has to be really extreme before it can be identified and dealt with by these methods. Some outlying sample values, provided they pass the survey edits (which typically only check for logical consistency), will then remain in the sample data and consequently impact on standard survey estimation methods. In other cases, the temptation to define too many units as outliers may be hard to resist, for example this may seem desirable in order to reduce the change in estimates compared to a previous iteration of the survey. If too many sample values have their contribution reduced, this could lead to highly biased estimators.

Strategy (c) does not require pre-identification of sample outliers. However, because the restriction on sample weights applies to all the sample units, this strategy can lead to a considerable loss of efficiency. Furthermore, it offers no protection against extreme sample values of Y. Strategy (d) also does not require pre-identification of sample outliers. It would be expected to be more efficient than strategy (c). However, the cutoff values need to be chosen in some way, usually based on analysis of previous surveys, and this is not at all straightforward.

There are alternative approaches one can take when there are sample outliers, however. In essence these all consist of replacing the optimal linear predictor under the working model by a slightly less efficient but considerably more robust (i.e. less outlier sensitive) predictor. In many cases such robust predictors are defined by using modern robust statistical methods to estimate model parameters. This in effect replaces a non-robust but efficient linear predictor (e.g. the BLUP) by a robust non-linear predictor that can 'accommodate' sample outliers.

All outlier robust estimation strategies involve a bias-variance trade-off, where some bias in the predictor is accepted in order to down-weight the influence of outliers. This down-weighting of outliers decreases the variance of the predictor and (hopefully) the mean squared error as well. However one then runs into problems with prediction interval estimation, as we shall see.

We begin by noting that there is now a well-developed theory of outlier robust estimation, dating from the seminal paper of Huber (1964), where outlier robust *M-estimators* for the parameters of statistical models were first introduced. These estimators are typically defined as solutions to estimating equations that have been modified to ensure that no one data value has undue influence on the estimate of a model parameter. They are motivated by the observation that the majority of the sample data follow the pattern of behaviour expected under the working model, but there a small number of sample outliers or contaminants that are inconsistent with this model. Ideally, therefore, these outlying values should be given zero weight in any inference about the parameters of the working model.

Chambers (1986) extended this idea to finite population prediction and introduced the concept of representative outliers. These are legitimate population units (i.e. their outlying values are *not* errors arising in data collection or data processing) and we have no guarantee that there are no further outliers in the non-sampled part of the population. Clearly it is inappropriate to zero-weight representative outliers. Non-representative outliers (errors or unique values) on the other hand should be either given unit weights (if they are unique) or should be corrected (if they are due to errors) or zero-weighted (i.e. discarded).

The basic problem is then how to deal with survey data containing representative outliers. Since there may well be (and usually are) more representative outliers in the non-sampled part of the population, it is inadequate just to isolate these outliers in the survey sample and, as in the post-stratification approach (a) described above, give them unit weights. Sample outliers provide limited information about non-sample outliers. However, there are few sample outliers (by definition), so any attempt to use standard weighting methods to extract this information is a recipe for disaster.

10.2 Robust Parametric Bias Correction

Provided representative sample outliers can be identified, robust bias correction is a compromise solution to this problem. The idea is straightforward. First predict the finite population total t_y as if the working model applies to all non-sample units. That is, either delete or give unit weights to all sample outliers when calculating this estimate. Clearly the working model implicit in this approach will generally be wrong (the evidence for this is the presence of outliers in the sample data!) so this predictor is biased if the sample outliers are representative. We therefore add a robust bias correction to this prediction, based on the information in the sample outliers.

To illustrate how this approach works we consider a simple *mixture model*. In particular suppose that our working model for the population assumes that the values of Y are independently drawn from a distribution with mean μ_1 and variance σ_1^2. However, the actual population values of Y are in fact independently drawn from the two-component mixture

$$y_i = \Delta_i(\mu_1 + \sigma_1 \varepsilon_{1i}) + (1 - \Delta_i)(\mu_2 + \sigma_2 \varepsilon_{2i}) \qquad (10.1)$$

where Δ_i is a zero-one variable denoting outlier/non-outlier status respectively, $\mu_2 >> \mu_1$ and $\sigma_2 >> \sigma_1$, and ε_{1i}, ε_{2i} are independent N(0,1) variables. It is assumed that $\theta_i = \Pr(\Delta_i = 1)$ is close to but strictly less than one, so there are few population (and hence sample) outliers.

Given a non-informative sample drawn from this mixture, we can use outlier robust methods to estimate μ_1. In particular, let $\hat{\mu}_1$ be an outlier-robust M-estimate of this parameter, defined by the estimating equation

$$\sum_s \psi_1 \left\{ \hat{\sigma}_1^{-1}(y_i - \hat{\mu}_1) \right\} = 0$$

where $\hat{\sigma}_1$ is a robust estimate of σ_1 (e.g. the median absolute deviation from the sample median of Y) and ψ_1 is a bounded skew-symmetric function that behaves like the identity function near the origin and drops away to zero for values far from the origin. It can be seen that a sample outlier will contribute little or nothing to the left hand side of this estimating equation since its value of $y_i - \hat{\mu}_1$ will be large in absolute value and so $\psi_1 \left\{ \hat{\sigma}_1^{-1}(y_i - \hat{\mu}_1) \right\}$ will be close to zero. As a consequence the value of $\hat{\mu}_1$ is determined almost entirely by the sample non-outliers and so will be (approximately) unbiased for μ_1. An example of ψ_1 is the bisquare function (see Beaton and Tukey, 1974),

$$\psi_1(t) = t \left(1 - t^2/k_1^2 \right)^2 I(-k_1 \leq t \leq k_1) \quad (10.2)$$

where $I(-k_1 \leq t \leq k_1)$ denotes the indicator function that takes the value one if $-k_1 \leq t \leq k_1$ and is zero otherwise. Here k_1 is a tuning constant (a typical value is 4.5). The smaller the value of k_1 the more outlier robust (and the more inefficient) is $\hat{\mu}_1$. The function ψ_1 is typically referred to as the influence function underlying $\hat{\mu}_1$. Note that we can write $\hat{\mu}_1 = \sum_s u_{i1} y_i / \sum_s u_{i1}$, where $u_{i1} = \left\{ \hat{\sigma}_1^{-1}(y_i - \hat{\mu}_1) \right\}^{-1} \psi_1 \left\{ \hat{\sigma}_1^{-1}(y_i - \hat{\mu}_1) \right\}$.

An initial predictor of t_y is then $\hat{t}_y^1 = \sum_s y_i + (N-n)\hat{\mu}_1$ (or the even simpler projection type predictor $\tilde{t}_y^1 = N\hat{\mu}_1$). This is very close (in spirit at least) to the commonly used post-stratification estimator. It assumes that there are no non-sample outliers in the population, and so predicts the sum of the non-sample Y-values on the assumption that these all follow the working model. That is, the sample outliers are non-representative. See Rivest and Rouillard (1991) and Gwet and Rivest (1992) for examples of this approach.

The predictor $\hat{t}_y^1$ is biased under the mixture model (10.1). In particular, under this model and assuming that $E(\hat{\mu}_1) = \mu_1$, we have

$$E\left(\hat{t}_y^1 - t_y\right) = (\mu_1 - \mu_2) \sum_r (1 - \theta_i).$$

To correct this bias we must estimate it in some way. Suppose the mixture probabilities are constant, that is $\theta_i = \theta$ for all i, then

$$E\left(\hat{t}_y^1 - t_y\right) = (N-n)(1-\theta)(\mu_1 - \mu_2) = -(N-n)E(\bar{r}),$$

where $\bar{r} = n^{-1} \sum_s (y_i - \hat{\mu}_1)$. A bias corrected version of $\hat{t}_y^1$ is therefore

$$\hat{t}_y^{1,adj} = \hat{t}_y^1 + (N-n)\bar{r} = \sum_s y_i + (N-n)(\hat{\mu}_1 + \bar{r}).$$

This modified predictor is unbiased under the mixture model (10.1). Unfortunately simple algebra also demonstrates that it is in fact just the highly non-robust expansion estimator, that is $\hat{t}_y^{1,adj} = N\bar{y}_s$!

The problem is the bias adjustment $\bar{r}$. This is computed as the mean of the raw residuals $r_{1i} = y_i - \hat{\mu}_1$. These residuals will be small for the well-behaved units in the sample, but will be large for the sample outliers. A more robust bias adjustment is the modified mean

$$\bar{e} = \hat{\sigma}_1 n^{-1} \sum_s \psi_2(r_{1i}/\hat{\sigma}_1)$$

where $\hat{\sigma}_1$ is the robust estimate of σ_1 defined earlier, and ψ_2 is a 'prediction' influence function that gives relatively more weight to the sample outliers than the 'estimation' influence function ψ_1 underlying $\hat{\mu}_1$. A natural choice is Huber's influence function (Huber, 1964),

$$\psi_2(t) = \begin{cases} k_2 & t > k_2 \\ t & |t| \leq k_2 \\ -k_2 & t < -k_2 \end{cases} \quad (10.3)$$

where the tuning constant k_2 is quite large, say $k_2 = 6$. A robust bias corrected predictor of the population total t_y is then

$$\hat{t}_y^{1*} = \hat{t}_y^1 + (N-n)\bar{e} = \sum_s y_i + (N-n)(\hat{\mu}_1 + \bar{e}). \quad (10.4)$$

As an aside, we note that (10.4) is equivalent to using the standard expansion estimator with modified Y-values:

$$y_i^* = \frac{n}{N} y_i + \left(1 - \frac{n}{N}\right)(\hat{y}_i + \bar{e})$$

where

$$\hat{y}_i = \left[\frac{\psi_1\{\hat{\sigma}_1^{-1}(y_i - \hat{\mu}_1)\}}{y_i - \hat{\mu}_1}\right] \left[n^{-1} \sum_{j \in s} \frac{\psi_1\{\hat{\sigma}_1^{-1}(y_j - \hat{\mu}_1)\}}{y_j - \hat{\mu}_1}\right]^{-1} y_i.$$

This can be useful in computation of (10.4).

Even though (10.4) includes a bias correction, it remains biased. This is because the bias correction is itself biased, due to the need to maintain robustness against extreme sample outliers. Consequently, standard large sample arguments for prediction interval estimation based on consistent estimation of the prediction variance of (10.4) do not apply, and we do not expect such prediction intervals to achieve (even asymptotically) their nominal coverage levels.

It is straightforward to extend the above approach to outlier robust prediction under a more general population model. In particular, suppose that population elements are uncorrelated and satisfy $E(y_i|z_i) = \mu(z_i;\omega)$ and $Var(y_i|z_i) =$

$\sigma^2(z_i; \omega)$. Here Z is an auxiliary variable (which can be multivariate) and μ and σ^2 are known functions of this auxiliary variable and a vector of unknown parameters ω. Let $\hat{\omega}_1$ denote an estimate of ω that effectively ignores all sample outliers. For example, $\hat{\omega}_1$ could be an M-estimate of ω based on an influence function like ψ_1 (defined earlier). This influence function assigns zero weight to sample outliers in estimation. A predictor of t_y that treats all sample outliers as non-representative is then

$$\hat{t}_y^M = \sum_s y_i + \sum_r \mu(z_i; \hat{\omega}_1).$$

This predictor will be biased if there are outliers in the sample data. A robust bias corrected version that allows for representative sample outliers is

$$\hat{t}_y^{M*} = \hat{t}_y^M + \sum_s w_i \sigma(z_i; \hat{\omega}_1) \psi_2 \left\{ \sigma^{-1}(z_i; \hat{\omega}_1)(y_i - \mu(z_i; \hat{\omega}_1)) \right\} \quad (10.5)$$

where ψ_2 is a robust influence function that 'accommodates' (rather than discards) extreme values and the weights w_i are $O\{n^{-1}(N-n)\}$. This latter condition ensures that the bias correction term in $\hat{t}_y^{M*}$ is $O(N-n)$ (i.e. has the right order of magnitude).

Chambers (1986) recommends that the weights w_i in (10.5) be chosen so that this predictor reduces to the BLUP when there are no sample outliers (i.e. when ψ_2 is equal to the identity function on the sample). We refer to this type of bias correction as a *robust parametric bias correction* since it essentially assumes that the distribution of the errors $y_i - \mu(z_i; \omega)$ is derived from the working model for the y_i.

Bias correction along the lines described above is essentially *twicing*, see Tukey (1977). The initial estimate $\hat{t}_y^M$ represents the first attempt to estimate the population total t_y. The difference between this estimate and t_y is then another population parameter, which is estimated by the bias correction term. Conceptually this process could continue indefinitely, since one could argue that there may be bias in the bias correction term that itself needs to be estimated, and so on. However, we only consider two-step (i.e. twiced) estimators.

10.3 Robust Non-parametric Bias Correction

The kernel-based non-parametric smoothers described in Section 8.4 can be modified to make them robust to outliers in the sample data. For example, outlier robust versions $a^M(z_0)$ and $b^M(z_0)$ of the weighted least squares coefficients $a(z_0)$ and $b(z_0)$ underpinning the non-parametric local linear fit (8.20) at z_0 can be defined by the weighted M-type estimating equations

$$\sum_s u_i(z_0) \psi_e \left(\frac{y_i - a^M(z_0) - b^M(z_0) z_i}{\sigma_0^M} \right) \begin{pmatrix} 1 \\ z_i \end{pmatrix} = \begin{pmatrix} 0 \\ 0 \end{pmatrix}.$$

Here ψ_e denotes an influence function suitable for parametric estimation (i.e. it ensures sample outliers have zero, or close to zero, impact on this estimating equation), $u_i(z_0)$ denotes the kernel-based local mean weights defined in (8.16) and σ_0^M is a robust estimate of the scale of the residuals from the robust fit (e.g. the median of the absolute deviations from this fit). In practice, these estimating equations are often solved via an iteratively reweighted least squares algorithm. This leads to modified weighted least squares estimates $a^M(z_0)$ and $b^M(z_0)$, defined by weights of the form

$$u_i^M(z_0) = u_i(z_0) \frac{\sigma_0^M \psi_e \left\{ (y_i - a^M(z_0) - b^M(z_0)z_i)/\sigma_0^M \right\}}{y_i - a^M(z_0) - b^M b^M(z_0) z_i}.$$

Such weights are routinely produced by software for outlier robust estimation (e.g. the *rlm* function in R software, see Venables and Ripley, 2002, Section 8.3). They can be substituted for $u_i(z_0)$ in (8.20) to define a robust local linear smooth at z_0. We can then predict t_y using (8.21).

The resulting predictor of t_y will have a bias of

$$B_y = \sum_r E\{y_i - \mu(z_i; \hat{\omega})\}. \tag{10.6}$$

We can calculate a robust non-parametric estimator of (10.6) using

$$\hat{B}_y^{RNP} = \sum_{j \in r} \sum_{i \in s} v_i^M(z_j) \{y_i - \mu(z_i; \hat{\omega})\} = \sum_s w_i^M \{y_i - \mu(z_i; \hat{\omega})\}$$

where $v_i^M(z_j)$ is the value of (8.20b) at $z_0 = z_j$ (with $u_i^M(z_0)$ substituted for $u_i(z_0)$) and $w_i^M = \sum_{j \in r} v_i^M(z_j)$. A robust non-parametrically bias corrected predictor of t_y is

$$\hat{t}_y^{RNP} = \hat{t}_y + \hat{B}_y^{RNP} = \hat{t}_y + \sum_s w_i^M \{y_i - \mu(z_i; \hat{\omega})\}. \tag{10.7}$$

10.4 Outlier Robust Design

In general it is not possible to use the sample design to provide protection against outliers in the same way as it can be used to provide protection against model misspecification. This is because we have no idea *a priori* where outliers will occur. However, a measure of outlier robustness is achieved by implementing a sample design where the weights w_i do not vary too much from one sample unit to another, since this minimises the opportunity for a sample outlier to team up with a large sample weight to destabilise the estimator.

Sample weights are typically functions of one or more auxiliary variables (Z), and sample designs where these weights do not vary (or at least vary little) are typically designs that are balanced with respect to these variables. Consequently sample designs that attempt to ensure such balance (e.g. by only choosing samples that have this property) can therefore be expected to be less outlier sensitive than designs that place no restrictions on the sample weights.

10.5 Outlier Robust Ratio Estimation: Some Empirical Evidence

In this section we use some numerical results extracted from a simulation study reported by Chambers and Dorfman (1994, hereafter referred to as CD) to illustrate the consequences of applying the robust prediction ideas sketched out in Sections 10.2 and 10.3 to the ratio estimator $\hat{t}_y^R$. Recollect that $\hat{t}_y^R$ is the BLUP under the ratio population model, that is when the population satisfies $E(y_i|z_i) = \beta z_i$ and $Var(y_i|z_i) = \sigma^2 z_i$, where the z_i are the values of a strictly positive auxiliary variable Z that is positively correlated with Y. Under this model, our optimal estimator of β is the ratio $\hat{\beta} = \sum_s y_i / \sum_s z_i$.

In order to robustify $\hat{t}_y^R$ against sample outliers, CD first followed the robust parametric bias calibration approach described at the end of Section 10.2. In particular, the version of (10.5) corresponding to $\hat{t}_y^R$ that they investigated was

$$\hat{t}_y^{RM} = \sum_s y_i + \hat{\beta}_1 \sum_r z_i + \hat{\sigma}_1 \left(\frac{N\bar{z}}{n\bar{z}_s} - 1 \right) \sum_s \sqrt{z_i}\psi_2 \left(\frac{y_i - \hat{\beta}_1 z_i}{\hat{\sigma}_1 \sqrt{z_i}} \right) \qquad (10.8)$$

where $\hat{\beta}_1$ satisfies the outlier robust estimating equation

$$\sum_s \frac{1}{\sqrt{z_i}} \psi_1 \left(\frac{y_i - \hat{\beta}_1 z_i}{\hat{\sigma}_1 \sqrt{z_i}} \right) = 0.$$

The influence functions ψ_1 and ψ_2 are the same as those specified in Section 10.2, that is ψ_1 is the bisquare function (10.2) which gives sample outliers zero, or close to zero, weight in estimation of β, while ψ_2 is the more moderate Huber specification (10.3), with $k_2 = 6$, that allows these outliers to contribute (in a controlled way) to a bias correction. The estimate $\hat{\sigma}_1$ is the so-called MAD estimate of the scale of the residuals $z_i^{-1/2}(y_i - \hat{\beta}_1 z_i)$, that is 1.4826 times the median of the absolute deviations of these residuals from their median.

CD also investigated the performance of the robust non-parametric bias corrected predictor (10.7). In the case of the ratio estimator this is

$$\hat{t}_y^{RNP} = \sum_s y_i + \hat{\beta} \sum_r z_i + \sum_s w_i^M (y_i - \hat{\beta} z_i) \qquad (10.9)$$

where the weights w_i^M are the same as those used in (10.7), with ψ_e chosen to equal the Huber influence function ψ_2 used in (10.8). Note that these weights combine the outlier robustness properties of the Huber influence function with local linear non-parametric smoothing. The smoother itself used the second order optimal Epanechnikov kernel function (Gasser et al., 1985, Table 10.1), and asymptotic results about optimal bandwidth sizes for kernel regression (Härdle, 1990) were used to justify a bandwidth

Table 10.1 Estimation performance in simulation study based on the beef farms' population.

Estimator	AVE	RMSE	MAE
$N^{-1}\hat{t}_y^R$	4.42	23.61	16.31
$N^{-1}\hat{t}_y^{RM}$	11.54	20.43	13.01
$N^{-1}\hat{t}_y^{RNP}$	1.89	15.19	9.61

$$b_j = \min\left(b_{j5}, \frac{3 \times (sample\ range\ of\ Z)}{4n^{1/5}}\right)$$

for the smooth at $z_0 = z_j$; $j \in r$. Here b_{j5} denotes the smallest possible bandwidth such that at least five sample units contribute positively to the value of the smooth at z_j.

The population for CD's study consisted of 430 beef cattle farms, with $Y =$ income from sale of cattle and $Z =$ number of cattle. Figure 10.1 shows the relationship between Y and Z for this population. It is clear that there is model misspecification if the ratio population model is assumed for this population. It is also clear that there are a number of outliers.

The simulation experiment carried out by CD consisted of selecting 500 independent simple random samples selected from this population. The sample size in each case was $n = 60$. For each sample, CD evaluated three prediction strategies for the population mean of Y on the basis of their behaviour over these 500 samples. These were the predictor $N^{-1}\hat{t}_y^R$ based on the ratio estimator, together with the misspecification robust estimator of its prediction variance due to Royall and Eberhardt (1975),

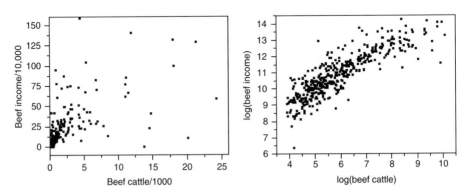

Fig. 10.1 The population of beef farms. The figure on the left shows the relationship between Y and Z in the raw scale, while the figure on the right shows this relationship on the logarithmic scale.

$$\hat{V}_D = \frac{1}{n}\left(1-\frac{n}{N}\right)\left(\frac{\bar{z}_r\bar{z}_U}{\bar{z}_s^2}\right)\left[1-\left\{\frac{s_z^2}{n\bar{z}_s^2}\right\}\right]^{-1}\frac{1}{n-1}\sum_s\left(y_i-\frac{\bar{y}_s}{\bar{z}_s}z_i\right)^2$$

(see Section 9.1), and the corresponding estimates defined by the robust parametric bias corrected version of the ratio estimator (10.8) and the robust non-parametric bias corrected version (10.9). Because explicit estimators are not available for the prediction variances of (10.8) and (10.9), bootstrap simulations were used to generate 95% prediction intervals for the actual population mean of Y when these predictors were calculated. Details of this procedure are set out in Section 12.3.

In order to evaluate the performances of the three different estimation strategies, CD calculated the average prediction error over the 500 samples (AVE), the root mean squared prediction error over the 500 samples (RMSE), and the median absolute prediction error over the 500 samples (MAE) for each predictor. Values for these measures, all expressed as a percentage of the true population mean $\bar{y}_U = 130441$, are set out in Table 10.1 below.

In terms of RMSE and MAE, the ratio estimator is substantially outperformed by the two robust bias corrected predictors. However, the parametric bias corrected predictor $N^{-1}\hat{t}_y^{RM}$ also has a relatively large value of AVE. This is not entirely surprising as far as this predictor is concerned, since it is based on the assumption that the majority of the survey population follow the simple ratio population model, which, as Fig. 10.1 clearly demonstrates, is clearly not the case for the beef population. The non-parametric bias corrected predictor $N^{-1}\hat{t}_y^{RNP}$ does not make this assumption, and so records the best performance of the three methods investigated in this study.

Table 10.2 shows the coverage performance of the prediction intervals generated by the three different strategies. In the case of the ratio estimator based strategy ($N^{-1}\hat{t}_y^R$ and $\hat{V}_D$), these prediction intervals were calculated using the standard large sample asymptotic normality assumption, that is as two estimated standard deviations either side of the estimate. For the two robust estimators however, prediction intervals were derived directly from the simulated bootstrap distribution.

All methods show undercoverage. In the case of the robust parametric bias corrected predictor $N^{-1}\hat{t}_y^{RM}$, we see that this undercoverage can be substantial. This is because of the bias of this predictor. In contrast, the robust non-parametric bias corrected predictor $N^{-1}\hat{t}_y^{RNP}$ has reasonably good all-round coverage performance. For the beef population at least, these results indicate that there are substantial gains to be had from being outlier robust as well as non-parametric in one's inference.

10.6 Practical Problems with Outlier Robust Estimators

The above promising results notwithstanding, it should be recognised that there are practical problems with adoption of the outlier robust predictors described

Table 10.2 Coverage performance in simulation study based on the beef farms' population.

Method	Nominal Coverage (%)			
	8 $Q = f(\bar{y}_1, \bar{y}_2, \ldots, \bar{y}_m)$ 0	90	95	98
	Actual Coverage (%)			
$N^{-1}\hat{t}_y^R + \hat{V}_D$	74.8	84.4	90.8	95.0
$N^{-1}\hat{t}_y^{RM}$ + bootstrap	62.6	73.2	80.8	87.8
$N^{-1}\hat{t}_y^{RNP}$ + bootstrap	76.4	88.2	94.0	96.8

above. One of the most important is caused by their intrinsic non-linearity. In particular, the robust predictors of the population totals of variables Y_A and Y_B may not sum to the value of a similarly derived robust predictor of their sum $Y = Y_A + Y_B$. Survey outputs are often hierarchically defined, and this aggregation consistency requirement is fundamental, limiting the effectiveness of the robust procedure at certain levels of categorisation or for certain variables. The problem can be overcome by imposing the aggregation constraint on the robust prediction procedure used, but this in turn requires that these constraints be identified *a priori*, which may not always be possible. It also adds considerable extra complication to the computation of the predictors themselves, since this now needs to be carried out in a multivariate manner. We are not aware of such a multivariate extension of the robust prediction methods described in this chapter, and so this remains an open problem.

A related practical problem is that, by definition, robust prediction strategies are variable specific. This means that they cannot be used in a survey processing system that assumes all variables are treated equally (e.g. they share the same sample weight). This can be overcome by representing the robust predictor as a standard linear predictor applied to a 'robustified' Y-value (see the example following (10.4)). However, this then suffers from the aggregation inconsistency problem discussed in the preceding paragraph, since there is no reason why the robustified values of Y_A and Y_B should sum to the robustified value of $Y = Y_A + Y_B$.

PART III
Applications of Model-Based Survey Inference

The emphasis in Parts 1 and 2 of this book is on model-based methods of inference for linear parameters of the target population. Such parameters are also the focus of traditional design-based sampling theory (see Cochran, 1977, or, more recently, Lohr, 1999). To a large extent this focus was well placed up to the 1980s, with surveys typically based on moderate to large probability-based samples and an emphasis on providing estimates of population totals, means and ratios. However, over the last quarter of a century, changes in social attitudes have led to a steady increase in the costs associated with collecting reliable data via traditional probability-based sample surveys. At the same time new developments in information technology have significantly enhanced our capacity to collect, process and analyse data, thereby contributing to an increased demand for more extensive and more coordinated data collections and more accurate and detailed inferences from these data. The combination of these two trends has resulted in survey methodologists developing procedures that allow efficient inference for a much wider range of population parameters from smaller and less well-structured samples. Model-based methods for sample survey inference have played a significant role in these new developments. In Part 3 of this book we illustrate this by describing how these methods are being applied in a number of these areas.

11 Inference for Non-linear Population Parameters

So far in this book we have focused on inferences about linear population parameters, that is linear functions of the population values of one or more survey variables (e.g. the population total or mean of a survey variable Y). We now consider the general situation where the population parameter of interest cannot be represented in this way.

11.1 Differentiable Functions of Population Means

Suppose that the population parameter of interest is $Q = f(\bar{y}_1, \ldots, \bar{y}_m)$, where f is a known differentiable function of the population means $\bar{y}_1, \bar{y}_2, \ldots, \bar{y}_m$ of m survey variables $Y_1, Y_2, \ldots, Y_m$. Let $\hat{\bar{y}}_1, \hat{\bar{y}}_2, \ldots, \hat{\bar{y}}_m$ denote suitable (i.e. unbiased) predictors of these population means. For example, they could be the BLUPs of $\bar{y}_1, \bar{y}_2, \ldots, \bar{y}_m$ under a model for the joint distribution of $Y_1, Y_2, \ldots, Y_m$ in the survey population. The natural predictor of Q is then the plug-in predictor $\hat{Q} = f(\hat{\bar{y}}_1, \hat{\bar{y}}_2, \ldots, \hat{\bar{y}}_m)$. We note that $\hat{Q}$ will be approximately unbiased in large samples provided the component predictors $\hat{\bar{y}}_1, \hat{\bar{y}}_2, \ldots, \hat{\bar{y}}_m$ are unbiased.

In order to estimate the prediction variance of $\hat{Q}$ we can use Taylor's Theorem (Serfling, 1980) to develop a first order approximation to the sample error $\hat{Q} - Q$. Let $\mu_k = E(\bar{y}_k) = E(\hat{\bar{y}}_k)$ and put $\theta = f(\mu_1, \mu_2, \ldots \mu_m)$. Linear approximations to $\hat{Q}$ and Q follow directly:

$$\hat{Q} \approx \theta + \sum_{a=1}^{m} \frac{\partial \theta}{\partial \mu_a}(\hat{\bar{y}}_a - \mu_a)$$

$$Q \approx \theta + \sum_{a=1}^{m} \frac{\partial \theta}{\partial \mu_a}(\bar{y}_a - \mu_a)$$

and so

$$\hat{Q} - Q \approx \sum_{a=1}^{m} \frac{\partial \theta}{\partial \mu_a}(\hat{\bar{y}}_a - \bar{y}_a). \tag{11.1}$$

A first order approximation to the variance of the sample error $\hat{Q} - Q$ is therefore

$$Var\left(\hat{Q} - Q\right) \cong \sum_{a=1}^{m}\sum_{b=1}^{m} \left(\frac{\partial \theta}{\partial \mu_a}\right)\left(\frac{\partial \theta}{\partial \mu_b}\right) Cov(\hat{\bar{y}}_a - \bar{y}_a, \hat{\bar{y}}_b - \bar{y}_b). \qquad (11.2)$$

Let $\hat{C}(\hat{\bar{y}}_a, \hat{\bar{y}}_b)$ denote an estimator of the covariance between the prediction errors of $\hat{\bar{y}}_a$ and $\hat{\bar{y}}_b$. An estimator of the right hand side of (11.2) is

$$\hat{V}(\hat{Q}) = \sum_{a=1}^{m}\sum_{b=1}^{m} \left(\frac{\partial \theta}{\partial \mu_a}\bigg|_{\hat{\mathbf{y}}}\right)\left(\frac{\partial \theta}{\partial \mu_b}\bigg|_{\hat{\mathbf{y}}}\right) \hat{C}(\hat{\bar{y}}_a, \hat{\bar{y}}_b) \qquad (11.3)$$

where $\frac{\partial \theta}{\partial \mu_a}\big|_{\hat{\mathbf{y}}}$ is the partial derivative of $\theta = f(\mu_1, \mu_2, \ldots \mu_m)$ with respect to μ_a, evaluated at $\hat{\mathbf{y}} = (\hat{\bar{y}}_1 \; \hat{\bar{y}}_2 \; \ldots \; \hat{\bar{y}}_m)'$. The estimator (11.3) is often referred to as the *Taylor Series Linearisation* variance estimator.

An important special case is where the estimators $\hat{\bar{y}}_1, \hat{\bar{y}}_2, \ldots, \hat{\bar{y}}_m$ are linear and use the same sample weights $\{w_i; i \in s\}$. That is, $\hat{\bar{y}}_a = N^{-1}\sum_s w_i y_{ai}$. From the representation (11.1), we see that we can then write $Var(\hat{Q}-Q) \approx Var(\hat{\bar{g}}_U - \bar{g}_U)$, where $\bar{g}_U$ is the population mean of the linearised variable whose value for the ith population element is

$$g_i = \sum_{a=1}^{m}\left(\frac{\partial \theta}{\partial \mu_a}\right) y_{ai} \qquad (11.4)$$

and

$$\hat{\bar{g}}_U = N^{-1}\sum_s w_i g_i = N^{-1}\sum_s w_i \left(\sum_{a=1}^{m}\left(\frac{\partial \theta}{\partial \mu_a}\right) y_{ai}\right).$$

That is, a first order approximation to the prediction variance of $\hat{Q}$ is the prediction variance of the predictor $\hat{\bar{g}}_U$ of the population mean of the linearised variable (11.4). This representation is due to Woodruff (1971). Note that (11.4) cannot be evaluated directly, since it depends on the partial derivatives $\partial\theta/\partial\mu_a$. In practice, we therefore replace it by

$$\hat{g}_i = \sum_{a=1}^{m}\left(\frac{\partial \theta}{\partial \mu_a}\bigg|_{\hat{\mathbf{y}}}\right) y_{ai}.$$

In order to illustrate this approach, we consider the situation where the parameter of interest is the ratio $R = \bar{y}_U/\bar{x}_U$ of the population means of two survey variables, Y and X. Suppose this population is such that distinct population elements are uncorrelated, with $E(y_i) = \mu_y$, $E(x_i) = \mu_x$, $Var(y_i) = \sigma_y^2$, $Var(x_i) = \sigma_x^2$ and $Cov(y_i, x_i) = \sigma_{yx}$. The predictor $\hat{R} = \bar{y}_s/\bar{x}_s$ is then the corresponding ratio of the BLUPs of the population means $\bar{y}_U$ and $\bar{x}_U$ (see Chapter 3). In order to estimate the prediction variance of $\hat{R}$, we adopt the Taylor series linearisation approach. In particular, in this case we have $\theta = \mu_y/\mu_x$, so $\partial\theta/\partial\mu_y = 1/\mu_x$ and $\partial\theta/\partial\mu_x = -\mu_y/\mu_x^2$. Hence the linearised variable (11.4) becomes

$$g_i = (1/\mu_x)y_i - (\mu_y/\mu_x^2)x_i = (y_i - \theta x_i)/\mu_x.$$

Since $E(g_i) = 0$ and $Var(g_i) = \mu_x^{-2} Var(y_i - \theta x_i) = \mu_x^{-2}(\sigma_y^2 - 2\theta\sigma_{yx} + \theta^2\sigma_x^2)$, it is clear that this variable, like Y and X themselves, can be modelled as second order homogeneous across the population of interest, and so our estimator of the prediction variance of $\hat{R}$ satisfies

$$\hat{V}(\hat{R}) = \hat{V}(\hat{g}_U) = n^{-1}(1 - nN^{-1})(n-1)^{-1} \sum_s (g_i - \bar{g}_s)^2.$$

Of course, g_i depends on unknown parameters and so cannot be calculated, leading us to replace it by $\hat{g}_i = (y_i - \hat{R}x_i)/\bar{x}_s$. It is easy to see that then

$$\hat{V}(\hat{R}) = n^{-1}\bar{x}_s^{-2}(1 - nN^{-1})(n-1)^{-1} \sum_s (y_i - \hat{R}x_i)^2.$$

11.2 Solutions of Estimating Equations

Many population parameters cannot be written down explicitly, and are instead defined as solutions to population level estimating equations. For example, the finite population median m_y of the variable Y is defined by

$$m_y = \min_m \left(N^{-1} \sum_U I(y_i \leq m) \geq 0.5 \right)$$

An alternative definition is

$$m_y^* = \begin{cases} y_{((N+1)/2)} & \text{if } N \text{ is odd} \\ (y_{([(N+1)/2])} + y_{([(N+1)/2]+1)})/2 & \text{if } N \text{ is even} \end{cases}$$

where $y_{(j)}$ is the j-th order statistic of the population values of Y, and $[u]$ refers to the integer part of u. The two definitions converge for large values of N.

In general, a population parameter Q is defined by a population level estimating equation if it is a solution to

$$H(Q) = \sum_U g(\mathbf{y}_i; Q) = 0. \tag{11.5}$$

where g is a function of both Q and $\mathbf{y}_i$, where the latter corresponds to the value of a vector of survey variables associated with population element i. For fixed Q the function $H(Q)$ in (11.5) is the population total of the derived variable $g(\mathbf{y}_i; Q)$. Given a set of sample weights $\{w_i; i \in s\}$ we can predict the value of $H(Q)$ by

$$\hat{H}(Q) = \sum_s w_i g(\mathbf{y}_i; Q).$$

We can then predict Q by $\hat{Q}$, where $\hat{Q}$ is a solution of

$$\hat{H}(\hat{Q}) = \sum_s w_i g(\mathbf{y}_i; \hat{Q}) = 0. \tag{11.6}$$

For the special case where g is a differentiable function of Q, estimation of the prediction variance of $\hat{Q}$ can be based on Taylor series linearisation. To illustrate this approach, we assume that there exists a value θ such that the random variables $g(\mathbf{y}_i; \theta)$ are uncorrelated with mean zero. We also assume that for large N and n the solutions Q and $\hat{Q}$ to (11.5) and (11.6) are unique. It follows that both Q and $\hat{Q}$ converge to θ, leading to the first order approximations

$$0 = H(Q) \cong H(\theta) + (Q - \theta) \sum_U \frac{\partial g(\mathbf{y}_i; \theta)}{\partial \theta}$$

$$0 = \hat{H}(\hat{Q}) \cong \hat{H}(\theta) + (\hat{Q} - \theta) \sum_s w_i \frac{\partial g(\mathbf{y}_i; \theta)}{\partial \theta}.$$

Therefore

$$Q \cong \theta - \left(\sum_U \frac{\partial g(\mathbf{y}_i; \theta)}{\partial \theta} \right)^{-1} H(\theta)$$

$$\hat{Q} \cong \theta - \left(\sum_s w_i \frac{\partial g(\mathbf{y}_i; \theta)}{\partial \theta} \right)^{-1} \hat{H}(\theta)$$

and hence

$$\hat{Q} - Q \cong \left(\sum_U \frac{\partial g(\mathbf{y}_i; \theta)}{\partial \theta} \right)^{-1} H(\theta) - \left(\sum_s w_i \frac{\partial g(\mathbf{y}_i; \theta)}{\partial \theta} \right)^{-1} \hat{H}(\theta).$$

Let $Var\{g(\mathbf{y}_i; \theta)\} = \gamma_i^2$. Then

$$Var\left\{ \hat{H}(\theta) \right\} = \sum_s w_i^2 \gamma_i^2$$

$$Var\left\{ H(\theta) \right\} = \sum_U \gamma_i^2$$

$$Cov\left\{ H(\theta), \hat{H}(\theta) \right\} = \sum_s w_i \gamma_i^2.$$

Put $d_i = \frac{\partial g(\mathbf{y}_i; \theta)}{\partial \theta}$, $t_d = \sum_U d_i$ and $t_{wd} = \sum_s w_i d_i$. Since $\hat{Q} - Q \cong t_d^{-1} H(\theta) - t_{wd}^{-1} \hat{H}(\theta)$, a large sample approximation to the prediction variance of $\hat{Q}$ is

$$Var(\hat{Q} - Q) \cong t_{wd}^{-2} \sum_s w_i^2 \gamma_i^2 - 2 t_{wd}^{-1} t_d^{-1} \sum_s w_i \gamma_i^2 + t_d^{-2} \sum_U \gamma_i^2.$$

Let $\hat{\gamma}_i^2$ be an estimator of γ_i^2, for example $\hat{\gamma}_i^2 = g(\mathbf{y}_i; \hat{Q})^2$ could be used. Then the prediction variance of $\hat{Q}$ can be estimated by

$$\hat{V}(\hat{Q}) = \hat{t}_{wd}^{-2} \left(\sum_s w_i^2 \hat{\gamma}_i^2 - 2 \sum_s w_i \hat{\gamma}_i^2 + \sum_U \hat{\gamma}_i^2 \right). \tag{11.7}$$

Here $\hat{t}_{wd} = \sum_s w_i \hat{d}_i$, with $\hat{d}_i = \left.\frac{\partial g(\mathbf{y}_i;\theta)}{\partial \theta}\right|_{\hat{Q}}$, i.e. the value of the derivative $\frac{\partial g(\mathbf{y}_i;\theta)}{\partial \theta}$ at $\theta = \hat{Q}$.

11.3 Population Medians

To illustrate the general approach described in Section 11.2, we will now consider the prediction of the population median more closely. We will suppose that the population of interest is homogenous and can be modelled via (3.1). In fact, we extend this model to assume that the population values of Y are independent and identically distributed realisations of a random variable with distribution function F and probability density function f. Suppose that we wish to predict the median Y-value m_y for this population. As noted earlier, for large values of N we (approximately) have $N^{-1} \sum_U I(y_i \leq M) = 0.5$, so our predictor $\hat{m}_y$ of this median satisfies

$$N^{-1} \sum_s w_i I(y_i \leq \hat{m}_y) = 0.5$$

to a similar degree of approximation. Here $w_i = N/n$ for every sampled element i. Note that in this case $g(\mathbf{y}_i;\theta) = I(y_i \leq \theta) - 0.5$, where θ satisfies $\Pr(y_i \leq \theta) = F(\theta) = 0.5$. It follows that

$$E\{g(\mathbf{y}_i;\theta)\} = 0$$

and

$$Var\{g(\mathbf{y}_i;\theta)\} = \frac{1}{4}.$$

Using (11.7), we can then write down a prediction variance estimator for $\hat{m}_y$:

$$\hat{V}(\hat{m}_y) = \hat{t}_{wd}^{-2}\left\{\frac{N^2}{4n} - \frac{2N}{4} + \frac{N}{4}\right\} = \frac{1}{4}\hat{t}_{wd}^{-2} N^2 n^{-1}(1 - nN^{-1}) \quad (11.8)$$

where

$$t_{wd} = \sum_s w_i \frac{\partial g(\mathbf{y}_i;\theta)}{\partial \theta} = \frac{\partial}{\partial \theta}\left\{\sum_s w_i g(\mathbf{y}_i;\theta)\right\}. \quad (11.9)$$

Since $g(\mathbf{y}_i;\theta)$ is an indicator function, the derivative in (11.9) is not strictly defined. However, for large sample sizes

$$\sum_s w_i g(\mathbf{y}_i;\theta) \cong (F(\theta) - 0.5)\sum_s w_i = N(F(\theta) - 0.5)$$

so that

$$t_{wd} \cong \frac{\partial}{\partial \theta}\{N(F(\theta) - 0.5)\} = Nf(\theta).$$

Let $\hat{f}$ denote an estimate of the (superpopulation) density f of Y. We can then approximate (11.8) by

$$\frac{1}{4}\hat{f}(\hat{m}_y)^{-2} n^{-1}(1 - nN^{-1}). \quad (11.10)$$

Table 11.1 Population counts of 64 cities (in 1000s) in 1930.

Y = 1930 population count			
900	**364**	**209**	113
822	**317**	183	**115**
781	**328**	163	123
805	302	253	154
670	**288**	232	140
1238	291	260	119
573	253	201	**130**
634	291	**147**	**127**
578	308	292	**100**
487	272	164	107
442	284	143	114
451	255	**169**	111
459	270	139	163
464	214	170	116
400	195	**150**	122
366	**260**	143	134

Let m' be the next highest sample value of Y above $\hat{m}_y$. A crude approximation to $\hat{f}(\hat{m}_y)$ which is sometimes used in surveys is

$$\hat{f}(\hat{m}_y) = \frac{\hat{F}(m') - \hat{F}(\hat{m}_y)}{m' - \hat{m}_y} = \frac{n^{-1}}{m' - \hat{m}_y} \qquad (11.11)$$

where

$$\hat{F}(t) = N^{-1} \sum_s w_i I(y_i \leq t). \qquad (11.12)$$

However (11.11) can be unstable, and sometimes m' and $\hat{m}_y$ will be equal so that the expression cannot be calculated. A better approach is to estimate $\hat{f}(\hat{m}_y)$ using a more sophisticated method such as kernel smoothing (Silverman, 1986).

To illustrate these ideas, suppose our population consists of the 1930s' city population counts shown in Table 4.1. The median of the population values can be readily calculated as 253. Suppose we take a simple random sample without replacement consisting of the 15 cities shown in bold (we will assume here that no stratifying information is available, unlike Chapter 4):

The sample median is equal to the 8th highest sample value, 209. To calculate the prediction variance using (11.10), we need an estimate of the probability density function at the median. Figure 11.1 shows a histogram of the 15 sampled values with the count in each range shown at the top of each bar. The kernel density estimate is shown as a smooth curve superimposed on this histogram. The

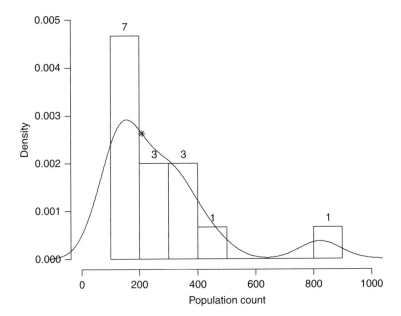

Fig. 11.1 Histogram of data from sampled cities with kernel density estimate superimposed.

estimated density at the median value of 209 is equal to 0.00262 and is marked by an asterisk. The kernel density estimate and the plots were produced using the *hist* and *density* functions in the R statistical environment (R Development Core Team, 2005).

Using (11.10), the estimated variance of the sample median is then

$$\hat{V}(\hat{m}_y) = \frac{1}{4 \times 15 \times 0.00262^2}\left(1 - \frac{15}{64}\right) = 1859.4$$

and so the estimated standard error of the sample median is the square root of this value, which is 43.1.

Although the preceding theory assumed a homogeneous population, we are usually interested in more complex populations. For example, suppose that our focus is prediction of the population median m_y for a population of H strata, where it is reasonable to assume that the values of Y in stratum h are independent and identically distributed realisations of a random variable with continuous distribution function F_h and probability density function f_h. This is a special case of the stratified population model (4.1) and so our predictor $\hat{m}_y$ of the stratified median m_y satisfies (up to a suitable degree of approximation),

$$N^{-1}\sum_s w_i I(y_i \leq \hat{m}_y) = 0.5 \qquad (11.13)$$

where now $w_i = N_h n_h^{-1}$ for sampled unit i in stratum h.

We still have $g(\mathbf{y}_i; \theta) = I(y_i \leq \theta) - 0.5$, where θ satisfies $\Pr(y_i \leq \theta) = 0.5$. Put

$$\pi_h = \Pr(y_i \leq \theta | i \in h) = F_h(\theta).$$

It follows that

$$E\{g(\mathbf{y}_i; \theta) | i \in h\} = \pi_h - 0.5$$
$$\mathrm{Var}\{g(\mathbf{y}_i; \theta) | i \in h\} = \pi_h(1 - \pi_h).$$

Using (11.7), we can write down a prediction variance estimator for $\hat{m}_y$:

$$\hat{V}(\hat{m}_y) = \hat{t}_{wd}^{-2} \sum_h N_h^2 n_h^{-1}(1 - n_h N_h^{-1})\hat{\pi}_h(1 - \hat{\pi}_h) \qquad (11.14)$$

where

$$t_{wd} = \sum_s w_i \frac{\partial g(\mathbf{y}_i; \theta)}{\partial \theta} = \frac{\partial}{\partial \theta}\left\{\sum_s w_i g(\mathbf{y}_i; \theta)\right\}$$

and

$$\hat{\pi}_h = n_h^{-1} \sum_{s_h} I(y_i \leq \theta).$$

As noted earlier, $g(\mathbf{y}_i; \theta)$ is an indicator function, and so the derivative defining t_{wd} above is not strictly defined. However, for large stratum sample sizes,

$$\sum_s w_i g(\mathbf{y}_i; \theta) = \sum_s w_i \left(I(y_i \leq \theta) - 0.5\right) \cong \sum_h N_h(\pi_h - 0.5).$$

A first approximation to t_{wd} is

$$t_{wd} \cong \frac{\partial}{\partial \theta}\left\{\sum_h N_h(\pi_h - 0.5)\right\} = \sum_h N_h \frac{\partial \pi_h}{\partial \theta} = \sum_h N_h \frac{\partial F_h(\theta)}{\partial \theta} = \sum_h N_h f_h(\theta).$$

We can calculate an estimator $\hat{f}_h$ of f_h using kernel smoothing, as discussed above, or by some other method. Our final approximation to t_{wd} is therefore

$$\sum_h N_h \hat{f}_h(\hat{m}_y). \qquad (11.15)$$

The estimated prediction variance of the stratified median predictor $\hat{m}_y$ defined by (11.13) is then calculated by substituting (11.15) into (11.14) to obtain

$$\hat{V}(\hat{m}_y) = \frac{\sum_h N_h^2 n_h^{-1}(1 - n_h N_h^{-1})\hat{\pi}_h(1 - \hat{\pi}_h)}{\left(\sum_h N_h \hat{f}_h(\hat{m}_y)\right)^2}.$$

12 Survey Inference via Sub-Sampling

Our next problem area relates to the increasing need for more complex inferences from sample survey data. Sometimes the definition of the population parameter of interest may be so complex that application of Taylor series linearisation methods for variance estimation (see Chapter 11) is difficult, if not impossible. An example is an annual 'chain ratio' index based on monthly business survey data, which is a function of the product of 12 ratios, based on 24 different estimates, computed using data collected in 12 different monthly 'waves' of the survey. Another is where the survey estimate is a complex function of the sample data, with this function in turn determined by a parameter that is estimated from the survey data. This is the case, for example, in a poverty survey where the parameter of interest is the median income of single adult households with children whose household incomes lie below the 10th percentile of the overall household income distribution. In both these examples, prediction of the parameter of interest can be achieved in a relatively straightforward manner via a plug-in approach (i.e. we compute the value of the target parameter, substituting sample-based predictions for all unknown population quantities), but then estimation of the prediction variance via Taylor series linearisation is difficult. In such cases we can use alternative variance estimation methods that are simple to implement, but are typically numerically intensive. Even where the Taylor series approach is feasible, it may be difficult to calculate or explain, and these alternatives may be preferable.

The basic idea behind these methods is straightforward. In effect, we use sub-sampling methods to estimate the variance of a statistic, by:

(i) making repeated draws from a distribution whose variance is related in a simple (and known) way to the variance of interest;
(ii) empirically estimating the variance of this distribution;
(iii) adjusting this variance estimate so that it is an estimate of the variance of interest.

In the following we use W to denote the (complex) population parameter of interest and $\hat{W}$ to denote our preferred predictor of W based on data from the full sample.

12.1 Variance Estimation via Independent Sub-Samples

This approach has a long history in sample survey theory. The simplest version of it leads to the *Random Groups Variance Estimator*, whose origin lies in the concept of *interpenetrating samples* (Mahalanobis, 1946; Deming, 1960). Here the actual sample selected is made up of G independent replicate or interpenetrating sub-samples. Each of these sub-samples is representative of the population, and is drawn according to the same design and with the same sample size $m = n/G$. In particular, it is assumed that each sub-sample is made up of distinct *primary sampling units* (PSUs) containing population elements that are mutually independent with respect to the survey variables whose population values define W, the (complex) population parameter of interest.

To motivate the approach, let $\hat{W}_g$ denote the plug-in predictor of W based on the data obtained in the gth replicate sample. Suppose also that our full sample predictor of W is $\hat{W} = G^{-1} \sum_g \hat{W}_g$. By construction, the $\hat{W}_g$ are independently distributed across the replicate samples. For example, if we have a model that observations from units in different PSUs are independent, then we can define the replicate samples to consist of non-overlapping groups of PSUs. If the $\hat{W}_g$ are also identically distributed then we can unbiasedly estimate the variance of their (common) distribution from their empirical variance around their average, $\hat{W}$. An unbiased estimator of the variance of $\hat{W}$ is this 'replicate variance' divided by the number of replicates, G. This is

$$\hat{V}_{rep}(\hat{W}) = \frac{1}{G(G-1)} \sum_g (\hat{W}_g - \hat{W})^2. \tag{12.1}$$

The estimator (12.1) is still unbiased for $Var(\hat{W})$ even if the replicate estimates are not identically distributed. All that is required is that they are independent of one another, and that they all have the same expected value, so $E(\hat{W}_g - \hat{W}) = 0$. To see this, we note that, given these assumptions

$$E(\hat{W}_g - \hat{W})^2 = Var(\hat{W}_g - \hat{W}) = \frac{(G-1)^2}{G^2} Var(\hat{W}_g) + \frac{1}{G^2} \sum_{h \neq g} Var(\hat{W}_h)$$

and so

$$E\left\{\hat{V}_{rep}(\hat{W})\right\} = \frac{1}{G(G-1)} \sum_g E(\hat{W}_g - \hat{W})^2 = \frac{1}{G^2} \sum_g Var(\hat{W}_g) = Var(\hat{W})$$

after some simplification.

Of course, in practice what interests us is the prediction variance $Var(\hat{W} - W) = Var\left(G^{-1} \sum_g e_g\right) = \frac{1}{G^2} \sum_g Var(e_g) + \frac{1}{G^2} \sum_g \sum_{h \neq g} Cov(e_g, e_h)$ where $e_g = \hat{W}_g - W$. Let $e = \hat{W} - W$. Then we can write

$$E\left\{\hat{V}_{rep}(\hat{W})\right\} = \frac{1}{G(G-1)} \sum_g E(e_g - e)^2$$

$$= \frac{1}{G(G-1)} \sum_g \left\{ \begin{array}{l} \frac{(G-1)^2}{G^2} Var(e_g) - 2\frac{(G-1)}{G^2} \sum_{h \neq g} Cov(e_g, e_h) \\ + \frac{1}{G^2} \left[\sum_{h \neq g} Var(e_h) + \sum_{h \neq g} \sum_{i \neq h \neq g} Cov(e_h, e_i) \right] \end{array} \right\}$$

$$= \frac{1}{G^2} \sum_g Var(e_g) + \frac{1}{G^2} \sum_g \sum_{h \neq g} Cov(e_g, e_h) - \frac{1}{G(G-1)} \sum_g \sum_{h \neq g} Cov(e_g, e_h)$$

We can see that $\hat{V}_{rep}(\hat{W})$ is generally biased for the prediction variance of $\hat{W}$, with

$$E\left\{\hat{V}_{rep}(\hat{W})\right\} - Var(\hat{W} - W) = -\frac{1}{G(G-1)} \sum_g \sum_{h \neq g} Cov(\hat{W}_g - W, \hat{W}_h - W).$$

This bias term is of the same order of magnitude as $Cov(\hat{W}_g - W, \hat{W}_h - W)$. Typically we expect this covariance to be of order m^{-1}, so the bias of $\hat{V}_{rep}(\hat{W})$ as an estimator of the prediction variance of $\hat{W}$ will be negligible provided that the sub-sample size, m, is large.

One issue with this approach is that the actual full sample predictor $\hat{W}$ used in (12.1) may not be the average of the $\hat{W}_g$. Using (12.1) then leads to a conservative variance estimator, that is it produces a larger estimate of variance than would be the case if $\hat{W}$ were the average of the $\hat{W}_g$.

12.2 Variance Estimation via Dependent Sub-Samples

A problem with the replication-based approach to variance estimation is the stability of these estimates. The more groups there are, the more stable these variance estimates are. However, the more groups there are, the less reliable is the approximation that $\hat{W} \cong G^{-1} \sum_g \hat{W}_g$. This leads naturally to the idea of using overlapping (i.e. dependent) groups.

There are essentially two approaches to using overlapping groups. The first is *Balanced Repeated Replication*, where groups are formed using experimental design methods so that covariances induced by the same unit belonging to different groups cancel out when we use (12.1). This can be quite difficult to accomplish in general. The method is typically restricted to certain types of multistage designs, with $G = 2$, and is rarely used in business surveys. See Wolter (1985) and Shao and Tu (1995).

The other, more commonly used, approach is the *Jackknife* (Quenouille, 1956; Tukey, 1958). Here again the sample is divided into G groups which are mutually independent, but this time the G predictions are computed by excluding each of the G groups from the sample in turn. The variability between the resulting dependent predictions is then used to estimate the variability of the overall predictor of W.

Let $\hat{W}_{(g)}$ be the predictor of W based on the sample excluding group g. The jackknife estimator of variance is given by

$$\hat{V}_{jack}(\hat{W}) = \frac{G-1}{G} \sum_g (\hat{W}_{(g)} - \hat{W})^2. \qquad (12.2)$$

There are two types of jackknife. The Type 1 jackknife has $\hat{W} = G^{-1} \sum_g \hat{W}_{(g)}$. The Type 2 jackknife has $\hat{W}$ equal to the 'full sample' estimate of W. It can easily be shown that the Type 2 jackknife will be more conservative (produce larger estimates of variance) than the Type 1 jackknife.

Like the random groups variance estimator (12.1), the jackknife variance estimator (12.2) is an estimator of $Var(\hat{W})$. However, it is not an unbiased estimator of this quantity in general. To see this, observe that, for a Type 1 jackknife,

$$E\left\{\hat{V}_{jack}(\hat{W})\right\} = \frac{G-1}{G} \sum_g E(\hat{W}_{(g)} - \hat{W})^2 = \frac{G-1}{G} \Big[\sum_g Var(\hat{W}_{(g)} - \hat{W})$$
$$+ \sum_g \left\{E(\hat{W}_{(g)} - \hat{W})\right\}^2\Big].$$

If for all values of g,

$$E\left(\hat{W}_{(g)} - \hat{W}\right) = 0 \qquad (12.3)$$

then

$$E\left\{\hat{V}_{jack}(\hat{W})\right\} = \frac{(G-1)^3}{G^3} \sum_g Var(\hat{W}_{(g)}) - 2\frac{(G-1)^2}{G^3} \sum_g \sum_{h \neq g} Cov(\hat{W}_{(g)}, \hat{W}_{(h)})$$
$$+ \frac{G-1}{G^3} \sum_g \sum_{h \neq g} Var(\hat{W}_{(h)}) + \frac{G-1}{G^3} \sum_g \sum_{h \neq g} \sum_{i \neq h \neq g} Cov\left(\hat{W}_{(h)}, \hat{W}_{(i)}\right)$$
$$= \left(\frac{G-1}{G}\right)^2 \sum_g Var(\hat{W}_{(g)}) - \frac{G-1}{G^2} \sum_g \sum_{h \neq g} Cov\left(\hat{W}_{(g)}, \hat{W}_{(h)}\right)$$

while

$$Var(\hat{W}) = \frac{1}{G^2} \sum_g Var(\hat{W}_{(g)}) + \frac{1}{G^2} \sum_g \sum_{h \neq g} Cov(\hat{W}_{(g)}, \hat{W}_{(h)}).$$

Comparing these two expressions, we see that they will be the same if

$$(G-2) \sum_g Var(\hat{W}_{(g)}) = \sum_g \sum_{h \neq g} Cov(\hat{W}_{(g)}, \hat{W}_{(h)}). \qquad (12.4)$$

That is, the Type 1 jackknife variance estimator (12.2) will be unbiased for the variance of $\hat{W}$ provided the identities (12.3) and (12.4) hold.

The identities (12.3) and (12.4) provide an insight into the reasoning behind the jackknife variance estimator. In particular this approach assumes that the groups are all approximately the same size and exchangeable, in the sense that excluding any particular group does not change the statistical properties of the predictor of W defined using the remaining sample data. This implies that the predictor $\hat{W}_{(g)}$ defined by excluding group g has (approximately) the same expected value and variance for every g. Furthermore, the method also assumes that the correlation between the predictors $\hat{W}_{(g)}$ and $\hat{W}_{(h)}$ defined by excluding two different groups g and h, is the same for any two distinct values of g and h and equals (at least approximately) the proportion $(G-2)(G-1)^{-1}$ of the sample they have in common.

Some important points about applying the jackknife method in practice are:

- The jackknife method does not specify how $\hat{W}_{(g)}$ is defined. However, it is assumed that this predictor uses the same auxiliary information as that used in the full sample predictor of W. To illustrate, suppose that the full sample predictor is an EB predictor, in the sense that it can be written

$$\hat{W} = \hat{E}(W | \text{ sample } s \text{ data, auxiliary population data})$$

where $\hat{E}$ denotes a full sample estimator of the conditional expectation. That is, it is the value of this conditional expectation when all unknown parameters are set equal to corresponding estimates calculated using data from the full sample s. It then follows that $\hat{W}_{(g)}$ is the EB predictor of W based on the sample s data excluding that obtained from group g. That is

$$\hat{W}_{(g)} = \hat{E}(W | \text{ sample} s(g) \text{ data, auxiliary population data}).$$

Here $s(g)$ denotes the sample s with group g excluded. Implicit in this interpretation is the requirement that all parameter estimates used in $\hat{W}_{(g)}$ must be re-estimated from the data in $s(g)$.
- The jackknife variance estimator (12.2) is typically computed at PSU level in multistage samples. That is, G equal sized groups of PSUs are formed and (12.2) calculated.
- The most common type of jackknife is when G is equal to the number of PSU's in the sample, that is one PSU is dropped from the sample each time a value of $\hat{W}_{(g)}$ is calculated. This is the so-called *drop-out-one jackknife*.
- Since the jackknife variance estimator (12.2) is an estimator of $Var(\hat{W})$, it is biased as an estimator of the prediction variance of $\hat{W}$. When the sample size is small relative to the population size, this is of no great concern. However, for business surveys the sampling fraction can be substantial in some strata. In

this case a standard modification is to multiply (12.2) by the finite population correction factor $1 - nN^{-1}$.

There can be a heavy computational burden when G is large (e.g. $G = n$) in the jackknife, so it is sometimes convenient to calculate an approximation that can be computed in one 'pass' of the sample data. This is the so-called *linearised jackknife* and is defined by essentially replacing $\hat{V}_{jack}(\hat{W})$ by a first order Taylor series approximation to it. Note that this assumes that $\hat{W}$ is a differential function of the sample data.

To simplify the presentation we assume a drop-out-one jackknife, a single stage sample of size n and a population model under which $E(y_i) = \mu_i$.

We first approximate $\hat{W}$ by

$$\hat{W} \cong \hat{W}(\mu_s) + \sum_s \left(\frac{\partial \hat{W}}{\partial y_j} \bigg|_{\mathbf{y}_s = \mu_s} \right) (y_j - \mu_j)$$

where μ_s is the n-vector of expected values for the sample Y-values $\mathbf{y}_s$ and $\hat{W}(\mu_s)$ is the value of $\hat{W}$ when these sample Y-values are replaced by μ_s. Similarly, we approximate $\hat{W}_{(i)}$ by

$$\hat{W}_{(i)} \approx \hat{W}_{(i)}(\mu_{s(i)}) + \sum_{s(i)} \left(\frac{\partial \hat{W}_{(i)}}{\partial y_j} \bigg|_{\mathbf{y}_{s(i)} = \mu_{s(i)}} \right) (y_j - \mu_j)$$

where $\mu_{s(i)}$ is μ_s with μ_i deleted, $\hat{W}_{(i)}$ is the predictor of W based on the sample excluding y_i and $\hat{W}_{(i)}(\mu_{s(i)})$ is $\hat{W}_{(i)}$ evaluated at $\mu_{s(i)}$. Finally, we need two extra assumptions:

$$\hat{W}(\mu_s) = \hat{W}_{(i)}(\mu_{s(i)}) = \theta_0 \qquad (12.5a)$$

and

$$\frac{\partial \hat{W}}{\partial y_j} \bigg|_{\mathbf{y}_s = \mu_s} = \left(\frac{n}{n-1} \right) \frac{\partial \hat{W}_{(i)}}{\partial y_j} \bigg|_{\mathbf{y}_{s(i)} = \mu_{s(i)}}. \qquad (12.5b)$$

Assumption (12.5a) is that $\hat{W}_{(i)}$ is a consistent predictor, which would almost always be the case. Assumption (12.5b) essentially states that the contribution of each observation to $\hat{W}$ is $n/(n-1)$ times the contribution of each observation to $\hat{W}_{(i)}$. This is also reasonable, because $\hat{W}$ and $\hat{W}_{(i)}$ are calculated from n and $n-1$ observations, respectively.

We can then replace the approximation to $\hat{W}_{(i)}$ above by

$$\hat{W}_{(i)} = \frac{n}{n-1} \left\{ \hat{W} - \left(\frac{\partial \hat{W}}{\partial y_i} \bigg|_{\mathbf{y}_s = \mu_s} \right) (y_i - \mu_i) \right\} - \frac{\theta_0}{n-1}.$$

This expression can be calculated for every unit in sample in one pass of the data. Its use in the jackknife variance estimator formula (12.2) leads to the linearised Type 1 jackknife variance estimator:

$$\hat{V}_{jack,lin}^{(1)}(\hat{W}) = \frac{n}{n-1} \sum_s \left\{ \left(\left. \frac{\partial \hat{W}}{\partial y_i} \right|_{\mathbf{y}_s=\hat{\mu}_s} \right) (y_i - \hat{\mu}_i) - \frac{1}{n} \sum_s \right.$$
$$\left. \left(\left. \frac{\partial \hat{W}}{\partial y_j} \right|_{\mathbf{y}_s=\hat{\mu}_s} \right) (y_j - \hat{\mu}_j) \right\}^2 \qquad (12.6)$$

where $\hat{\mu}_s$ is the full sample estimate of μ.

12.3 Variance and Interval Estimation via Bootstrapping

The variance estimation methods described in Sections 12.1 and 12.2 are just that – methods that attempt to directly estimate the variance of $\hat{W}$. Prediction intervals and other methods of inference for W are then based on the usual large sample assumption of an underlying normal distribution for $\hat{W}$. Alternatively, a t-distribution with G degrees of freedom could be used, but this is also based on a large sample assumption, albeit a better one than the Central Limit Theorem. The *bootstrap method* (Efron, 1982; Hall, 1992) is more ambitious. Its aim is to recreate the actual distribution of either $\hat{W}$ or $\hat{W} - W$ by a process of simulation. This is particularly useful for many sample designs where sample sizes are too small for central limit behaviour to be applicable (e.g. fine strata containing relatively few units) or situations where the distribution of the sample error may be quite non-normal. Bootstrapping is then a way of estimating the distribution of $\hat{W}$ or $\hat{W} - W$ directly. Once this estimated distribution is obtained, variance estimates and prediction intervals can be 'read off' it quite easily.

To simplify presentation, we assume a single level population, with the parameter of interest, W, defined in terms of the values of a single Y-variable with distribution specified by a slightly more restrictive version of the general regression model considered in Section 10.2. Under this model we assume that there exists a conditional mean function $\mu(z; \omega)$ and a strictly positive conditional variance function $\sigma^2(z; \omega)$, both completely specified except for an unknown vector of parameters ω, such that the population values satisfy

$$\varepsilon_i = \frac{y_i - \mu(z_i; \omega)}{\sigma(z_i; \omega)} \sim IID(0,1). \qquad (12.7)$$

Here the Z-values correspond to an auxiliary variable known for all population units and $IID(0,1)$ denotes independently and identically distributed with zero mean and unit variance. As usual, we assume non-informative sampling given Z.

The first bootstrap method we describe is extremely simple and effectively assumes that the population and sample distributions of Z are the same. It is also non-parametric, in the sense that the model (12.7) forms no part of the bootstrap procedure (although this model will typically influence the definition of $\hat{W}$). This is the *naïve unconditional bootstrap*, which is defined by the steps in the following algorithm:

A1. Given a sample s of n units from the population, we draw a sub-sample s^* of size n randomly and with replacement from s. The values of Y and Z defined by the units (not all distinct) in s^* are denoted $\mathbf{y}_{s^*}$ and $\mathbf{z}_{s^*}$ respectively. Let r denote the non-sampled population units. We also randomly resample $N - n$ times with replacement from the non-sample Z-values $\mathbf{z}_r$ to define a bootstrap realisation $\mathbf{z}_{r^*}$ of this quantity. The bootstrap realisation of $\hat{W}$ is the value of this statistic computed using $\mathbf{y}_{s^*}$, $\mathbf{z}_{s^*}$ and $\mathbf{z}_{r^*}$. We denote this value by $\hat{W}^*$.

A2. The process described in step 1 is independently repeated a large number of times, to generate a bootstrap distribution of $\hat{W}^*$ values.

A3. The bootstrap estimate of the variance of $\hat{W}$ is the (empirical) variance of this bootstrap distribution. Similarly, a bootstrap 95% prediction interval for W can be defined by the (empirical) 2.5 and 97.5 percentiles of this distribution.

The other bootstrap procedure we describe is more complex, and depends on correct specification of the model (12.7). Let $\hat{\omega}$ denote an efficient estimator of ω based on fitting (12.7) to the sample data, and let $r_i(\hat{\omega}) = (y_i - \mu(z_i; \hat{\omega}))/\sigma(z_i; \hat{\omega})$ for $i \in s$ be the resulting set of *studentised* residuals generated by the sample data under this model. For large samples, they satisfy $E\{r_i(\hat{\omega})\} \cong 0$ and $Var\{r_i(\hat{\omega})\} \cong 1$ under (12.7). The steps in this second bootstrap procedure are then

B1. Generate N bootstrap residuals $\{r_j^*; j \in U\}$ by sampling at random and with replacement N times from the n studentised residuals $\{r_i(\hat{\omega}); i \in s\}$.

B2. Generate a bootstrap realisation $\{y_j^* = \mu(z_j; \hat{\omega}) + \sigma(z_j; \hat{\omega})r_j^*; j \in U\}$ of the population Y-values.

B3. Compute the value W^* of W for this bootstrap population.

B4. Compute the bootstrap estimate $\hat{W}^*$ of W^* based on the values $\{y_j^*; i \in s\}$. The bootstrap realisation of the sample error is then $\hat{W}^* - W^*$.

B5. Repeat steps B1–B4 above a large number of times thus generating a distribution of bootstrap sample errors $E^* = \hat{W}^* - W^*$. Denote the (empirical) mean of this bootstrap distribution by $\bar{E}^*$ and its (empirical) variance by V^*.

B6. The final bootstrap estimate of the distribution of prediction errors $\hat{W} - W$ is then defined by the values $E^* - \bar{E}^*$. By construction, this distribution has a zero mean and variance V^*.

B7. Let $L_{boot}(\alpha/2)$ and $U_{boot}(1 - \alpha/2)$ denote the $\alpha/2$ and $1 - \alpha/2$ (empirical) percentiles respectively of the bootstrap estimate of the distribution of prediction errors $\hat{W} - W$. A $100(1 - \alpha)\%$ prediction interval for W is then $\hat{W} - U_{boot}(1 - \alpha/2) \leq W \leq \hat{W} - L_{boot}(\alpha/2)$.

The procedure (B1)–(B7) above is an example of the *percentile bootstrap* (Hall 1992). It has been called *Bootstrap World* by some authors, since it simulates the distribution of the prediction error by first using the working model to simulate the target population, and then 'reads off' this error from the simulated population and sample values. Since it conditions on the actual sample and non-sample distributions of Z, an alternative name is the *model-based conditional bootstrap*.

Unfortunately, the variance V^* of the bootstrap estimate of the distribution of prediction errors generated by (B1)–(B7) is typically an underestimate of the actual prediction variance of $\hat{W}$. This is because it does not take into account the error in $\hat{\omega}$ when generating the population values. If we have a good estimate of this prediction variance, we can rescale the bootstrap distribution defined by the values $E^* - \bar{E}^*$ in step (B6) above to recover this estimated prediction variance. However, this is a complex process that introduces further difficulties which we will not go into. Another issue is that, as we have already seen in Chapter 9, misspecification of the variance function $\sigma^2(z;\omega)$ can lead to biased estimation of prediction variances. So even if we did have an estimate of the prediction variance of $\hat{W}$, and used it to rescale the bootstrap estimate of the distribution of prediction errors, there is no guarantee that the result will lead to valid prediction intervals for W.

In order to get around these problems, Chambers and Dorfman (1994, 2003) take a similar approach to that underpinning the robust prediction variance methods described in Chapter 9 and base the conditional bootstrap on the raw residuals $\{y_i - \mu(z_i; \hat{\omega}); i \in s\}$ rather than the studentised residuals $\{r_i(\hat{\omega}); i \in s\}$. The justification for their approach is Liu (1988), who shows that bootstrap simulation based on independent but not identically distributed data is still valid, provided the data all have essentially the same location. Here this means steps (B1) and (B2) of the conditional bootstrap need to be replaced by

B1a. Generate N bootstrap residuals $\{r_j^*; j \in U\}$ by sampling at random and with replacement N times from the n raw residuals $\{y_i - \mu(z_i; \hat{\omega}); i \in s\}$.

B2a. Generate a bootstrap realisation $\{y_j^* = \mu(z_j; \hat{\omega}) + r_j^*; j \in U\}$ of the population Y-values.

Steps (B3)–(B7) of the algorithm are unchanged.

This modified conditional bootstrap method was used to compute interval estimates for the outlier robust versions of the ratio estimator discussed in Section 10.5. See Table 10.2. In particular, using the same notation as there, the raw residuals used to define the bootstrap procedure for $\hat{t}_y^{RM}$ were the robust ratio residuals $\{y_i - \hat{\beta}_1 z_i; i \in s\}$, while those used to define the bootstrap procedure for $\hat{t}_y^{RNP}$ were the standard ratio residuals $\{y_i - \hat{\beta} z_i; i \in s\}$.

For a recent review of resampling methods in surveys, see Gershunskaya *et al.* (2009), particularly Section 6.

13 Estimation for Multipurpose Surveys

Much of the theory developed in Parts 1 and 2 of this book assumes a scalar auxiliary variable Z, although in Chapter 7 this was generalised to the case of a vector-valued auxiliary variable. In general, however, we have assumed that there is only a single survey variable of interest, Y. In this chapter we allow Y to be multivariate, reflecting the reality that surveys are typically multipurpose, that is they collect data on many variables simultaneously. We therefore focus on the realistic situation where the survey variable is vector valued, with each component Y_j related to the vector-valued auxiliary variable $\mathbf{Z}$ in a different way.

To start, we assume that it is reasonable to model these different relationships linearly. Let $\mathbf{y}_{Uj}$ denote the N-vector of population values $\{y_{ij}; i \in U\}$ of Y_j, and suppose we wish to predict the corresponding population total $t_j = \sum_U y_{ij}$. Let $\mathbf{Z}_U = [z_{ik}; i \in U, k = 1, \ldots, p]$ denote the $N \times p$ matrix of values of p auxiliary variables. This matrix is assumed to be of full rank. As usual, we assume non-informative sampling and full response (or non-informative non-response).

Our working model therefore is the general linear model (7.5), defined by

$$\mathbf{y}_{Uj} = \mathbf{Z}_U \beta_j + \mathbf{e}_{Uj} \tag{13.1}$$

where β_j is an unknown vector of regression parameters of dimension p, $E(\mathbf{e}_{Uj}) = \mathbf{0}_N$ and $Var(\mathbf{e}_{Uj}) = \sigma_j^2 \mathbf{V}_U$. Here $\mathbf{0}_N$ is the zero vector of dimension N, σ_j^2 is an unknown positive scaling constant and $\mathbf{V}_U = \mathbf{V}(\mathbf{Z}_U)$ is a known positive definite matrix. As usual, we assume that both $\mathbf{Z}_U$ and $\mathbf{V}_U$ can be partitioned conformably into sample and non-sample sub-matrices as

$$\mathbf{Z}_U = \begin{bmatrix} \mathbf{Z}_s \\ \mathbf{Z}_r \end{bmatrix} \quad \text{and} \quad \mathbf{V}_U = \begin{bmatrix} \mathbf{V}_{ss} & \mathbf{V}_{sr} \\ \mathbf{V}_{rs} & \mathbf{V}_{rr} \end{bmatrix}.$$

An important consideration in multipurpose surveys is the *aggregation consistency* of the estimation methods used. That is, the sum of the predicted values of the population totals of two different Y-variables should be the same as the predicted value of the derived variable defined by the sum of these two Y-variables. More generally, it is usually required that the predicted value of the population total of any linear combination of Y-variables should be the same as the corresponding linear combination of the predicted values of the individual population totals of these variables.

This consistency is typically achieved by means of linear estimation based on a **fixed** set of sample weights. For arbitrary j we therefore consider a general linear estimator for the population total t_j of Y_j based on sample weights that are fixed for the sample units (i.e. they do not vary for different Y-variables). This is of the form

$$\hat{t}_j^w = \sum_s w_i y_{ij} = \mathbf{w}_s' \mathbf{y}_{sj} \qquad (13.2)$$

where $\mathbf{y}_{sj} = \{y_{ij}; i \in s\}$ denotes the vector of sample values of Y_j and $\mathbf{w}_s = \{w_i; i \in s\}$ denotes the vector of fixed sample weights. Note that although these weights do not vary from one Y-variable to the next, they usually will depend on the auxiliary information in $\mathbf{Z}$ and the sample s of selected population units.

13.1 Calibrated Weighting via Linear Unbiased Weighting

A common requirement for such general-purpose sample weights is that they are *calibrated* on $\mathbf{Z}_U$, or more accurately, on the auxiliary variables that define $\mathbf{Z}_U$. That is, when we calculate (13.2) with the sample values of y_{ij} replaced by the corresponding sample values of the component variables that make up $\mathbf{Z}_U$, we actually recover the known population totals of these auxiliary variables. More formally, we require

$$\hat{\mathbf{t}}_Z^w = \mathbf{Z}_s' \mathbf{w}_s = \mathbf{Z}_U' \mathbf{1}_N = \mathbf{t}_Z \qquad (13.3)$$

where $\mathbf{Z}_s$ denotes the restriction of $\mathbf{Z}_U$ to the sample, $\mathbf{1}_N$ denotes an N-vector of ones and $\mathbf{t}_Z$ denotes the p-vector made up of the population totals of the component variables in $\mathbf{Z}_U$.

Arguments for calibration are often practical, rather than theoretical. Typically, the totals in $\mathbf{t}_Z$ constitute knowledge about the population of interest, often referred to as *benchmark* information, and it is desirable that the prediction method defined by (13.2) gets them 'right'. Other arguments relate to consistency with other surveys, with much larger sample sizes, where the totals in $\mathbf{t}_Z$ are estimates derived from these surveys and we want predictions derived using (13.2) to be consistent with these more 'precise' external estimates.

From a model-based perspective, a compelling theoretical argument for calibration is not hard to find. Essentially (13.3) is a sufficient condition for (13.2) to be unbiased under (13.1), since then

$$E\left(\hat{t}_j^w - t_j\right) = \mathbf{w}_s' \mathbf{Z}_s \beta_j - \mathbf{1}_N' \mathbf{Z}_U \beta_j = (\mathbf{w}_s' \mathbf{Z}_s - \mathbf{1}_N' \mathbf{Z}_U) \beta_j = 0.$$

That is, any set of weights that ensure (13.2) is unbiased under (13.1) will be calibrated on $\mathbf{Z}_U$. Conversely, any set of weights that is calibrated on $\mathbf{Z}_U$ will define an unbiased predictor under (13.1).

Since unbiasedness (i.e. calibration) is a rather weak condition, it makes sense to consider using efficient unbiased weights. In Section 7.2 we derived the weights that define the best linear unbiased predictor (BLUP) of t_j under (13.1). These are

$$\mathbf{w}_s^{BLUP} = \mathbf{1}_n + \left\{\mathbf{H}_L'\mathbf{Z}_r' + (\mathbf{I}_n - \mathbf{H}_L'\mathbf{Z}_s')\mathbf{V}_{ss}^{-1}\mathbf{V}_{sr}\right\}\mathbf{1}_{N-n} \qquad (13.4)$$

where $\mathbf{I}_n$ is the identity matrix of order n, $\mathbf{1}_n$ is a n-vector of one's, $\mathbf{1}_{N-n}$ is a $(N-n)$-vector of one's and $\mathbf{H}_L = \left(\mathbf{Z}_s'\mathbf{V}_{ss}^{-1}\mathbf{Z}_s\right)^{-1}\mathbf{Z}_s'\mathbf{V}_{ss}^{-1}$. These weights can readily be shown to be calibrated on $\mathbf{t}_Z$.

More generally, we can consider a family of weights that define linear unbiased predictors under (13.1). In order to do so we observe that $\mathbf{H}_L\mathbf{Z}_s = \mathbf{I}_p$, where $\mathbf{I}_p$ is the identity matrix of order p. Let $\mathbf{H}$ be any p by n matrix that satisfies $\mathbf{H}\mathbf{Z}_s = \mathbf{I}_p$. Weights defined by (13.4) but with $\mathbf{H}_L$ replaced by $\mathbf{H}$, that is

$$\mathbf{w}_s^H = \mathbf{1}_n + \mathbf{H}'\left(\mathbf{Z}_U'\mathbf{1}_N - \mathbf{Z}_s'\mathbf{1}_n\right) + \left(\mathbf{I}_n - \mathbf{H}'\mathbf{Z}_s'\right)\mathbf{V}_{ss}^{-1}\mathbf{V}_{sr}\mathbf{1}_{N-n} \qquad (13.5)$$

will be referred to as linear unbiased (LU) weights. The matrix $\mathbf{H}$ itself will be referred to as a *LU matrix*. It is straightforward to see that the LU weights (13.5) are calibrated on $\mathbf{Z}_U$ for any $\mathbf{V}_U$ and so define an unbiased estimator of t_j under (13.1). When $\mathbf{V}_U$ is diagonal, these LU weights take the form

$$\mathbf{w}_s^H = \mathbf{1}_n + \mathbf{H}'\left(\mathbf{Z}_U'\mathbf{1}_N - \mathbf{Z}_s'\mathbf{1}_n\right) \qquad (13.6)$$

in which case we refer to them as diagonal LU weights. These weights are also calibrated on $\mathbf{Z}_U$.

13.2 Calibration of Non-parametric Weights

The linear model assumption (13.1) in the previous section may be going too far. In such a situation, some form of non-parametric weighting may be preferable. See Section 8.4. In particular we could consider non-parametric smoothing against one of the components of $\mathbf{Z}_U$, leading to the non-parametric predictor (8.18) or (8.21), with associated non-parametric weights $\mathbf{w}_s^{NP} = \{w_i^{NP}; i \in s\}$. Note that provided each Y_j is smoothed against the same $\mathbf{Z}_U$ with the same bandwidth, the non-parametric weights w_i^{NP} satisfy the requirement of being the same for each Y_j.

Unfortunately, non-parametric estimators are usually not calibrated on $\mathbf{Z}_U$, so they are biased under the general linear model (13.1). There are two general approaches to remedying this situation.

The first approach (Deville and Särndal, 1992) is to choose sample weights that are close to the non-parametric weights but which at the same time define a predictor that is unbiased under the general linear model (13.1) (i.e. calibrated on $\mathbf{Z}_U$). In order to develop this approach we require a metric for 'closeness'.

We use the Euclidean metric $Q = \left(\mathbf{w}_s - \mathbf{w}_s^{NP}\right)' \mathbf{\Omega}_s \left(\mathbf{w}_s - \mathbf{w}_s^{NP}\right)$, where $\mathbf{\Omega}_s$ is a positive definite diagonal matrix of order n.

Minimising Q with respect to $\mathbf{w}_s$ subject to the calibration constraint (13.3) leads to calibrated weights that are as close as possible to the original non-parametric weights. Using standard Lagrangian arguments, it can be shown that these closest calibrated weights are:

$$\mathbf{w}_s^{CAL} = \mathbf{w}_s^{NP} + \mathbf{H}'_\Omega \left(\mathbf{Z}'_U \mathbf{1}_N - \mathbf{Z}'_s \mathbf{w}_s^{NP}\right) \tag{13.7}$$

where $\mathbf{H}_\Omega$ is the LU matrix $\mathbf{H}_\Omega = \left(\mathbf{Z}'_s \mathbf{\Omega}_s^{-1} \mathbf{Z}_s\right)^{-1} \mathbf{Z}'_s \mathbf{\Omega}_s^{-1}$.

The second approach tackles the problem from the other end. The idea here is that we want to use a set of diagonal LU weights, $\mathbf{w}_s^\Omega = \mathbf{1}_n + \mathbf{H}'_\Omega \left(\mathbf{Z}'_U \mathbf{1}_N - \mathbf{Z}'_s \mathbf{1}_n\right)$ defined by the LU matrix $\mathbf{H}_\Omega$ above. See (13.6). Such weights are calibrated on $\mathbf{Z}_U$ and hence define an unbiased estimator of t_j under (13.1). However, suppose this model does not actually fit our data. Can we protect ourselves against bias due to potential model misspecification?

A solution to this problem has already been explored in Section 10.3. The basic idea is to add a non-parametric bias correction term to the linear predictor that uses the diagonal LU weights $\mathbf{w}_s^\Omega$. See (10.7). This correction term corresponds to a non-parametric predictor of the bias of the linear predictor, and is given by a non-parametrically weighted sum of the sample residuals under (13.1). Under LU weighting defined by $\mathbf{H}_\Omega$, the fitted values of Y_j for the sampled units are $\hat{\mathbf{y}}_{sj} = \mathbf{Z}_s \mathbf{H}_\Omega \mathbf{y}_{sj} = \{\hat{y}_{ij}; i \in s\}$, with residuals $\mathbf{r}_{sj} = (\mathbf{I}_n - \mathbf{Z}_s \mathbf{H}_\Omega) \mathbf{y}_{sj} = \{r_{ij}; i \in s\}$. If we use the same non-parametric weights $\mathbf{w}_s^{NP}$ to predict the value of this bias, then the correction term is given by

$$\hat{B}_j^{NP} = \sum_s \left(w_i^{NP} - 1\right)(y_{ij} - \hat{y}_{ij}) = \left(\mathbf{w}_s^{NP} - \mathbf{1}_n\right)' \mathbf{r}_{sj}$$
$$= \left(\mathbf{w}_s^{NP} - \mathbf{1}_n\right)' (\mathbf{I}_n - \mathbf{Z}_s \mathbf{H}_\Omega) \mathbf{y}_{sj}.$$

The final predictor (i.e. the one defined by the sum of the original linear predictor and this non-parametric bias correction term) is then also a linear predictor, with weights given by

$$\begin{aligned}\mathbf{w}_s^{\Omega NP} &= \mathbf{w}_s^\Omega + (\mathbf{I}_n - \mathbf{H}'_\Omega \mathbf{Z}'_s)\left(\mathbf{w}_s^{NP} - \mathbf{1}_n\right) \\ &= \mathbf{1}_n + \mathbf{H}'_\Omega \left(\mathbf{Z}'_U \mathbf{1}_N - \mathbf{Z}'_s \mathbf{1}_n\right) + (\mathbf{I}_n - \mathbf{H}'_\Omega \mathbf{Z}'_s)\left(\mathbf{w}_s^{NP} - \mathbf{1}_n\right) \\ &= \mathbf{w}_s^{NP} + \mathbf{H}'_\Omega (\mathbf{Z}'_U \mathbf{1}_N - \mathbf{Z}'_s \mathbf{w}_s^{NP}) \\ &= \mathbf{w}_s^{CAL}.\end{aligned}$$

That is, we end up using the *same* calibrated weights as under the first approach. In other words, parametrically bias calibrating a non-parametric predictor is the same as non-parametrically bias correcting a parametric predictor.

Note that this equivalence result only holds when the parametric predictor is defined by diagonal LU weights, that is where $\mathbf{V}_U$ is diagonal or the population values of Y_j are uncorrelated given $\mathbf{Z}_U$.

13.3 Problems Associated With Calibrated Weights

Since calibration is equivalent to unbiasedness under the general linear model (13.1) specified by $\mathbf{Z}_U$, it is immediately clear that the introduction of too many calibration constraints (i.e. overspecification of $\mathbf{Z}_U$) will lead to a loss in efficiency. The evidence for this is an increase in the variability of the sample weights as the number of constraints increases. Thus, suppose $\mathbf{Z}_U$ contains an intercept term and $\mathbf{V}_U$ is the identity matrix. Then adding an extra column $\mathbf{z}_{p+1}$ to $\mathbf{Z}_U$ (i.e. changing $\mathbf{Z}_U$ to $[\mathbf{Z}_U \; \mathbf{z}_{p+1}]$) increases the sample variance of the BLUP weights (13.4) by $G_1^2 \left(\mathbf{G}_2'\mathbf{G}_2\right)^{-1}$, where this sample variance is defined by

$$\left\{ \sum_{i=1}^{n} w_i^2 - n^{-1} \left(\sum_{i=1}^{n} w_i \right)^2 \right\} / (n-1)$$

and

$$G_1 = \mathbf{1}_r' \left[\mathbf{Z}_r (\mathbf{Z}_s'\mathbf{Z}_s)^{-1} \mathbf{Z}_s' \mathbf{z}_{(p+1)s} - \mathbf{z}_{(p+1)r} \right]$$
$$\mathbf{G}_2 = \mathbf{z}_{(p+1)s} - \mathbf{Z}_s (\mathbf{Z}_s'\mathbf{Z}_s)^{-1} \mathbf{Z}_s' \mathbf{z}_{(p+1)s}.$$

That is, the greater the 'size' p of the linear model (13.1), the greater the variability of a set of LU weights based on that model. Equivalently, the more calibration constraints one imposes, the higher the variability in the resulting set of sample weights. This increased variability can lead to extreme weights, even weights that are substantially negative. This raises the possibility of negative estimates for strictly positive quantities, especially in domain analysis. It also results in larger standard errors.

This problem has been extensively researched. Huang and Fuller (1978) describe an algorithm that numerically searches for strictly positive calibrated weights. In contrast, Deville and Särndal (1992) suggest replacing the Euclidean metric Q by alternative metrics that guarantee positive weights. However, we then lose the natural interpretability of Q. Also, there is then no finite sample theory for this approach. Bankier, Rathwell and Majkowski (1992) adopt a more pragmatic approach, reducing p (removing calibration constraints), until all (calibrated) weights are strictly positive.

Other approaches have focused on minimum mean squared error of predictors of a population total, rather than minimum variance. Silva and Skinner (1997) and Clark and Chambers (2008) search for lower mean squared error by using the sample data to suggest appropriate variables to include in $\mathbf{Z}_U$ rather than by including all possible benchmark variables in this matrix (smaller p – less likely to get negative weights). However, this approach has the disadvantage of requiring that each survey variable have its own set of sample weights.

Bardsley and Chambers (1984) take a different approach. Assuming a diagonal $\mathbf{V}_U$, these authors use a ridge regression approach to 'ridge' the $\mathbf{Z}_U$-based

BLUP weights (13.4) in order to obtain strictly positive weights. This reduces the mean squared error of the resulting linear predictor at the expense of introducing some bias, since these ridged weights are not exactly calibrated (i.e. $\mathbf{Z}'_U \mathbf{1}_N - \mathbf{Z}'_s \mathbf{w}_s = 0$ does not hold exactly). Their approach was extended by Chambers (1996) to allow calculation of weights that lead to predictions with smaller mean squared error and at the same time are close to an initial set of non-parametric weights, thus generalising (13.7).

In order to describe this extension, we assume, as in the development leading up to (13.7), that one starts with an initial set of non-parametric weights $\mathbf{w}_s^{NP}$. The aim then is to derive a modified set of weights $\mathbf{w}_s$ that are close to $\mathbf{w}_s^{NP}$ but possess better calibration properties. Rather than minimising the Euclidean distance between $\mathbf{w}_s^{NP}$ and $\mathbf{w}_s$ subject to exact calibration, this approach minimises the penalised Euclidean metric

$$Q_\lambda = \left(\mathbf{w}_s - \mathbf{w}_s^{NP}\right)' \Omega_s \left(\mathbf{w}_s - \mathbf{w}_s^{NP}\right) + \frac{1}{\lambda} \left(\mathbf{Z}'_U \mathbf{1}_N - \mathbf{Z}'_s \mathbf{w}_s\right)' \mathbf{C} \left(\mathbf{Z}'_U \mathbf{1}_N - \mathbf{Z}'_s \mathbf{w}_s\right).$$

Here λ is a positive scalar parameter (often referred to as the ridge parameter) and $\mathbf{C}$ is a diagonal matrix of order p whose entries reflect:

(i) the relative importance attached to each of the p calibration constraints;
(ii) the different scales of measurement for the benchmark variables that define $\mathbf{Z}_U$.

The solution to this (unconstrained) optimisation problem is a vector of ridged weights

$$\mathbf{w}_s^{ridge}(\lambda, \mathbf{C}) = \mathbf{w}_s^{NP} + \mathbf{G}'_\Omega(\lambda, \mathbf{C}) \left(\mathbf{Z}'_U \mathbf{1}_N - \mathbf{Z}'_s \mathbf{w}_s^{NP}\right) \qquad (13.9)$$

where $\mathbf{G}_\Omega(\lambda, \mathbf{C}) = \left(\lambda \mathbf{C}^{-1} + \mathbf{Z}'_s \Omega_s^{-1} \mathbf{Z}_s\right)^{-1} \mathbf{Z}'_s \Omega_s^{-1}$ is a ridged LU matrix. As $\lambda \downarrow 0$, these ridged weights become standard calibrated weights based on the LU matrix $\mathbf{H}_\Omega = \left(\mathbf{Z}'_s \Omega_s^{-1} \mathbf{Z}_s\right)^{-1} \mathbf{Z}'_s \Omega_s^{-1}$, while as $\lambda \uparrow \infty$, the ridged weights reduce to the non-parametric (and uncalibrated) weights $\mathbf{w}_s^{NP}$ provided all elements of $\mathbf{C}$ are strictly positive and finite. Note that choosing $\mathbf{w}_s^{NP} = \mathbf{1}_n$ leads to the ridged generalisation of the BLUP weights (13.4) described in Bardsley and Chambers (1984).

Zeroing some components of $\mathbf{C}^{-1}$ is equivalent to forcing calibration on the corresponding components of $\mathbf{Z}_U$. In this case, as $\lambda \uparrow \infty$ the ridge weights defined by (13.9) interpolate smoothly between calibration weights under the 'large' version of (13.1) defined by $\mathbf{Z}_U$ and calibration weights under a 'small' version of this model where $\mathbf{Z}_U$ only contains these zeroed components. Thus ridging can be interpreted as a smooth reduction in the dimension of the linear model (13.1).

The calibration weights under the small model above will typically have much smaller variation than those under the large model. Provided these small model weights are all greater than one, there will exist a smallest value of λ, say λ_{min}, such that the ridged weights (13.9) are all at least one. The value λ_{min} can be found by inspection of the behaviour of the weights (13.9) as λ increases. Following Bardsley and Chambers (1984) one can substitute λ_{min} in (13.9) to define strictly positive multipurpose weights that are exactly calibrated on the important components of $\mathbf{Z}_U$ and approximately calibrated on the rest.

13.4 A Simulation Analysis of Calibrated and Ridged Weighting

In order to illustrate the behaviour of the various calibrated and ridged weighting methods described so far, we reproduce selected results from a simulation study reported in Chambers (1996). The target population for this study consisted of 904 cropping, livestock and dairy farms that participated in an economic survey of farm business performance in Australia in the 1980s. Table 13.1 provides details of four important economic variables measured in the survey as well as four associated production outputs that were also measured for these farms. Scatterplots of the distribution of each economic variable against its relevant production output variable are provided in Fig. 13.1. Note that DSE (Dry Sheep Equivalent) refers to an overall size measure for a farm. This measure was calculated as a linear combination of the four production outputs of a farm (wheat, sheep, beef and dairy).

In addition to the above variables, each farm was also classified into one of seven agricultural industries using the Australian Standard Industry Classification (ASIC) code (see Table 13.2) and into one of 39 geographically defined regions (Region).

The simulation study involved a number of different sampling methods, designed to mimic actual sampling reality. In all cases 1000 samples were drawn independently according to each design. These designs are described in Table 13.3.

Table 13.1 Key farm economic and production variables.

Variable	Definition
Wheat income	Annual income from sale of wheat
Beef income	Annual income from sale of beef cattle
Sheep income	Annual income from sale of wool and sheep
Dairy income	Annual income from sale of milk products
Total income	Annual income from all four activities above
Wheat area	Area (hectares) sown to wheat during the year
Beef number	Number of beef cattle on the farm at the end of the year
Sheep number	Number of sheep on the farm at the end of the year
Dairy number	Number of dairy cattle on the farm at the end of the year

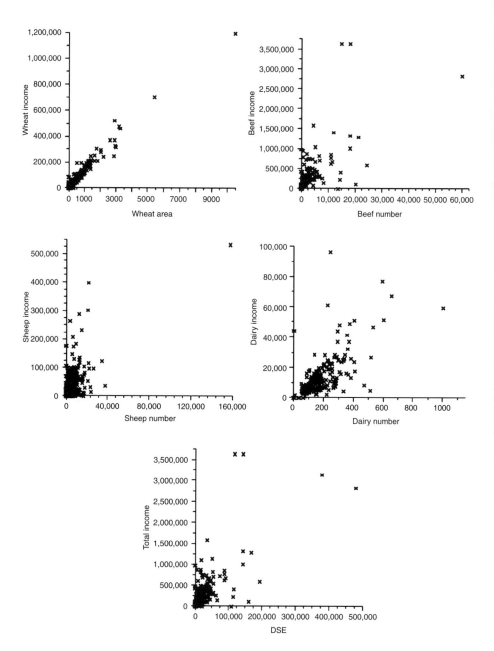

Fig. 13.1 Economic performance vs. production performance for $N = 904$ Australian farms.

Table 13.2 ASIC Industry groups for the farms.

ASIC Number	Industry Description
181	Cropping specialist
182	Mixed crops and sheep production
183	Mixed crops and beef cattle production
184	Mixed livestock production
185	Sheep specialist
186	Beef cattle specialist
187	Dairy farming

Table 13.3 Sample designs used in simulation study.

Design	Description
Simple Random Sampling	Random sample of size $n = 100$ taken without replacement from the population of $N = 904$ farms. A sample was rejected if it did not contain one or more farms from each of the seven ASIC industries, or was without production in one of the four farm outputs (wheat, sheep, beef or dairy).
Size Stratification + Compromise Allocation	Independent random samples taken from four size strata, defined by values of the size variable DSE. Equal aggregate stratification was used (see Section 4.10), based on DSE. Stratum allocations were defined by averaging proportional and Neyman allocation (based on DSE), resulting in the design: | Stratum	DSE Range	N_h	n_h
---	---	---	---
1	200–9499	665	50
2	9500–24999	166	25
3	25000–99999	52	18
4	100000+	12	7	 Nine farms with DSE < 200 were excluded from selection. A sample was rejected if it did not contain one or more farms from each of the seven ASIC industries, or was without wheat, sheep, beef or dairy production.
Size Stratification + Optimal Allocation	Same stratification and sample rejection rule as for size stratification with compromise allocation, but with Neyman allocation based on DSE and with stratum 4 completely enumerated. | Stratum	DSE Range	N_h	n_h
---	---	---	---
1	200–9499	665	30
2	9500–24999	166	29
3	25000–99999	52	29
4	100000+	12	12	

Table 13.4 Prediction methods evaluated in the simulation study.

Method	Description
RATIO	Separate ratio estimators that use Z = production output for a commodity when predicting Y = income from that commodity.
S/BLUP	BLUP weighting (13.4) based on model S.
S/RIDGE	Ridged BLUP weighting (13.9 with $\mathbf{w}_s^{NP} = \mathbf{1}_n$) based on model S with $C_k = 1000$ for each of the four production benchmarks in the model.
S/D3	Non-parametrically bias corrected and ridged BLUP weighting (13.9) based on model S with non-parametric weights $\mathbf{w}_s^{NP}$ defined by local mean smoothing against Z = DSE. Same C-values as S/RIDGE.
S/DAR3	Non-parametrically bias corrected and ridged BLUP weighting (13.9) based on model S, with non-parametric weights $\mathbf{w}_s^{NP}$ defined by local mean smoothing against Z_1 = DSE within poststrata defined by Z_2 = ASIC and Z_3 = Region. Same C-values as S/RIDGE.
L/BLUP	BLUP weighting (13.4) based on model L.
L/RIDGE	Ridged BLUP weighting (13.9 with $\mathbf{w}_s^{NP} = \mathbf{1}_n$) based on model L with $C_k = 1000$ for each of the four production benchmarks in the model, and $C_k = 100000$ for each of the seven industry benchmarks in the model. Because of differences in scale, this actually puts a lower cost on not meeting the industry calibration constraints.
L/D3	Non-parametrically bias corrected and ridged BLUP weighting (13.9) based on model L with non-parametric weights $\mathbf{w}_s^{NP}$ defined by local mean smoothing against Z = DSE. Same C-values as L/RIDGE.
L/DAR3	Non-parametrically bias corrected and ridged BLUP weighting (13.9) based on model L, with non-parametric weights $\mathbf{w}_s^{NP}$ defined by local mean smoothing against Z_1 = DSE within poststrata defined by Z_2 = ASIC and Z_3 = Region. Same C-values as L/RIDGE.

Finally, in Table 13.4, we describe the prediction methods investigated in the study. These methods assumed two working models, both variations on (13.1). The first, denoted model S, defined $\mathbf{Z}_U$ as an intercept column plus four more columns corresponding to the four production outputs specified in Table 13.1 (so $p = 5$). The second, denoted model L, replaced the intercept column in $\mathbf{Z}_U$ by indicator columns for the seven industry groups (so $p = 11$). Both models assumed $\mathbf{V}_U$ was the diagonal matrix defined by the population values of the overall size measure DSE. Three approaches to ridging, labelled /RIDGE, /D3 and /DAR3, were implemented for each of the two models. In each case, ridge weights were calculated using (13.9) and the ridge parameter λ was chosen to be the smallest value such that all weights for the sample were greater than or equal to 1.

As expected, a number of the simulated samples generated negative weights for the non-ridged (i.e. calibrated) weighting methods. Table 13.5 shows the percentages of samples that generated negative weights under BLUP weighting

Table 13.5 Percentages of samples with at least one negative weight. The numbers in parentheses are the average number of sample units with a negative weight in samples containing at least one negative weight.

Weighting Method	Simple Random Sampling	Size Stratification/ Compromise Allocation	Size Stratification/ Optimal Allocation
S/BLUP	44 (4.58)	6 (1.33)	48 (1.81)
L/BLUP	77 (14.11)	20 (1.83)	93 (5.87)

Table 13.6 Root mean squared errors of the different prediction methods under simple random sampling (expressed as percentages of the population total of interest). Minimum value for each variable shown in boldface, and methods within 5% of the minimum are shaded.

Method	Wheat income	Beef income	Sheep income	Dairy income	Total income
RATIO	14.7	28.9	19.1	**14.4**	16.7
S/BLUP	14.0	27.4	17.2	15.6	17.8
S/RIDGE	15.8	24.2	16.3	20.4	15.8
S/D3	15.1	22.2	16.1	18.1	14.5
S/DAR3	14.4	22.6	15.9	17.3	14.7
L/BLUP	**13.6**	26.1	17.0	15.0	17.3
L/RIDGE	15.7	23.6	16.0	17.1	15.7
L/D3	15.0	**22.1**	15.9	17.5	14.6
L/DAR3	14.5	22.4	**15.6**	17.0	14.7

(13.4). The increase in the incidence of negative weights with the increase in the 'size' of the model is clear.

The efficiencies of the different predictors investigated in the simulation study were measured by computing their empirical root mean squared errors over the simulations. These are shown in Tables 13.6–13.8.

Examination of Tables 13.6–13.8 leads to a number of interesting observations.

- Losses in efficiency from adopting a multipurpose weighting strategy seem minimal. The RATIO weighting method, which is variable specific, did not perform significantly better overall than the multipurpose weighting methods. In fact, it recorded the best mean squared error for just one variable (Dairy income).

Table 13.7 Root mean squared errors of the different prediction methods under size stratification with compromise allocation (expressed as percentages of the population total of interest). Minimum value for each variable shown in boldface, and methods within 5% of the minimum are shaded.

Method	Wheat income	Beef income	Sheep income	Dairy income	Total income
RATIO	10.0	11.6	15.5	**19.2**	8.3
S/BLUP	10.8	14.5	14.8	25.2	10.2
S/RIDGE	11.4	14.5	14.8	25.2	10.2
S/D3	10.1	11.8	13.9	19.6	**8.1**
S/DAR3	**9.9**	12.1	13.8	19.9	8.2
L/BLUP	10.8	12.8	14.3	20.5	8.9
L/RIDGE	13.2	13.1	15.6	23.1	9.8
L/D3	10.5	**11.5**	14.1	19.8	**8.1**
L/DAR3	10.5	11.6	14.1	19.7	**8.1**

Table 13.8 Root mean squared errors of the different prediction methods under size stratification with optimal allocation (expressed as percentages of the population total of interest). Minimum value for each variable shown in boldface, and methods within 5% of the minimum are shaded.

Method	Wheat income	Beef income	Sheep income	Dairy income	Total income
RATIO	10.1	10.1	15.9	**25.7**	7.9
S/BLUP	**9.1**	11.1	14.8	34.2	8.3
S/RIDGE	12.6	10.7	15.7	37.2	8.7
S/D3	11.5	9.8	**14.3**	29.2	7.4
S/DAR3	11.5	9.6	14.4	29.7	**7.2**
L/BLUP	11.9	11.1	16.4	32.1	8.0
L/RIDGE	23.5	9.6	21.3	47.8	11.9
L/D3	12.5	9.1	15.6	30.7	7.3
L/DAR3	12.9	**8.9**	15.7	31.5	7.3

- The RIDGE procedure, which combines the linear model (13.1) with strictly positive weights, is typically less efficient than the BLUP method that also assumes (13.1) but allows negative weights. This shows the price imposed by constraining weights to be strictly positive under a linear model assumption.
- There are gains (relative to RIDGE) from weakening the linear model assumption for this population, that is by using a non-parametric bias correction, particularly for the variables Beef income, Sheep income and Total income. This is in spite of the restriction to positive weights. Most of these gains occur with simple one-dimensional smoothing (/D3) against the overall size measure. There are further gains to be had from post-stratifying the smoother (/DAR3), but these are not consistent.
- Weighting based on the model L does best under simple random sampling. Under more complex stratified designs the model S does better, and overall prediction performance improves generally under the stratified designs. This provides some evidence for a claim that more complex models are required with simple designs, while more complex designs allow the use of simpler models.

Overall, these results suggest that combining a stratified sampling strategy with a ridge weighting strategy and non-parametric bias correction is a good approach to prediction of economic variables for this population.

13.5 The Interaction Between Sample Weighting and Sample Design

Suppose one has the choice about which sample to select, but calibration is a requirement no matter what sample is selected. Should this influence the way one selects the sample? In particular, selection of the sample so that the calibration constraints are automatically satisfied for a pre-specified method of sample weighting is an alternative way of ensuring that a linear predictor remains unbiased under (13.1). That is, we can achieve calibration by choosing an appropriate sample rather than by modifying sample weights. In Subsection 8.2.6 we defined a $\mathbf{W}$-balanced sample as one that satisfies $\mathbf{1}'_{N-n}\mathbf{W}\mathbf{Z}_s = \mathbf{1}'_{N-n}\mathbf{Z}_r$, where $\mathbf{W}$ is an arbitrary $(N-n) \times n$ matrix. It immediately follows that in such a sample $\mathbf{Z}'_s(\mathbf{1}_n + \mathbf{W}'\mathbf{1}_{N-n}) = \mathbf{Z}'_U\mathbf{1}_N$. That is, a linear predictor based on the sample weights $\mathbf{w}_s = \mathbf{1}_n + \mathbf{W}'\mathbf{1}_{N-n}$ will be calibrated, that is unbiased under (13.1), in a $\mathbf{W}$-balanced sample.

At the design stage of a survey we therefore have two options:

(i) Select a $\mathbf{W}$-balanced sample, then use the sample weights $\mathbf{w}_s$ as defined above.

(ii) Select the sample according to other, more convenient, criteria, then use a calibrated version of $\mathbf{w}_s$.

To illustrate the trade-off between these two options, Chambers (1997) considered two sampling strategies when there is a single auxiliary variable Z. In both cases, calibration to N and the population total t_z of Z is required, indicating that unbiasedness with respect to a linear model with an intercept is the aim. The two strategies are:

1. Ratio estimation with balanced sampling. In general, the ratio estimator achieves calibration to t_z but not to N. If the sample is balanced so that $\bar{z}_s = \bar{z}_U$, then calibration to N is also achieved.
2. Regression estimation with simple random sampling.

Which is the better approach to achieve calibration on N and t_z? Strategy 1 achieves calibration partly through the use of balanced sampling. In contrast, Strategy 2 does not use a balanced sample, and achieves calibration purely through unbiased prediction. (Note that the two strategies also use different variance models. To be fair, a third strategy should therefore perhaps be added, based on the BLUP under a linear population model with variance proportional to z_i.)

It seems sensible to choose the option that leads to a smaller mean squared error. Since both options lead to an unbiased predictor under (13.1), this is equivalent to choosing the option that leads to the smaller prediction variance. That is, if the values of the Calibration Efficiency Ratio (*CER*)

$$CER = \frac{Var\left(\hat{t}_y - t_y \mid \text{unbalanced sample, calibrated weights}\right)}{Var\left(\hat{t}_y - t_y \mid \text{balanced sample, uncalibrated weights}\right)}.$$

are generally greater than one then option (i) is preferable, otherwise option (ii) is preferable.

Consider the popular sampling scenario defined by simple random sampling without replacement with ratio estimation based on a scalar auxiliary variable Z. Note that the ratio weights (i.e. the weights defining the ratio estimator) are $\mathbf{w}_s^R = (N\bar{z}_U/n\bar{z}_s)\mathbf{1}_n$. Clearly, these weights are calibrated on Z. However they are not calibrated on the population size N. Let $\tilde{\mathbf{Z}}_s = [\mathbf{1}_n \; \mathbf{Z}_s]$ and $\mathbf{Z}_s$ be the n-vector of sample values of Z. Substituting $\mathbf{w}_s^R$ and $\mathbf{V}_{ss} = diag(\mathbf{Z}_s)$ for $\mathbf{w}_s^{NP}$ and Ω_s in (13.7) we obtain modified ratio weights that are calibrated on N and Z,

$$\mathbf{w}_s^{RCAL} = \{w_i^{RCAL}; i \in s\} = \frac{N\bar{z}_U}{n\bar{z}_s}\mathbf{1}_n + \mathbf{V}_{ss}^{-1}\tilde{\mathbf{Z}}_s\left(\tilde{\mathbf{Z}}_s'\mathbf{V}_{ss}^{-1}\tilde{\mathbf{Z}}_s\right)^{-1}\begin{pmatrix} N\left(1 - \bar{z}_U/\bar{z}_s\right) \\ 0 \end{pmatrix}.$$

Here $\tilde{\mathbf{Z}}_s = [\mathbf{1}_n \; \mathbf{Z}_s]$, $\mathbf{Z}_s$ is the n-vector of sample values of Z and $\mathbf{V}_{ss} = diag(\mathbf{Z}_s)$. Conversely, if we select a first order balanced sample, that is one such that $\bar{z}_s = \bar{z}_U$, then $\mathbf{w}_s^R = (N/n)\mathbf{1}_n$ and these weights are calibrated on both N and Z in this sample (and this type of sample only).

Under balanced sampling the prediction variance of the ratio estimator under the ratio population model is $\sigma^2 n^{-1} N(N-n)\bar{z}_U$ (see Subsection 8.2.3). On the other hand, for arbitrary s the prediction variance of the linear estimator based on the calibrated weights $\mathbf{w}_s^{RCAL}$ defined above is

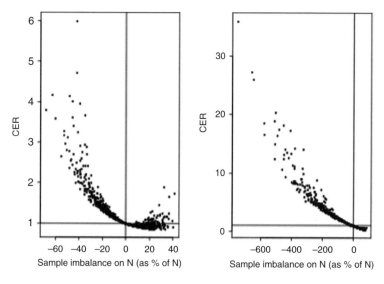

Fig. 13.2 Estimated CER values for samples of size $n = 10$ under the SRS/RATIO sampling strategy. Results for SUGAR are shown in the left pane and those for BEEF in the right pane.

$\sigma^2 \left[\sum_s (w_i^{RCAL} - 1)^2 z_i + \sum_r z_i \right]$ under the same model. The estimated value of CER for an arbitrarily selected sample is therefore

$$C\hat{E}R = \frac{\sum_s \left(w_i^{RCAL} - 1 \right)^2 z_i + \sum_r z_i}{n^{-1} N(N-n) \bar{z}_U}. \tag{13.10}$$

Consequently, by repeatedly sampling using SRSWOR from a population that follows the ratio population model, we can estimate the distribution of CER values from the resulting values of (13.10).

In an empirical study of the distribution of the estimated CER values (13.10), Chambers (1997) considers repeated sampling from two real populations. The first population (SUGAR) is defined by the sample of 338 sugarcane farms introduced in Section 5.1 (see Figure 5.2). Here Z = area assigned for cane growing. As we saw there, the ratio population model fits this population quite well. The second population (BEEF) is the sample of 430 Australian beef cattle farms that was described in Section 10.5 (see Figure 10.1). Here Z = number of beef cattle at the end of the financial year. The distribution of Z in the BEEF population is much more skewed than in the SUGAR population.

Figures 13.2 to 13.4 are taken from Chambers (1997) and show estimated CER values calculated via (13.10) for the SUGAR and BEEF populations plotted against sample imbalance on Z for three different sample sizes ($n = 10$, 30 and 100). These values were obtained by repeatedly sampling via SRS from these populations. Note that sample imbalance is defined as $(1 - \bar{z}_U/\bar{z}_s) \times 100$, so negative values of sample imbalance correspond to samples made up of population

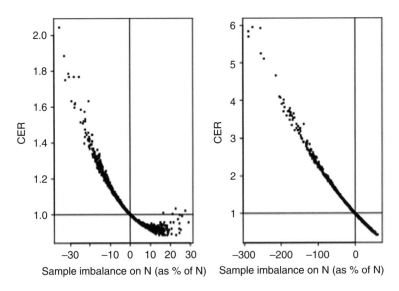

Fig. 13.3 Estimated *CER* values for samples of size $n = 30$ under the SRS/RATIO sampling strategy. Results for SUGAR are shown in the left pane and those for BEEF in the right pane.

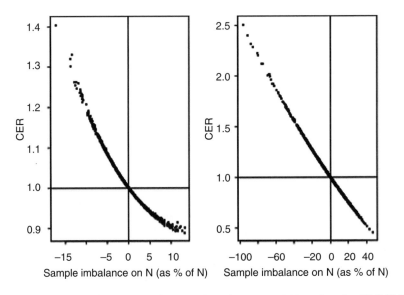

Fig. 13.4 Estimated *CER* values for samples of size $n = 100$ under the SRS/RATIO sampling strategy. Results for SUGAR are shown in the left pane and those for BEEF in the right pane.

units with smaller Z-values, while positive values correspond to samples made up of population units with larger Z-values.

Inspection of these plots indicates that for these populations calibrating ratio weights to sum to the population size N is always more inefficient than choosing a balanced sample when $\bar{z}_s < \bar{z}_U$. Conversely, it can be more efficient than choosing a balanced sample when $\bar{z}_s > \bar{z}_U$, provided the sample size is large enough. However, for smaller sample sizes weight calibration can be less efficient than balancing. For populations with positively skewed Z distributions, we expect that samples with $\bar{z}_s < \bar{z}_U$ will occur more than 50% of the time under SRS. Consequently, we have a 50:50 chance at best of improving on balanced sampling by combining SRS with calibrated ratio weights. A risk-averse strategy would therefore seek to calibrate via balancing rather than via weight modification when sampling from these two populations.

Before closing this discussion, it should be noted that the preceding results are based on complete response. In practice this almost never happens. Consequently, even if a sample that balanced on N and Z is initially selected, it will often end up unbalanced because of sample non-response. In such cases we have no option but to calibrate the sample weights to correct for the imbalance caused by this non-response.

14 Inference for Domains

A domain is a subgroup of the target population for which a separate estimate of the total of Y (or mean etc.) is required. For example, in many business surveys the sample frame is out of date, so the industry and size classifications of many units on the frame do not agree with their 'current' industry and size classifications. After the survey is carried out, estimates are required for the current industry by size classes, which can be determined from the information provided by the sampled businesses. These classes then correspond to domains of interest.

As the example above demonstrates, a basic assumption is that domain membership is observable on the sample. Consequently, we can define a domain membership variable D with value d_i for population unit i, such that $d_i = 1$ if unit i is in the domain and is zero otherwise. The number of population units in the domain is the population total of D, $N_d = \sum_U d_i$, and the population total of Y for the domain is $t_{dy} = \sum_U d_i y_i$. The domain total of interest is therefore the population total of the derived variable DY.

14.1 Unknown Domain Membership

The most common situation in domain estimation is that membership of the domain is only observable on the sample. We do not know how the non-sampled units are distributed over the domains. In particular, we do not know how many of them there are in the domain identified by D. In this case we need to treat the domain membership variable D on the same footing as the survey variable Y and model their joint distribution.

To illustrate, consider a working model for the distribution of Y values in the domain that is a simple extension of the homogeneous population model (3.1). Specifically, we assume that domain membership (D) can be modelled as N independent and identically distributed realisations of a Bernoulli random variable, and, conditional on D, the population values of Y are uncorrelated with constant mean and variance. Under this working model we therefore have

$$E(y_i|d_i = 1) = \mu_d \qquad (14.1\text{a})$$
$$Var(y_i|d_i = 1) = \sigma_d^2 \qquad (14.1\text{b})$$

Unknown Domain Membership

$$Cov(y_i, y_j | d_i, d_j; i \neq j) = 0 \quad (14.1c)$$
$$E(d_i) = \theta_d \quad (14.1d)$$
$$Var(d_i) = \theta_d(1 - \theta_d) \quad (14.1e)$$
$$Cov(d_i, d_j | i \neq j) = 0. \quad (14.1f)$$

As always we assume non-informative sampling, that is sample inclusion is independent of the values of the variables of interest. Sample inclusion and domain membership are therefore independent of one another, so we can estimate θ_d from the sample data. This will be true if the sample is randomly chosen, for example.

Under (14.1) we then have

$$E(d_i y_i) = E(y_i | d_i = 1) pr(d_i = 1) = \theta_d \mu_d \quad (14.2a)$$
$$Var(d_i y_i) = E\left(y_i^2 | d_i = 1\right) pr(d_i = 1) - E^2(d_i y_i)$$
$$= \theta_d \sigma_d^2 + \theta_d(1 - \theta_d)\mu_d^2 \quad (14.2b)$$
$$Cov(d_i y_i, d_j y_j | i \neq j) = 0. \quad (14.2c)$$

This is just a special case of the homogeneous population model (3.1), and so the BLUP for t_{dy} is the expansion estimator for the product DY,

$$\hat{t}_{dy}^E = \frac{N}{n} \sum_s d_i y_i = \frac{N}{n} \sum_{s_d} y_i = \frac{N n_d}{n} \bar{y}_{sd} = N \hat{\theta}_d \bar{y}_{sd} \quad (14.3)$$

where s_d denotes the set of n_d sample units in domain d, $\hat{\theta}_d = n^{-1} n_d$ is the unbiased estimator of θ_d and $\bar{y}_{sd}$ is the unbiased estimator of μ_d defined by the mean of the sample Y-values from the domain. As a consequence, the prediction variance of (14.3) is

$$Var\left(\hat{t}_{dy}^E - t_{dy}\right) = \frac{N^2}{n}\left(1 - \frac{n}{N}\right)\{\theta_d \sigma_d^2 + \theta_d(1-\theta_d)\mu_d^2\}. \quad (14.4)$$

We can estimate this variance unbiasedly using (3.6), with DY substituted for Y, leading to

$$\hat{V}\left(\hat{t}_{dy}^E\right) = \frac{N^2}{n}\left(1 - \frac{n}{N}\right) s_{dy}^2 \quad (14.5)$$

where

$$s_{dy}^2 = (n-1)^{-1} \sum_s (d_i y_i - \hat{\theta}_d \bar{y}_{sd})^2.$$

Setting

$$\hat{\sigma}_d^2 = (n_d - 1)^{-1} \sum_{s_d} (y_i - \bar{y}_{sd})^2$$

it is then straightforward to show that (14.5) can equivalently be written

$$\hat{V}\left(\hat{t}_{dy}^E\right) = \frac{N^2}{n}\left(1-\frac{n}{N}\right)\frac{n}{n-1}\left\{\frac{n_d-1}{n_d}\hat{\theta}_d\hat{\sigma}_d^2 + \hat{\theta}_d\left(1-\hat{\theta}_d\right)\bar{y}_{sd}^2\right\}.$$

This is the 'plug-in' estimator of (14.4), if one ignores second order terms.

Domain sample sizes are often small, which means that confidence intervals based on the Central Limit Theorem may perform poorly. Casady, Dorfman and Wang (1998) compare alternative confidence intervals in this case.

The above development can be easily extended to more complex models for the distribution of D and the distribution of Y within the domain. See exercise E.33.

14.2 Using Information about Domain Membership

Suppose that we know which population units belong to the domain of interest; call this information $\mathbf{D_U}$. Thus we can specify our model conditional on $\mathbf{D_U}$. Model (14.1) becomes:

$$E(y_i) = \mu_d \qquad (14.6\text{a})$$

$$Var(y_i) = \sigma_d^2 \qquad (14.6\text{b})$$

$$Cov(y_i, y_j | i \neq j) = 0 \qquad (14.6\text{c})$$

for all units i and j in the domain d, where the expected values, variances and covariances are conditional on $\mathbf{D_U}$. For each domain, this is just the homogenous model, so we know from Chapter 3 that an EB estimator of t_{dy} is

$$\hat{t}_{dy}^E = N_d\bar{y}_{sd} = N_d\frac{\sum_{s_d}y_i}{n_d} = N_d\frac{\sum_s d_i y_i}{\sum_s d_i} \qquad (14.7)$$

where s_d contains the n_d sample units in the domain. An unbiased estimator of the prediction variance of $\hat{t}_{dy}^E$ is then:

$$\hat{V}\left(\hat{t}_{dy}^E\right) = N_d^2 n_d^{-1}\left(1 - n_d/N_d\right)\hat{\sigma}_d^2. \qquad (14.8)$$

Neither $\hat{t}_{dy}^E$ nor $\hat{V}\left(\hat{t}_{dy}^E\right)$ require knowledge of the domain membership of the non-sampled units, provided that N_d is known. This is convenient in practice, because sometimes, although we do not know which non-sampled units are in the domain of interest, we do know how many there are. For example, suppose the domain of interest in a social survey is a particular age by gender group in the population. Here, although we do not know the age and gender of the individuals making up this population, we know from demographic analysis how many people there are in each age by gender group.

14.3 The Weighted Domain Estimator

Suppose that we have calculated a set of weights incorporating auxiliary information on a vector-valued variable Z, for example we may have used the BLUP weights from Section 7.1. Suppose that the weights satisfy the calibration property (13.3). As discussed in Chapter 13, any weights satisfying this property will give unbiased estimates of t_y for any variable Y generated by the general linear model (13.1). In practice a single set of weights $\{w_i : i \in s\}$ are typically used for many variables of interest in a survey. It is also a common practice to use the same set of weights to estimate domain totals in the same way, using:

$$\hat{t}_{dy}^w = \sum_s w_i d_i y_i = \sum_{s_d} w_i y_i. \tag{14.8}$$

This has many desirable practical properties, for example the sum of all of the estimates for a set of domains partitioning the population will add to the estimate of the population total.

Is (14.8) a good estimator? It will be an unbiased estimator of t_{dy} when

$$E\left(\hat{t}_{dy}^w - t_{dy}\right) = 0$$

or equivalently

$$\sum_s w_i E\left(d_i y_i\right) = \sum_U E\left(d_i y_i\right). \tag{14.9}$$

We will suppose that (13.1) holds so that $E\left(y_i | \mathbf{z}_i\right) = \mathbf{z}_i' \beta$. Equation (14.9) will then be satisfied whenever

$$E\left(d_i y_i | \mathbf{z}_i\right) = \mathbf{z}'_i \gamma \tag{14.10}$$

for some vector γ, that is, $E(DY)$ is linear in Z. Two scenarios in which this will occur are:

a. When domain membership D is unrelated to Y given Z, that is

$$E\left(d_i | y_i, \mathbf{z}_i\right) = E\left(d_i | \mathbf{z}_i\right) = \theta.$$

In this case (14.10) is satisfied since

$$E\left(d_i y_i | \mathbf{z}_i\right) = E\left\{E\left(d_i y_i | y_i, \mathbf{z}_i\right) | \mathbf{z}_i\right\} = E\left(\theta y_i | \mathbf{z}_i\right) = \theta \mathbf{z}' \beta_i.$$

b. The auxiliary information consists of stratum membership only. Then (14.10) must hold as it is the most general possible model in this case.

Even if (a) or (b) do not apply, (14.10) may still sometimes be a reasonable approximation.

An even better approach would be to simultaneously model the domain membership variable D and the survey variable Y as a mixture (i.e. treat the population distribution of Y as made up of a finite mixture of domain-specific distributions for this variable) and use EB prediction. This is equivalent to separately modelling $\Pr(d_i = 1 | \mathbf{z}_i)$ and $E(y_i | d_i = 1, \mathbf{z}_i)$, for example using a logistic regression for the former and a linear model for the latter, and then combining these models in the EB predictor. However, the resulting estimator of t_{dy} is then not linear in the variable Y, which would be inconvenient in surveys with many variables and domains.

15 Prediction for Small Areas

Large national samples are often used to produce estimates for subnational domains with small sample sizes. Such domains are typically referred to as 'small areas' because in many cases they are defined geographically. We adopt this convention throughout this chapter, even though the domains of interest can correspond to very large areas. As has already been noted, their defining characteristic, however, is that the sample size in many of these domains is small. When the meaning is clear, we will often leave out the adjective 'small' and just refer to areas.

Small sample sizes make small area prediction problematic, because predictions based on the sample data from any particular area can be unstable. For example, suppose that a sample s is taken from a target population U and a linear predictor based on weights $\{w_i; i \in s\}$ is used to make inferences about population level quantities using data from the sample units. Suppose also that the population can be divided up into $d = 1, \ldots, D$ areas and we wish to use the sample data to predict the average value of a survey variable Y in each of them. If there are sample units from area d then the average $\bar{y}_d$ of Y in this area can be predicted by

$$\hat{\bar{y}}_d^{Ha} = \left(\sum\nolimits_{s_d} w_i\right)^{-1} \left(\sum\nolimits_{s_d} w_i y_i\right) \qquad (15.1\text{a})$$

or, if the population size N_d of the area is known, by

$$\hat{\bar{y}}_d^{HT} = N_d^{-1} \left(\sum\nolimits_{s_d} w_i y_i\right). \qquad (15.1\text{b})$$

Here s_d denotes the set of sampled units in area d. Both versions of (15.1) are sometimes referred to as *direct predictors* of $\bar{y}_d$. More precisely, we refer to (15.1a) as the *Hajek* form of the direct predictor, and (15.1b) as the *Horvitz–Thompson* (HT) form of the direct predictor. These names refer to alternative approaches to estimating finite population means in the design-based sampling literature (see Sections 3.9 and 3.10 of Tillé, 2001). Irrespective of which version of (15.1) is used, however, it is easy to see that its prediction variance is $O\left(n_d^{-1}\right)$ and so the prediction error can be large when the area sample size n_d is small.

This small sample problem can be resolved provided an auxiliary variable Z (typically multivariate) is available to 'strengthen' the limited sample data from area d. Values of Z can be the values of the survey variable Y from the same area in the past and/or the values of other variables that are related to the variable of interest. A key feature of the approach is the assumption that some of the parameters of a model for the relationship between Y and Z in any particular area are the same in every area, and so our inference for any particular area can *borrow strength* from other areas by using data from the entire sample to estimate these common parameters. In this chapter we describe some methods for doing this.

15.1 Synthetic Methods

There are essentially two classes of models in use in small area inference. The first corresponds to the situation where the model that is assumed to hold in any particular area actually holds in every area, and so holds for the population as a whole. That is, since the same fitted model is used everywhere, we don't need to know the area in which a particular population unit is located (or even to have sample units in the area) in order to predict its value of Y. The second corresponds to the situation where a different model holds in each small area, and so any corresponding population model is defined by the accumulation of these area level models. Here we need to know the area in which a population unit is located in order to predict its value of Y and, preferably, to have sample units in the area on which to base this predicted value. In this case small area inference usually proceeds by restricting the number of model parameters that can vary from area to area, with parameters common to all areas estimated using data from the entire sample. We consider models of this type in Section 15.2 below.

The assumption that the population level model and the area level model are the same is often referred to as the *synthetic assumption*. It is equivalent to assuming that between-area variability in the values of Y is explained entirely in terms of between-area variability in the values of the auxiliary variable Z. Hence a working model for the population that predicts Y using Z also applies in each area. To illustrate, suppose that our working model for the relationship between Y and Z in the population is the general linear model (7.1), which, using the same notation as in Section 7.1, we write as

$$\mathbf{y}_U = \mathbf{Z}_U \beta + \mathbf{e}_U. \tag{15.2}$$

Given an estimate $\hat{\beta}$ of β, we can then predict the average $\bar{y}_d$ of Y in area d assuming that (15.2) also holds in area d. This leads to the *synthetic* predictor of $\bar{y}_d$ based on the linear model (15.2),

$$\hat{\bar{y}}_d^{SYN} = N_d^{-1} \left(\sum\nolimits_{s_d} y_i + \sum\nolimits_{r_d} \mathbf{z}_i' \hat{\beta} \right) = \bar{\mathbf{z}}_d' \hat{\beta} + N_d^{-1} n_d \left(\bar{y}_{sd} - \bar{\mathbf{z}}_{sd}' \hat{\beta} \right) \tag{15.3a}$$

where $\mathbf{z}_i$ denotes the (vector) value of Z for population unit i and r_d denotes the non-sampled units in the same area. From now on we denote restriction of a vector or matrix to values from area d by adding a subscript of d to the notation for that vector or matrix. Thus, the area d population mean of Z is denoted $\bar{\mathbf{z}}_d$ above, while the corresponding sample means of Y and Z are denoted $\bar{y}_{sd}$ and $\bar{\mathbf{z}}_{sd}$ respectively. Note that calculation of (15.3) requires that, at a minimum, area affiliation is known for all sample units, and that the values of N_d and $\bar{\mathbf{z}}_d$ are known for area d. Observe also that if there is no sample in area d, that is $n_d = 0$, then a synthetic predictor under (15.2) can still be computed, since (15.3a) reduces in this case to

$$\hat{\bar{y}}_d^{SYN} = N_d^{-1} \sum_{U_d} \mathbf{z}_i' \hat{\beta} = \bar{\mathbf{z}}_d' \hat{\beta}. \tag{15.3b}$$

An alternative approach would be to calculate the BLUP weights from model (15.2), and then use these weights in (15.1b). This is not equivalent to (15.3a). To see this, suppose that the variance-covariance matrix of the regression error vector $\mathbf{e}_U$ in (15.2) is proportional to the identity matrix of order N and $\hat{\beta}$ is the BLUE of the regression parameter β in (15.2). In this case the vector of BLUP weights used in (15.1b) is (see Section 7.2)

$$\mathbf{w}_s^{BLUP} = \mathbf{1}_n + \mathbf{Z}_s \left(\mathbf{Z}_s' \mathbf{Z}_s \right)^{-1} \mathbf{Z}_r' \mathbf{1}_{N-n}.$$

With this choice, (15.1b) is of the form

$$N_d^{-1} \sum_{s_d} w_i y_i = N_d^{-1} \left\{ \sum_{s_d} y_i + \left(\sum_g \mathbf{1}_{N_g - n_g}' \mathbf{Z}_{rg} \right) \left(\mathbf{Z}_s' \mathbf{Z}_s \right)^{-1} \mathbf{Z}_{sd}' \mathbf{y}_{sd} \right\}.$$

where we use g to index the areas represented in the sample. This estimator is biased for the average of Y in area d under (15.2) since

$$E\left(\sum_{s_d} w_i y_i - \sum_{i \in d} y_i \right) = \left\{ \left(\sum_g \mathbf{1}_{N_g - n_g}' \mathbf{Z}_{rg} \right) (\mathbf{Z}_s' \mathbf{Z}_s)^{-1} \mathbf{Z}_{sd}' \mathbf{Z}_{sd} \right.$$
$$\left. - \mathbf{1}_{N_d - n_d}' \mathbf{Z}_{rd} \right\} \beta \neq 0.$$

In contrast, (15.3a) cannot be represented as a linear combination of the sample Y-values in area d, but is instead a linear combination of **all** the Y-values in the sample. That is, it is of the form $N_d^{-1} \mathbf{w}_{sd}' \mathbf{y}_s$, with $\mathbf{w}_{sd} = \Delta_{sd} + \mathbf{Z}_s \left(\mathbf{Z}_s' \mathbf{Z}_s \right)^{-1} \mathbf{Z}_{rd}' \mathbf{1}_{N_d - n_d}$. Here Δ_{sd} denotes the n-vector with ones for every sample unit in area d and zeros everywhere else. In this case we do have unbiasedness under (15.2) since

$$E\left(\mathbf{w}_{sd}' \mathbf{y}_s - \sum_{i \in d} y_i \right) = \mathbf{1}_{N_d - n_d}' \mathbf{Z}_{rd} \left\{ (\mathbf{Z}_s' \mathbf{Z}_s)^{-1} \mathbf{Z}_s' \mathbf{Z}_s - \mathbf{I}_p \right\} \beta = 0.$$

An important special case of (15.3a) is where the population level model is the homogeneous strata population model, with the strata 'cutting across' the small areas. If we index these strata by h, then (15.3a) becomes

$$\hat{\bar{y}}_d^{SYN} = N_d^{-1} \sum_h N_{hd} \left\{ \bar{y}_{sh} + f_{hd} \left(\bar{y}_{shd} - \bar{y}_{sh} \right) \right\} \tag{15.4}$$

where $\bar{y}_{sh}$ denotes the stratum h sample mean of Y and $f_{hd} = N_{hd}^{-1} n_{hd}$ is the stratum h sampling fraction in area d. Predictors like (15.4) were the original synthetic predictors investigated in the small area literature. See Gonzalez and Hoza (1978). They are valid provided the synthetic assumption holds. That is, the expected value of an arbitrarily chosen population unit depends on its stratum affiliation, rather than the area in which it is located. If this is true then (15.4) typically has a much smaller prediction variance than the corresponding direct predictor

$$\hat{\bar{y}}_d^{HT} = N_d^{-1} \sum_h N_{hd} \bar{y}_{shd} \tag{15.5}$$

where $\bar{y}_{shd}$ now denotes the sample mean of Y in the (d, h) cell of the area by stratum crossclassification. However, if the synthetic assumption fails, that is the homogeneous strata population model does not hold for a randomly selected unit from area d, then (15.4) can have a large bias, and so its mean squared error can in fact be larger than that of (15.5). In this case the small area estimation method might be considered as having 'borrowed weakness' rather than 'borrowed strength'.

15.2 Methods Based on Random Area Effects

Models like (15.2) account for differences in the distribution of Y between small areas purely in terms of corresponding differences in the distribution of Z. In many cases this is inadequate, in the sense that although its fit to the data from the entire sample is acceptable, its fit within any particular small area is not. Consequently it is now standard to add area specific terms to small area prediction models to allow them to better account for the between area variability in the distribution of Y. In particular, such models characterise the differences between the values of Y in different small areas in terms of the different Z-distributions for the different areas and the values of unobserved random area effects. That is, any between-area variability that remains after we condition on Z is modelled by the introduction of random area effects. An immediate consequence is that the relationship between Y and Z at the small area level is no longer the same as the relationship between these variables in the population as a whole.

When the population is made up of a large number of similar small areas, the simplest and most common assumption consistent with the above reasoning is that the random area effects correspond to independent and identically distributed realisations of a random variable with zero mean. This is the so-called *random intercepts* model, since the random effects then define different area-specific intercepts under a linear model for the regression of Y on Z in the population. See Battese *et al.* (1988) for an early development of this idea. A more sophisticated version of this model allows the regression parameters of the population level linear model to vary randomly between areas. This is sometimes referred to as a *random coefficients*, or *random slopes*, linear model.

Let D denote the total number of areas making up the target population. A quite general specification for a linear model with random area effects is the so-called *linear mixed model* specification, where a random effects term is added to the linear specification (15.2), leading to

$$\mathbf{y}_U = \mathbf{Z}_U \boldsymbol{\beta} + \mathbf{G}_U \mathbf{u} + \mathbf{e}_U. \tag{15.6a}$$

Here $\mathbf{G}_U$ is a $N \times qD$ matrix of fixed known constants and $\mathbf{u}' = [\mathbf{u}'_1 \mathbf{u}'_2 \ldots \mathbf{u}'_D]$ is a vector of dimension qD made up of D independent realisations $\{\mathbf{u}_d; d = 1, \ldots, D\}$ of a q-dimensional random area effect with zero mean vector and covariance matrix $\sigma_u^2 \Gamma$. The unit level error vector $\mathbf{e}_U$ is assumed to be such that $E(\mathbf{e}_U) = \mathbf{0}_N$ and with $Var(\mathbf{e}_U) = \sigma_e^2 \Sigma_U$, where Σ_U is a known diagonal matrix of order N. It is also assumed that $\mathbf{u}$ is distributed independently of $\mathbf{e}_U$. Typically Γ is itself a known function of a vector of parameters that we will call φ, so we write it in the form $\Gamma(\varphi)$. By construction $Var(\mathbf{u}) = \sigma_u^2 \Omega$, where $\Omega = diag\{\Gamma(\varphi); d = 1, \ldots, D\}$, that is the block diagonal matrix with main diagonal made up of D repetitions of $\Gamma(\varphi)$. The parameters σ_e^2, σ_u^2 and φ are often referred to as the *variance components* of the model. We observe that (15.6a) reduces to (15.2) when $\sigma_u^2 = 0$.

Model (15.6a) can also be written in the following unit-specific form, which is perhaps easier to interpret:

$$y_i = \mathbf{z}'_i \boldsymbol{\beta} + \mathbf{g}'_i \mathbf{u}_d + e_i \tag{15.6b}$$

where $\mathbf{z}_i$ and $\mathbf{g}_i$ are vectors of auxiliary variables for unit i that define the ith rows of $\mathbf{Z}_U$ and and $\mathbf{G}_U$ respectively and e_i is a unit specific random error with variance $\sigma_e^2 \Sigma_{ii}$, where Σ_{ii} is the ith diagonal element of Σ_U. The random area effect is defined by the q-vector $\mathbf{u}_d$. The components of this vector can be correlated among themselves within an area, but are assumed to be independent from one area to another. The D values of $\mathbf{u}_d$ are therefore modelled as independently and identically distributed realisations of an area specific q-dimensional random error with covariance matrix $\sigma_u^2 \Gamma(\varphi)$.

We now specify the form of the BLUP for the area d mean $\bar{y}_d$ of Y under (15.6). To start, observe that the restriction to a linear predictor under a BLUP-type approach implicitly assumes that the area population size N_d is known, as are all second order moments up to a constant of proportionality. Consequently, application of this approach under (15.6) requires that the variance components σ_e^2, σ_u^2 and φ be known. For the time being, therefore, we make this assumption. We also note that (15.6) is a special case of the general linear population model (7.1), with

$$Var\left(\mathbf{y}_U | \mathbf{Z}_U\right) = \mathbf{V}_U = \sigma_e^2 \Sigma_U + \sigma_u^2 \mathbf{G}_U \Omega \mathbf{G}'_U.$$

Furthermore, since Ω is block diagonal, $\mathbf{V}_U$ is also block diagonal, and can be written $\mathbf{V}_U = diag(\mathbf{V}_d; d = 1, \ldots, D)$, where $\mathbf{V}_d = Var(\mathbf{y}_d | \mathbf{Z}_d) = \sigma_e^2 \Sigma_d + \sigma_u^2 \mathbf{G}_d \Gamma(\varphi) \mathbf{G}'_d$. Here $\mathbf{y}_d$ is the vector of N_d values of Y for area d and Σ_d, $\mathbf{G}_d$ are the area d components of Σ_U, $\mathbf{G}_U$ respectively.

Given a non-informative sample from the target population, and assuming that the variance components in this model are known, we can then use (7.2) to write down the BLUP for the population mean in area d under (15.6). This is

$$\hat{\bar{y}}_d^{BLUP} = N_d^{-1}\left[\mathbf{1}'_{n_d}\mathbf{y}_{sd} + \mathbf{1}'_{N_d-n_d}\left\{\mathbf{Z}_{rd}\hat{\beta}^{BLUE} + \mathbf{V}_{rsd}\mathbf{V}_{ssd}^{-1}\left(\mathbf{y}_{sd} - \mathbf{Z}_{sd}\hat{\beta}^{BLUE}\right)\right\}\right] \tag{15.7}$$

where

$$\hat{\beta}^{BLUE} = (\mathbf{Z}'_s\mathbf{V}_{ss}^{-1}\mathbf{Z}_s)^{-1}(\mathbf{Z}'_s\mathbf{V}_{ss}^{-1}\mathbf{y}_s) = \left(\sum_d \mathbf{Z}'_{sd}\mathbf{V}_{ssd}^{-1}\mathbf{Z}_{sd}\right)^{-1}\left(\sum_d \mathbf{Z}'_{sd}\mathbf{V}_{ssd}^{-1}\mathbf{y}_{sd}\right)$$

is the best linear unbiased estimator (BLUE) of β. Here $\mathbf{V}_{rs}$ and $\mathbf{V}_{ss}$ are defined by the sample/non-sample decomposition of the covariance matrix $\mathbf{V}_U$ while $\mathbf{V}_{rsd}$ and $\mathbf{V}_{ssd}$ are defined by the corresponding decomposition of the area d specific covariance matrix $\mathbf{V}_d$.

Under the linear mixed model (15.6a),

$$\mathbf{V}_U = diag\left(\mathbf{V}_d; d = 1, \ldots, D\right) = diag\left\{\sigma_e^2\Sigma_d + \sigma_u^2\mathbf{G}_d\Gamma(\varphi)\mathbf{G}'_d; d = 1, \ldots, D\right\}, \tag{15.8}$$

so

$$\mathbf{V}_{ssd} = \sigma_e^2\Sigma_{ssd} + \sigma_u^2\mathbf{G}_{sd}\Gamma(\varphi)\mathbf{G}'_{sd}$$

and, since Σ_U is a diagonal matrix,

$$\mathbf{V}_{rsd} = \sigma_u^2\mathbf{G}_{rd}\Gamma(\varphi)\mathbf{G}'_{sd}.$$

It follows that

$$\mathbf{V}_{rsd}\mathbf{V}_{ssd}^{-1} = \sigma_u^2\mathbf{G}_{rd}\Gamma(\varphi)\mathbf{G}'_{sd}\left\{\sigma_e^2\Sigma_{ssd} + \sigma_u^2\mathbf{G}_{sd}\Gamma(\varphi)\mathbf{G}'_{sd}\right\}^{-1}$$

and we can write (15.7) in the form

$$\hat{\bar{y}}_d^{BLUP} = N_d^{-1}\left[\mathbf{1}'_{n_d}\mathbf{y}_{sd} + \mathbf{1}'_{N_d-n_d}\right.$$
$$\left.\left\{\begin{matrix}\mathbf{Z}_{rd}\hat{\beta}^{BLUE}\\ +\mathbf{G}_{rd}\left[\sigma_u^2\mathbf{G}_{rd}\Gamma(\varphi)\mathbf{G}'_{sd}\left\{\sigma_e^2\Sigma_{ssd}+\sigma_u^2\mathbf{G}_{sd}\Gamma(\varphi)\mathbf{G}'_{sd}\right\}^{-1}\right]\end{matrix}\right\}\right].$$

On the other hand, given (15.6b), an EB predictor of the population mean in area d is of the general form

$$\hat{\bar{y}}_d^{EB} = N_d^{-1}\left[\mathbf{1}'_{n_d}\mathbf{y}_{sd} + \mathbf{1}'_{N_d-n_d}\left\{\mathbf{Z}_{rd}\hat{\beta} + \mathbf{G}_{rd}\hat{\mathbf{u}}_d\right\}\right]$$

where $\hat{\beta}$ is an efficient estimator of β and $\hat{\mathbf{u}}_d$ is an efficient predictor of $\mathbf{u}_d$. Comparing $\hat{\bar{y}}_d^{BLUP}$ and $\hat{\bar{y}}_d^{EB}$ above, one can see that they are identical when $\hat{\beta} = \hat{\beta}^{BLUE}$ and $\hat{\mathbf{u}}_d = \hat{\mathbf{u}}_d^{BLUP}$, where

$$\hat{\mathbf{u}}_d^{BLUP} = \sigma_u^2 \Gamma(\varphi) \mathbf{G}'_{sd} \left\{ \sigma_e^2 \Sigma_{ssd} + \sigma_u^2 \mathbf{G}_{sd} \Gamma(\varphi) \mathbf{G}'_{sd} \right\}^{-1} \left(\mathbf{y}_{sd} - \mathbf{Z}_{sd} \hat{\beta}^{BLUE} \right). \tag{15.9}$$

An alternative representation of the BLUP (15.7) is therefore as the EB predictor of the area d mean of Y under the model (15.6) defined by the BLUE of β and the BLUP of $\mathbf{u}_d$, and which can be written in the form

$$\begin{aligned}\hat{\bar{y}}_d^{BLUP} &= N_d^{-1} \left\{ \mathbf{1}'_{n_d} \mathbf{y}_{sd} + \mathbf{1}'_{N_d - n_d} \left(\mathbf{Z}_{rd} \hat{\beta}^{BLUE} + \mathbf{G}_{rd} \hat{\mathbf{u}}_d^{BLUP} \right) \right\} \\ &= N_d^{-1} \left\{ n_d \bar{y}_{sd} + (N_d - n_d) \left(\bar{\mathbf{z}}'_{rd} \hat{\beta}^{BLUE} + \bar{\mathbf{g}}'_{rd} \hat{\mathbf{u}}_d^{BLUP} \right) \right\}\end{aligned} \tag{15.10}$$

where $\bar{\mathbf{g}}'_{rd} = (N_d - n_d)^{-1} \mathbf{1}'_{N_d - n_d} \mathbf{G}_{rd}$.

To illustrate, consider the *Random Means* (RM) model. Under this model, population Y-values in area d satisfy

$$y_i = \mu + u_d + e_i \tag{15.11}$$

where the random area effects $\{u_d; d = 1, \ldots, D\}$ are independently and identically distributed with $E(u_d) = 0$ and $Var(u_d) = \sigma_u^2$, and the random individual effects $\{e_i; i = 1, \ldots, N\}$ are also independently and identically distributed with $E(e_i) = 0$ and $Var(e_i) = \sigma_e^2$. Area effects and individual effects are assumed to be independent of one another.

The sample means $\{\bar{y}_{sd}; d = 1, \ldots, D\}$ of Y in the small areas are then independently distributed, with $E(\bar{y}_{sd}) = \mu$ and $Var(\bar{y}_{sd}) = \sigma_u^2 + n_d^{-1} \sigma_e^2 = \sigma_e^2 (\lambda + n_d^{-1})$ where $\lambda = \sigma_u^2 / \sigma_e^2$. Since (15.11) is the special case of (15.6b) where $z_i = g_i = 1$, substitution in the expression for $\hat{\beta}^{BLUE}$ immediately following (15.7), followed by simplification, shows that the BLUE of μ is

$$\hat{\mu}^{BLUE} = \left(\sum_d (\lambda + n_d^{-1})^{-1} \right)^{-1} \sum_d (\lambda + n_d^{-1})^{-1} \bar{y}_{sd}.$$

Also, under this model,

$$\mathbf{V}_{rsd} = \sigma_u^2 \mathbf{1}_{N_d - n_d} \mathbf{1}'_{n_d}$$

and

$$\mathbf{V}_{ssd} = \sigma_e^2 \mathbf{I}_{n_d} + \sigma_u^2 \mathbf{1}_{n_d} \mathbf{1}'_{n_d}$$

where $\mathbf{1}_{n_d}$, $\mathbf{1}_{N_d - n_d}$ denote unit vectors of dimension n_d and $N_d - n_d$ respectively, and $\mathbf{I}_{n_d}$ denotes the identity matrix of order n_d. It can then be shown (exercise E.33) that the BLUP of the area d mean of Y under this model is

$$\hat{\bar{y}}_d^{BLUP} = N_d^{-1}\left(n_d\bar{y}_{sd} + (N_d - n_d)\left\{\left(\frac{1}{1+n_d\lambda}\right)\hat{\mu}^{BLUE} + \left(\frac{n_d\lambda}{1+n_d\lambda}\right)\bar{y}_{sd}\right\}\right)$$
$$= N_d^{-1}\left(n_d\bar{y}_{sd} + (N_d - n_d)\left\{\hat{\mu}^{BLUE} + \left(\frac{n_d\lambda}{1+n_d\lambda}\right)(\bar{y}_{sd} - \hat{\mu}^{BLUE})\right\}\right)$$
$$= N_d^{-1}\left(n_d\bar{y}_{sd} + (N_d - n_d)\{\hat{\mu}^{BLUE} + \hat{u}_d^{BLUP}\}\right).$$

When $n_d \ll N_d$ the BLUP of this area mean under (15.11) is essentially a weighted average of the BLUE for the expected value of Y over the population and the area d sample mean of Y. In particular, when $\sigma_u^2 \ll \sigma_e^2$ (so $\lambda \cong 0$) and n_d is small relative to N_d this BLUP is approximately the population mean estimate $\hat{\mu}^{BLUE}$, while if n_d is large and λ is bounded away from zero it is approximately the area d sample mean. Note also that the BLUP $\hat{u}_d^{BLUP}$ of the area d effect is then

$$\hat{u}_d^{BLUP} = \left(\frac{n_d\lambda}{1+n_d\lambda}\right)(\bar{y}_{sd} - \hat{\mu}^{BLUE}),$$

which is a scaled down version of the average sample residual in area d. This so-called 'shrinkage' is a well-known property of predicted random effects.

So far, we have assumed that the variance components of (15.6) are known. In reality, this is almost never the case. However, under non-informative sampling (i.e. where the process used to select the sample is independent of both area effects and individual effects) it is possible to estimate these components from the sample data and then substitute these estimates in the previous development. This leads to the so-called *empirical BLUP* or EBLUP.

Thus, the EBLUP of the area d mean of Y under (15.6) is

$$\hat{\bar{y}}_d^{EBLUP} = N_d^{-1}\left\{\mathbf{1}'_{n_d}\mathbf{y}_{sd} + \mathbf{1}'_{N_d-n_d}\left(\mathbf{Z}_{rd}\hat{\beta}^{EBLUE} + \mathbf{G}_{rd}\hat{\mathbf{u}}_d^{EBLUP}\right)\right\} \quad (15.12)$$

where

$$\hat{\mathbf{u}}_d^{EBLUP} = \hat{\sigma}_u^2\Gamma(\hat{\varphi})\mathbf{G}'_{sd}\left\{\hat{\sigma}_e^2\Sigma_{ssd} + \hat{\sigma}_u^2\mathbf{G}_{sd}\Gamma(\hat{\varphi})\mathbf{G}'_{sd}\right\}^{-1}\left(\mathbf{y}_{sd} - \mathbf{Z}_{sd}\hat{\beta}^{EBLUE}\right)$$

is the EBLUP of the area d random effect, and

$$\hat{\beta}^{EBLUE} = \left(\sum_d \mathbf{Z}'_{sd}\left\{\hat{\sigma}_e^2\Sigma_{ssd} + \hat{\sigma}_u^2\mathbf{G}_{sd}\Gamma(\hat{\varphi})\mathbf{G}'_{sd}\right\}^{-1}\mathbf{Z}_{sd}\right)^{-1}$$
$$\left(\sum_d \mathbf{Z}'_{sd}\left\{\hat{\sigma}_e^2\Sigma_{ssd} + \hat{\sigma}_u^2\mathbf{G}_{sd}\Gamma(\hat{\varphi})\mathbf{G}'_{sd}\right\}^{-1}\mathbf{y}_{sd}\right)$$

is the estimated BLUE (or EBLUE) of the regression parameter β.

A variety of estimation methods can be used for the variance components of (15.6). The most common are *maximum likelihood* (ML) and *restricted maximum likelihood* (REML). See Searle et al. (2006) for an exposition of these methods. In both cases it is standard to assume area effects are normally distributed. Most popular statistical software packages have the capacity to compute these estimates. Estimation via the method of moments (Henderson, 1953) is sometimes done, although in this case it is possible to end up with negative estimates for one of the variance components.

There is an unspoken assumption in the preceding development that there are sample units in every small area. If area d contains no sample units, then the assumption that population units from different areas are uncorrelated means that our best prediction of the value of the area d effect is zero. In this case the EBLUP for $\bar{y}_d$ is the *mixed-synthetic* predictor $\hat{\bar{y}}_d^{EBLUP} = N_d^{-1}\mathbf{1}'_{N_d}\mathbf{Z}_d\hat{\boldsymbol{\beta}}^{EBLUE}$. Like all synthetic predictors, this predictor is model dependent. However, since there are no sampled units in area d, the fit of the model (15.6) underlying this mixed-synthetic predictor in area d can never be checked. As a consequence, small area estimates for such 'non-sample' areas should always be treated with considerable caution.

For a review of the use of mixed models for small area estimation, see Datta (2009).

15.3 Estimation of the Prediction MSE of the EBLUP

So far, we have focused on prediction of small area quantities, and not inference. In order to carry out the latter we need to be able to estimate the prediction MSE of the small area predictor. When this predictor is based on a mixed model with estimated variance components, estimation of its prediction MSE can be quite complex. This is because we need to allow for the contribution to this MSE from estimation of the variance components. Rao (2003, Subsection 6.2.5) discusses this issue in some detail. Here we provide motivation for the general approach by focusing on the linear mixed model (15.6) and the EBLUP (15.12). An estimator of the prediction MSE of (15.12) under (15.6) is then constructed by assuming that the random effects are normally distributed and following the steps described below.

To start, we note that under (15.6) the area d non-sample total of Y can be written $t_{rd} = \tau_{rd} + \mathbf{1}'_{N_d-n_d}\mathbf{e}_{rd}$, where $\tau_{rd} = \mathbf{1}'_{N_d-n_d}(\mathbf{Z}_{rd}\boldsymbol{\beta} + \mathbf{G}_{rd}\mathbf{u}_d)$. By construction, the EBLUP of τ_{rd} is $\hat{\tau}_{rd}^{EBLUP} = \mathbf{1}'_{N_d-n_d}\left(\mathbf{Z}_{rd}\hat{\boldsymbol{\beta}}^{EBLUE} + \mathbf{G}_{rd}\hat{\mathbf{u}}_d^{EBLUP}\right)$, and it is not difficult to see that the prediction MSE of the EBLUP (15.12) is then N_d^{-2} times the prediction MSE of $\hat{\tau}_{rd}^{EBLUP}$ plus $Var\left\{\mathbf{1}'_{N_d-n_d}\mathbf{e}_{rd}\right\}$. The basic idea therefore is to decompose the prediction MSE of $\hat{\tau}_{rd}^{EBLUP}$ into a number of components, each one of which can be estimated separately. Our first step in developing this decomposition is to suppose the regression vector β and the variance components σ_e^2, σ_u^2 and Σ are known. The minimum mean squared error predictor (MMSEP) of τ_{rd} is then

$$\hat{\tau}_{rd}^{MMSEP} = E\left(\tau_{rd}|\mathbf{y}_s, \mathbf{Z}_U\right) = \mathbf{1}'_{N_d-n_d}\left\{\mathbf{Z}_{rd}\boldsymbol{\beta} + \mathbf{G}_{rd}E\left(\mathbf{u}_d|\mathbf{y}_{sd}, \mathbf{Z}_U\right)\right\}$$

where, given normally distributed random effects,

$$E\left(\mathbf{u}_d | \mathbf{y}_{sd}, \mathbf{Z}_U\right) = \sigma_u^2 \Gamma(\varphi) \mathbf{G}'_{sd} \left\{\sigma_e^2 \Sigma_{ssd} + \sigma_u^2 \mathbf{G}_{sd} \Gamma(\varphi) \mathbf{G}'_{sd}\right\}^{-1} (\mathbf{y}_{sd} - \mathbf{Z}_{sd}\beta).$$

Put $\mathbf{T}_{sd} = \sigma_u^2 \Gamma(\varphi) \mathbf{G}'_{sd} \left\{\sigma_e^2 \Sigma_{ssd} + \sigma_u^2 \mathbf{G}_{sd} \Gamma(\varphi) \mathbf{G}'_{sd}\right\}^{-1}$. Then

$$\begin{aligned}
E &\left(\hat{\tau}_{rd}^{MMSEP} - \tau_{rd} | \mathbf{Z}_U\right)^2 \\
&= Var\left(\hat{\tau}_{rd}^{MMSEP} - \tau_{rd} | \mathbf{Z}_U\right) \\
&= \mathbf{1}'_{N_d-n_d} \mathbf{G}_{rd} \left\{ Var\left(\mathbf{T}_{sd}\mathbf{y}_{sd} - \mathbf{u}_d | \mathbf{Z}_U\right)\right\} \mathbf{G}'_{rd} \mathbf{1}_{N_d-n_d} \\
&= \mathbf{1}'_{N_d-n_d} \mathbf{G}_{rd} \left\{ Var\left[\mathbf{T}_{sd}(\mathbf{Z}_{sd}\beta + \mathbf{G}_{sd}\mathbf{u}_d + \mathbf{e}_{sd}) - \mathbf{u}_d | \mathbf{Z}_U\right]\right\} \mathbf{G}'_{rd} \mathbf{1}_{N_d-n_d} \\
&= \mathbf{1}'_{N_d-n_d} \mathbf{G}_{rd} \left\{ Var\left[\mathbf{T}_{sd}(\mathbf{G}_{sd}\mathbf{u}_d + \mathbf{e}_{sd}) - \mathbf{u}_d | \mathbf{Z}_U\right]\right\} \mathbf{G}'_{rd} \mathbf{1}_{N_d-n_d} \\
&= \mathbf{1}'_{N_d-n_d} \mathbf{G}_{rd} \left\{ Var\left[(\mathbf{T}_{sd}\mathbf{G}_{sd} - \mathbf{I}_q)\mathbf{u}_d + \mathbf{T}_{sd}\mathbf{e}_{sd} | \mathbf{Z}_U\right]\right\} \mathbf{G}'_{rd} \mathbf{1}_{N_d-n_d} \\
&= \mathbf{1}'_{N_d-n_d} \mathbf{G}_{rd} \left\{ \sigma_u^2 (\mathbf{T}_{sd}\mathbf{G}_{sd} - \mathbf{I}_q) \Gamma(\varphi) (\mathbf{T}_{sd}\mathbf{G}_{sd} - \mathbf{I}_q)' + \sigma_e^2 \mathbf{T}_{sd}\Sigma_{ssd}\mathbf{T}'_{sd}\right\} \\
&\quad \mathbf{G}'_{rd} \mathbf{1}_{N_d-n_d} \\
&= \mathbf{1}'_{N_d-n_d} \mathbf{G}_{rd} \left\{ \sigma_u^2 \Gamma(\varphi) - \sigma_u^4 \Gamma(\varphi) \mathbf{G}'_{sd} \left\{\sigma_e^2 \Sigma_{ssd} + \sigma_u^2 \mathbf{G}_{sd}\Gamma(\varphi)\mathbf{G}'_{sd}\right\}^{-1} \mathbf{G}_{sd}\Gamma(\varphi)\right\} \\
&\quad \mathbf{G}'_{rd} \mathbf{1}_{N_d-n_d} \\
&= g_{1d}.
\end{aligned}$$

Under the random means model (15.11) it can be shown that

$$g_{1d} = (N_d - n_d)^2 \sigma_e^2 n_d^{-1} \gamma_d$$

where $\gamma_d = \left(n_d^{-1}\sigma_e^2 + \sigma_u^2\right)^{-1} \sigma_u^2$.

The next step, still assuming the variance components are known, is to replace β by its BLUE

$$\hat{\beta}^{BLUE} = \left(\sum_d \mathbf{Z}'_{sd} \left\{\sigma_e^2 \Sigma_{ssd} + \sigma_u^2 \mathbf{G}_{sd}\Gamma(\varphi)\mathbf{G}'_{sd}\right\}^{-1} \mathbf{Z}_{sd}\right)^{-1}$$
$$\left(\sum_d \mathbf{Z}'_{sd} \left\{\sigma_e^2 \Sigma_{ssd} + \sigma_u^2 \mathbf{G}_{sd}\Gamma(\varphi)\mathbf{G}'_{sd}\right\}^{-1} \mathbf{y}_{sd}\right).$$

The BLUP of τ_{rd} is then

$$\begin{aligned}
\hat{\tau}_{rd}^{BLUP} &= \mathbf{1}'_{N_d-n_d} \left(\mathbf{Z}_{rd}\hat{\beta}^{BLUE} + \mathbf{G}_{rd}\hat{\mathbf{u}}_d^{BLUP}\right) \\
&= \mathbf{1}'_{N_d-n_d} \left\{\mathbf{Z}_{rd}\hat{\beta}^{BLUE} + \mathbf{G}_{rd}\mathbf{T}_{sd}\left(\mathbf{y}_{sd} - \mathbf{Z}_{sd}\hat{\beta}^{BLUE}\right)\right\}
\end{aligned}$$

and so

$$\begin{aligned}
\hat{\tau}_{rd}^{BLUP} - \hat{\tau}_{rd}^{MMSEP} &= \mathbf{1}'_{N_d-n_d} \left\{\mathbf{Z}_{rd}\hat{\beta}^{BLUE} + \mathbf{G}_{rd}\mathbf{T}_{sd}\left(\mathbf{y}_{sd} - \mathbf{Z}_{sd}\hat{\beta}^{BLUE}\right)\right\} \\
&\quad - \mathbf{1}'_{N_d-n_d} \left\{\mathbf{Z}_{rd}\beta + \mathbf{G}_{rd}\mathbf{T}_{sd}(\mathbf{y}_{sd} - \mathbf{Z}_{sd}\beta)\right\} \\
&= \mathbf{1}'_{N_d-n_d} \left\{\mathbf{Z}_{rd} - \mathbf{G}_{rd}\mathbf{T}_{sd}\mathbf{Z}_{sd}\right\} \left(\hat{\beta}^{BLUE} - \beta\right).
\end{aligned}$$

The prediction error of this BLUP can be decomposed as

$$\hat{\tau}_{rd}^{BLUP} - \tau_{rd} = \left(\hat{\tau}_{rd}^{BLUP} - \hat{\tau}_{rd}^{MMSEP}\right) + \left(\hat{\tau}_{rd}^{MMSEP} - \tau_{rd}\right)$$
$$= \left\{\mathbf{1}'_{N_d - n_d}\left(\mathbf{Z}_{rd} - \mathbf{G}_{rd}\mathbf{T}_{sd}\mathbf{Z}_{sd}\right)\left(\hat{\beta}^{BLUE} - \beta\right)\right\}$$
$$+ (\hat{\tau}_{rd}^{MMSEP} - \tau_{rd}).$$

It can be shown that $Cov\left(\hat{\beta}^{BLUE}, \hat{\tau}_{rd}^{MMSEP} - \tau_{rd}|\mathbf{Z}_U\right) = \mathbf{0}_p$, so the prediction MSE of the BLUP $\hat{\tau}_{rd}^{BLUP}$ contains two terms

$$E\left\{\left(\hat{\tau}_{rd}^{BLUP} - \tau_{rd}\right)^2 |\mathbf{Z}_U\right\} = Var\left(\hat{\tau}_{rd}^{BLUP} - \tau_{rd}|\mathbf{Z}_U\right) = g_{1d} + g_{2d}$$

where

$$g_{2d} = \mathbf{1}'_{N_d - n_d}\left(\mathbf{Z}_{rd} - \mathbf{G}_{rd}\mathbf{T}_{sd}\mathbf{Z}_{sd}\right) Var\left(\hat{\beta}^{BLUE}|\mathbf{Z}_U\right)\left(\mathbf{Z}_{rd} - \mathbf{G}_{rd}\mathbf{T}_{sd}\mathbf{Z}_{sd}\right)' \mathbf{1}_{N_d - n_d}$$
$$= \mathbf{1}'_{N_d - n_d}(\mathbf{Z}_{rd} - \mathbf{G}_{rd}\mathbf{T}_{sd}\mathbf{Z}_{sd})\left(\sum_h \mathbf{Z}'_{sh}\left\{\sigma_e^2 \Sigma_{ssh} + \sigma_u^2 \mathbf{G}_{sh}\Gamma(\varphi)\mathbf{G}'_{sh}\right\}^{-1} \mathbf{Z}_{sh}\right)^{-1}$$
$$(\mathbf{Z}_{rd} - \mathbf{G}_{rd}\mathbf{T}_{sd}\mathbf{Z}_{sd})' \mathbf{1}_{N_d - n_d}.$$

Under the random means model (15.11),

$$g_{2d} = (N_d - n_d)^2 \sigma_e^2 n^{-1}(1 - \gamma_d)^2 \left(1 - n^{-1}\sum_h n_h \gamma_h\right)^{-1}.$$

Comparing g_{1d} with g_{2d}, we see that the former is of order $n_d^{-1}(N_d - n_d)^2$, while the latter is of order $n^{-1}(N_d - n_d)^2$. That is, we expect g_{1d} to dominate g_{2d} in most situations.

The third, and final, step is to consider the EBLUP case, with the variance components estimated either via ML or REML, and with $\hat{\beta}^{BLUE}$ replaced by $\hat{\beta}^{EBLUE}$. Then

$$E\left(\hat{\tau}_{rd}^{EBLUP} - \tau_{rd}\right)^2 = E\left(\hat{\tau}_{rd}^{EBLUP} - \hat{\tau}_{rd}^{BLUP} + \hat{\tau}_{rd}^{BLUP} - \tau_{rd}\right)^2$$
$$= Var\left(\hat{\tau}_{rd}^{BLUP} - \tau_{rd}\right) + E\left(\hat{\tau}_{rd}^{EBLUP} - \hat{\tau}_{rd}^{BLUP}\right)^2$$
$$+ 2E\left(\hat{\tau}_{rd}^{BLUP} - \tau_{rd}\right)\left(\hat{\tau}_{rd}^{EBLUP} - \hat{\tau}_{rd}^{BLUP}\right).$$

The first term on the right hand side above has already been shown to be $g_{1d} + g_{2d}$. A naive estimator of the prediction variance of $\hat{\tau}_{rd}^{EBLUP}$ is therefore defined by disregarding the last two terms on the right hand side above and replacing the unknown variance components in $g_{1d} + g_{2d}$ by suitable estimators. However, Kacker and Harville (1984) point out that although the third term on the right hand side in the preceding expression is negligible (in fact, under normal random effects this cross product term is zero), the second is not. They also derive a plug in approximation to this second term. Prasad and Rao (1990) show that the prediction variance estimator based on this approximation underestimates the true prediction variance, and introduce a second order approximation to the prediction variance,

$$E\left(\hat{\tau}_{rd}^{EBLUP} - \tau_{rd}\right)^2 \cong g_{1d} + g_{2d} + g_{3d}$$

where

$$g_{3d} = E\left(\hat{\tau}_{rd}^{EBLUP} - \hat{\tau}_{rd}^{BLUP}\right)^2 = trace\left\{\left(\frac{\partial \mathbf{b}'_d}{\partial \theta}\right)\mathbf{V}_{ssd}\left(\frac{\partial \mathbf{b}'_d}{\partial \theta}\right)' Var(\hat{\theta})\right\}$$

with $\theta' = (\sigma_e^2, \sigma_u^2, \varphi')$, $\mathbf{b}'_d = \sigma_u^2 \mathbf{1}'_{N_d-n_d} \mathbf{G}_{rd} \Gamma(\varphi) \mathbf{G}'_{sd} \mathbf{V}_{ssd}^{-1}$ and $Var(\hat{\theta})$ is the asymptotic covariance matrix of the variance components. Under ML estimation, this is obtained from the relevant components of the inverse of the observed information matrix $I(\hat{\theta})$.

In most applications g_{3d} is of the same order of magnitude as g_{2d} and hence is also dominated by g_{1d}. For the random means model (15.11), one can show that

$$g_{3d} = (N_d - n_d)^2 n_d \left(\sigma_e^2 + n_d \sigma_u^2\right)^{-3} h(\hat{\theta})$$

with $h(\hat{\theta}) = \sigma_e^4 Var\left(\hat{\sigma}_u^2\right) + \sigma_u^4 Var\left(\hat{\sigma}_e^2\right) - 2\sigma_e^2 \sigma_u^2 Cov\left(\hat{\sigma}_u^2, \hat{\sigma}_e^2\right)$. When the variance components are estimated via ML their asymptotic covariance matrix is

$$Var\begin{pmatrix}\hat{\sigma}_u^2 \\ \hat{\sigma}_e^2\end{pmatrix} = \frac{2}{T}\begin{bmatrix}\frac{n-D}{\sigma_e^4} + \sum_{h=1}^{D}\left(n_h \sigma_u^2 + \sigma_e^2\right)^{-2} & -\sum_{h=1}^{D} n_h \left(n_h \sigma_u^2 + \sigma_e^2\right)^{-2} \\ -\sum_{h=1}^{D} n_h \left(n_h \sigma_u^2 + \sigma_e^2\right)^{-2} & \sum_{h=1}^{D} n_h^2 \left(n_h \sigma_u^2 + \sigma_e^2\right)^{-2}\end{bmatrix}$$

where

$$T = \frac{n-D}{\sigma_e^4} \sum_{h=1}^{D} n_h^2 \left(n_h \sigma_u^2 + \sigma_e^2\right)^{-2}$$
$$+ \left\{\sum_{h=1}^{D}\left(n_h \sigma_u^2 + \sigma_e^2\right)^{-2}\right\}\left\{\sum_{h=1}^{D} n_h^2 \left(n_h \sigma_u^2 + \sigma_e^2\right)^{-2}\right\}$$
$$- \left\{\sum_{h=1}^{D} n_h \left(n_h \sigma_u^2 + \sigma_e^2\right)^{-2}\right\}^2.$$

The corresponding expression when the variance components are estimated via REML is considerably more complicated. See McCulloch and Searle (2001, Section 2.2).

Collecting results, we see that the prediction MSE of the EBLUP (15.12) under (15.6) is

$$E\left(\hat{\bar{y}}_d^{EBLUP} - \bar{y}_d\right)^2 = N_d^{-2}\left\{E\left(\hat{\tau}_{rd}^{EBLUP} - \tau_{rd}\right)^2 + Var\left(\mathbf{1}'_{N_d-n_d}\mathbf{e}_{rd}\right)\right\}$$
$$\cong N_d^{-2}\left(g_{1d} + g_{2d} + g_{3d} + \sigma_e^2 \mathbf{1}'_{N_d-n_d}\Sigma_{rd}\mathbf{1}_{N_d-n_d}\right)$$
$$= N_d^{-2}\left(g_{1d} + g_{2d} + g_{3d} + g_{4d}\right).$$

Each term in this expression can be estimated by substituting the estimated values of the variance components. That is, our estimator of the prediction MSE of the EBLUP (15.12) is

$$\hat{V}\left(\hat{\bar{y}}_d^{EBLUP}\right) = N_d^{-2}\left(\hat{g}_{1d} + \hat{g}_{2d} + 2\hat{g}_{3d} + \hat{g}_{4d}\right) \tag{15.13}$$

where a 'hat' on a quantity denotes a corresponding plug-in estimator of that quantity. Note that the factor of 2 for the $\hat{g}_{3d}$ term in (15.13) arises because the plug-in estimator of g_{1d} is biased low, with $E(\hat{g}_{1d}) \cong g_{1d} - g_{3d}$. See Rao (2003, page 104).

A final issue arises when the variance components are estimated using maximum likelihood, in which case $E(\hat{g}_{1d}) \cong g_{1d} - g_{3d} + g_{5d}$, where the last term on the right hand side of this approximation has a somewhat complicated form. For details, see Datta and Lahiri (2000). In this case (15.13) is replaced by

$$\hat{V}\left(\hat{\bar{y}}_d^{EBLUP}\right) = N_d^{-2}\left(\hat{g}_{1d} + \hat{g}_{2d} + 2\hat{g}_{3d} + \hat{g}_{4d} + \hat{g}_{5d}\right).$$

15.4 Direct Prediction for Small Areas

There is a substantial disconnect between the mixed model-based small area estimation theory developed in the previous two sections and the approach to inference for domains described in Chapter 14. Since small areas are just special types of domains, it seems valid to ask why this is the case.

The answer to this question rests on the distinction between *direct* prediction methods and *indirect* prediction methods for small area characteristics. The direct approach to small area prediction was briefly described at the start of this chapter, see (15.1). It is also the approach that was used for prediction of domain characteristics in Chapter 14. In both cases, it was assumed that the characteristic of interest for area d is a well-defined function of the population values of the variables D and DY, where D is a zero-one indicator of area d exclusion/inclusion. A plug-in predictor of this function based on the sample values of D and DY is then used. Thus, if the mean of Y in area d,

$$\hat{\bar{y}}_d = \sum_U d_i y_i \bigg/ \sum_U d_i, \tag{15.14}$$

is of interest, then plug-in prediction of this mean based on weighted linear prediction of the numerator and the denominator on the right hand side of (15.14) leads to the Hajek-type predictor (15.1a), that is

$$\hat{\bar{y}}_{wd} = \sum_s w_i d_i y_i \bigg/ \sum_s w_i d_i = \sum_{s_d} w_i y_i \bigg/ \sum_{s_d} w_i. \tag{15.15}$$

Here d_i takes the value 1 if unit i is in area d and is zero otherwise, and w_i is a sample weight. Note that (15.15) is an aggregation consistent approach to prediction, since it is the standard weighted mean predictor of the population mean of Y when the domain is the entire population. Predictors like (15.15) are often said to be direct because they ostensibly depend only on the sample data from area d. However, this is not true in general. For example, the sample weight w_i can depend on data from the entire sample.

In contrast, the mixed model-based predictors that have been the focus of the previous two sections cannot be written as population level predictors applied to area-specific variables. For example, suppose $n_d > 0$. The EBLUP of the area d mean of Y under the Random Means model (15.11) is then

$$\hat{\bar{y}}_d^{EBLUP} = N_d^{-1}\left[n_d\bar{y}_{sd} + (N_d - n_d)\left\{\left(\frac{1}{1+n_d\hat{\lambda}}\right)\frac{\sum_h(\hat{\lambda}+n_h^{-1})^{-1}\bar{y}_{sh}}{\sum_h(\hat{\lambda}+n_h^{-1})^{-1}}\right.\right.$$
$$\left.\left.+\left(\frac{n_d\hat{\lambda}}{1+n_d\hat{\lambda}}\right)\bar{y}_{sd}\right\}\right] \quad (15.16)$$

where the summations on the right hand side are over all the small areas h represented in the sample, and $\hat{\lambda} = \hat{\sigma}_u^2/\hat{\sigma}_e^2$ is the sample-based estimator of the ratio of the between- to within-area variation of Y. When $n_d = 0$ this predictor is still defined, being

$$\hat{\bar{y}}_d^{EBLUP} = \frac{\sum_h\left(\hat{\lambda}+n_h^{-1}\right)^{-1}\bar{y}_{sh}}{\sum_h\left(\hat{\lambda}+n_h^{-1}\right)^{-1}}. \quad (15.17)$$

Note that both (15.16) and (15.17) cannot be written as functions of the sample values of D and DY. We therefore refer to them as indirect predictors of the area d mean of Y.

An important point that should be made here is that the lack of any sample data in area d means that we do not have the data necessary to check whether (15.11) holds in this area. As we have already noted at the end of Section 15.2, the validity of an indirect predictor like (15.17) for any non-sampled area is completely dependent on (15.11) holding in *every* area, whether sampled or not.

An attractive property of both (15.16) and (15.7) is that the weighted average of their values over all areas represented in the sample is the value of the EBLUP of the population mean of Y under the random means model (15.11), where the weights used in this average are the population proportions in each area. This is also true of the EBLUP (15.12) under the more general linear mixed model (15.6). This aggregation property is a consequence of the fact that if a model with random area effects holds for each area, then it must also hold for the population. However, it is often the case that the population level model underlying the weight w_i used in (15.15) is not the same as the model (15.11) underlying (15.16).

In this situation the population weighted average of the values of (15.16) over all the sampled small areas will not in general equal the value of the estimate $\hat{\bar{y}}_w = \sum_s w_i y_i / \sum_s w_i$ of the population mean of Y defined by these weights. Since population level estimates are (usually) more important than area level estimates, this inconsistency is typically resolved in favour of $\hat{\bar{y}}_w$ by scaling the area EBLUPs to ensure that they aggregate appropriately.

This scaling has implications for indirect predictors like (15.16), since it could substantially change the final predicted value of the area d mean of Y relative to its EBLUP value (15.16) – in which case one has to wonder about the efficiency of the final result. We illustrate this via a simple, quite artificial, example. Suppose that the EBLUE of the overall mean μ under (15.11) based on the data from the sampled areas is $\hat{\mu}^{EBLUE} = 3$, with the corresponding EBLUPs for these sampled areas varying around this figure. By definition, the EBLUPs for the non-sampled areas are all the same, and equal to $\hat{\mu}^{EBLUE}$. Suppose also that the sample size in each sampled area is the same, and is small relative to the population sizes of these areas, all of which are also the same. The EBLUP of the population mean of Y under (15.11) is then approximately equal to $\hat{\mu}^{EBLUE}$. Finally, suppose that the weights used for estimation of population level quantities lead to an estimate of 3.3 for the population mean of Y. Then scaling the EBLUPs for the different areas so that they add up to an estimate of 3.3 for the population mean of Y means that we increase the EBLUPs for the sampled areas by 10%, which can be large relative to their estimated standard errors. Furthermore, the absolute change in areas with large predicted area effects will be larger than in areas with small predicted area effects. As an aside, we note that since for sample area d, $\hat{\bar{y}}_d^{EBLUP} \approx \hat{\mu}^{EBLUE} + \hat{u}_d^{EBLUP}$, with $\sum_{h|n_h>0} \hat{u}_h^{EBLUP} = 0$ in this case, an alternative to scaling might be to predict the area d mean using $\hat{\bar{y}}_d = \hat{\bar{y}}_w + \hat{u}_d^{EBLUP}$, where $\hat{\bar{y}}_w = \sum_s w_i y_i / \sum_s w_i$ is the preferred estimate of the population mean of Y, and $\hat{u}_d^{EBLUP} = 0$ if $n_d = 0$.

Direct predictors like (15.15) are easy to interpret and to build into survey processing systems. They also do **not** permit prediction for areas where there are no sample data, which, in light of the comments at the end of Section 15.2, may be considered to be a good thing. In a certain sense they are also robust. This is because (15.15) is an unbiased estimator of the expected value of Y in area d (assuming non-informative sampling within this area) irrespective of the correctness or not of the working model that was used to derive the sample weights. However, as noted at the start of this chapter, depending on choice of the weights used in (15.15), direct estimation can be quite inefficient. Consequently, we will now construct a direct predictor which is an analogue of the indirect EBLUP (15.12). We will do this by identifying appropriate weights to use in (15.15).

Consider the linear mixed model (15.6) that underlies much of the development in the preceding part of this chapter. We have already seen in (15.7) the form of the BLUP for the area d mean of Y. The corresponding EBLUP of the population mean of Y under (15.6) is therefore

$$\hat{\bar{y}}^{EBLUP} = N^{-1}\left[\mathbf{1}'_n\mathbf{y}_s + \mathbf{1}'_{N-n}\left\{\mathbf{Z}_r\hat{\boldsymbol{\beta}}^{EBLUE} + \hat{\mathbf{V}}_{rs}\hat{\mathbf{V}}_{ss}^{-1}\left(\mathbf{y}_s - \mathbf{Z}_s\hat{\boldsymbol{\beta}}^{EBLUE}\right)\right\}\right] \tag{15.18}$$

where the notation is the same as in Section 15.2. Without loss of generality, we assume that the matrix $\mathbf{Z}_U$ of auxiliary variables contains an intercept. An equivalent representation of (15.18) is then the weighted mean

$$\hat{\bar{y}}^{EBLUP} = \sum_s w_i^{EBLUP} y_i \Big/ \sum_s w_i^{EBLUP}$$

where the EBLUP weights $\mathbf{w}_s^{EBLUP} = \{w_i^{EBLUP}; i \in s\}$ satisfy

$$\mathbf{w}_s^{EBLUP} = \mathbf{1}_n + \mathbf{H}'_L\left(\mathbf{Z}'_U\mathbf{1}_N - \mathbf{Z}'_s\mathbf{1}_n\right) + (\mathbf{I}_n - \mathbf{H}'_L\mathbf{Z}'_s)\hat{\mathbf{V}}_{ss}^{-1}\hat{\mathbf{V}}_{sr}\mathbf{1}_{N-n} \tag{15.19}$$

with $\mathbf{H}_L = \left(\mathbf{Z}'_s\hat{\mathbf{V}}_{ss}^{-1}\mathbf{Z}_s\right)^{-1}\mathbf{Z}'_s\hat{\mathbf{V}}_{ss}^{-1}$. These weights are calibrated on $\mathbf{Z}_U$ (see Section 13.1), so $\sum_s w_i^{EBLUP} = N$, and only depend on the variance-covariance structure of (15.6) via the estimates $\hat{\mathbf{V}}_{ss}$ and $\hat{\mathbf{V}}_{sr}$. In turn, these estimated variance-covariance matrices are defined by substituting estimates of unknown variance components – obtained by fitting (15.6) to the sample data – in the expressions shown below (15.8). This naturally leads us to defining the *model-based direct* (MBD) predictor of the area d mean of Y based on the population model (15.6) and the resulting EBLUP weights (15.19) as

$$\hat{\bar{y}}_d^{MBD} = \sum_s w_i^{EBLUP} d_i y_i \Big/ \sum_s w_i^{EBLUP} d_i = \sum_{s_d} w_i^{EBLUP} y_i \Big/ \sum_{s_d} w_i^{EBLUP}. \tag{15.20}$$

Note that (15.20) is **not** the same as the indirect EBLUP (15.12), although

$$N^{-1}\sum_h N_h \hat{\bar{y}}_h^{EBLUP} = N^{-1}\sum_h N_h \hat{\bar{y}}_h^{MBD} = \hat{\bar{y}}^{EBLUP}.$$

Also, if the linear mixed model (15.6) is actually *true*, then (15.12) can be expected to dominate (15.20). To illustrate, suppose the random means model (15.11) holds. Then, under the direct approach, the weight associated with sampled unit i in area d is

$$w_i = \frac{N}{n}\left[1 + \frac{1}{1 + n_d\hat{\lambda}}\left\{(N_d - n_d)\hat{\lambda} + \frac{\bar{N} - \bar{n}}{\bar{n}}\right\}\right]$$

where $\hat{\lambda} = \hat{\sigma}_u^2/\hat{\sigma}_e^2$, $\bar{N} = \sum_d N_d(1 + n_d\hat{\lambda})^{-1}/\sum_d(1 + n_d\hat{\lambda})^{-1}$ and $\bar{n}$ is defined similarly. That is, (15.20) reduces to the area d sample mean $\bar{y}_{sd}$, which has high variability in small samples. In contrast, (15.12) is then a linear combination of the overall sample mean and the area d sample mean, and has much less variability.

Applications of the MBD approach to small area estimation are described in more detail in Chandra and Chambers (2009, 2011) and Salvati et al. (2010).

15.5 Estimation of Conditional MSE for Small Area Predictors

The approach to MSE estimation outlined in Section 15.3 is unconditional. That is, it takes no account of the actual value of the area effect in any particular area, and how this could affect the uncertainty about the true value of the area mean, and so is driven essentially by the variability of area effects across all areas. It can be argued that this is not necessarily the best way to measure the uncertainty in the estimate for any particular area since this uncertainty is then being measured relative to situations where the effect for the area of interest is completely different, with a different sign for example. What is required is a measure of uncertainty conditional on the actual values of the area effects in the different small areas, rather than an unconditional measure that averages over these effects. In this section we describe an approach to MSE estimation that has this property.

In what follows we use a subscript of $\mathbf{u}$ to denote conditioning on area effects, observing that under the linear mixed model specification (15.6) this leads to the conditional model with expected value

$$E_\mathbf{u}(y_i | \mathbf{z}_i) = E(y_i | i \in d, \mathbf{z}_i) = \mathbf{z}_i' \beta + \mathbf{g}_i' \mathbf{u}_d, \qquad (15.21)$$

and where the Y-values in small area d are mutually uncorrelated with variance $Var_\mathbf{u}(y_i | \mathbf{z}_i) = \sigma_e^2$. An obvious estimator of $E_\mathbf{u}(y_i | \mathbf{z}_i)$ is $\hat{E}(y_i | i \in d) = \mathbf{z}_i' \hat{\beta}^{EBLUE} + \mathbf{g}_i' \hat{\mathbf{u}}_d^{EBLUP}$, where $\hat{\mathbf{u}}_d^{BLUP}$ is defined by (15.10). Unfortunately, even when the variance components are assumed known, this estimator is biased for $E_\mathbf{u}(y_i | \mathbf{z}_i)$ due to the shrinkage property of the BLUP of a random effect referred to earlier. An estimator that can be used instead is

$$\hat{y}_i = \mathbf{z}_i' \hat{\beta}^{EBLUE} + \mathbf{g}_i' \left(\mathbf{G}_{sd}' \mathbf{G}_{sd} \right)^{-1} \mathbf{G}_{sd}' \left(\mathbf{y}_{sd} - \mathbf{Z}_{sd} \hat{\beta}^{EBLUE} \right). \qquad (15.22)$$

This estimator is approximately unbiased in large samples since

$$\begin{aligned} E_\mathbf{u}(\hat{y}_i | \mathbf{z}_i) &\approx \mathbf{z}_i' \beta + \mathbf{g}_i' \left(\mathbf{G}_{sd}' \mathbf{G}_{sd} \right)^{-1} \mathbf{G}_{sd}' \{ E(\mathbf{y}_{sd} | \mathbf{u}_d) - \mathbf{Z}_{sd} \beta \} \\ &= \mathbf{z}_i' \beta + \mathbf{g}_i' \left(\mathbf{G}_{sd}' \mathbf{G}_{sd} \right)^{-1} \mathbf{G}_{sd}' \mathbf{G}_{sd} \mathbf{u}_d \\ &= \mathbf{z}_i' \beta + \mathbf{g}_i' \mathbf{u}_d \\ &= E_\mathbf{u}(y_i | \mathbf{z}_i). \end{aligned}$$

Consider estimation of the conditional MSE of the MBD predictor (15.20). As we now show, we can calculate this MSE estimate using a straightforward generalisation of the robust approach to prediction variance estimation discussed in Section 9.2. This can be compared with the rather more complicated estimator (15.13) of the (unconditional) MSE of the EBLUP (15.12) described in Section 15.3.

To start, we write down a first order approximation to the prediction variance of (15.20), based on treating the EBLUP weights (15.19) as fixed constants, that is we effectively replace $\hat{\mathbf{V}}_{sr}$ and $\hat{\mathbf{V}}_{ss}$ by $\mathbf{V}_{sr}$ and $\mathbf{V}_{ss}$ in (15.19). This leads to

$$Var_{\mathbf{u}}\left(\hat{\bar{y}}_d^{MBD} - \bar{y}_d\right) \cong N_d^{-2}\left\{\sum_{s_d} \theta_i^2 Var_{\mathbf{u}}(y_i) + \sum_{r_d} Var_{\mathbf{u}}(y_i)\right\} \quad (15.23)$$

where

$$\theta_i = \left(\sum_{s_d} w_j^{EBLUP}\right)^{-1}\left(N_d w_i^{EBLUP} - \sum_{s_d} w_j^{EBLUP}\right).$$

In order to use the robust prediction variance estimation methods described in Section 9.2 we need to replace $Var_{\mathbf{u}}(y_i)$ in the first term on the right hand side of (15.23) by a corresponding squared residual. In turn, this requires us to define an appropriate fitted value $\hat{y}_i$ for a sample value y_i. Unlike the situation considered in Section 9.2, however, $\hat{y}_i$ is not an estimate of the expected value $\mathbf{z}_i'\beta$ of y_i under (15.6) since this makes no allowance for the fact that we **know** that y_i is from area d. Instead, we use the estimator (15.22) suggested by the conditional model (15.21).

In what follows we therefore define the area d sample residuals as $\{y_i - \hat{y}_i; i \in s_d\}$ where $\hat{y}_i$ is defined by (15.22), and hence write down a robust estimator of the leading (i.e. first) term of the prediction variance (15.23) by replacing $Var_{\mathbf{u}}(y_i)$ there with the corresponding squared residual $(y_i - \hat{y}_i)^2$. As in Section 9.2, we refine this estimator by adjusting this squared residual so that it is exactly unbiased under (15.21). In order to do this, observe that when the variance components are known (so we can replace $\hat{\beta}^{EBLUE}$ by $\hat{\beta}^{BLUE}$), (15.22) is a linear combination of the sample Y-values, that is $\hat{y}_i = \sum_{j \in s} \phi_{ij} y_j$, where the ϕ_{ij} are fixed weights. In this case $y_i - \hat{y}_i = (1 - \phi_{ii})y_i - \sum_{j \in s(-i)} \phi_{ij} y_j$, where $s(-i)$ denotes the sample s with unit i excluded, and so $E_{\mathbf{u}}(y_i - \hat{y}_i)^2 = \sigma_e^2 \left\{(1-\phi_{ii})^2 + \sum_{j \in s(-i)} \phi_{ij}^2\right\}$. A robust estimator of (15.23) is then

$$\hat{V}_d = N_d^{-2} \sum_{s_d} \delta_i (y_i - \hat{y}_i)^2 \quad (15.24)$$

where $\delta_i = \left\{(1-\phi_{ii})^2 + \sum_{j \in s(-i)} \phi_{ij}^2\right\}^{-1} \left\{\theta_i^2 + (N_d - n_d)n_d^{-1}\right\}$.

In order to estimate the prediction MSE of $\hat{\bar{y}}_d^{MBD}$ we also need to estimate its squared prediction bias under (15.21). Here we note that

$$E_{\mathbf{u}}\left(\hat{\bar{y}}_d^{MBD} - \bar{y}_d\right) \cong (\bar{\mathbf{z}}_{wd} - \bar{\mathbf{z}}_d)'\beta + (\bar{\mathbf{g}}_{wd} - \bar{\mathbf{g}}_d)'\mathbf{u}_d$$

under (15.21), where $\bar{\mathbf{z}}_{wd}$ denotes the weighted average of the sample $\mathbf{z}_i$ in area d and $\bar{\mathbf{g}}_{wd}$ denotes the corresponding weighted average of the sample values of $\mathbf{g}_i$. A simple estimator of this bias is

$$\hat{B}_d = \left(\sum_{s_d} w_i^{EBLUP}\right)^{-1} \sum_{s_d} w_i^{EBLUP} \hat{y}_i - N_d^{-1} \sum_{i \in d} \hat{y}_i \quad (15.25)$$

where $\hat{y}_i$ is defined by (15.22). Our suggested estimator $\hat{M}_d$ of the conditional MSE of $\hat{\bar{y}}_d^{MBD}$ combines (15.24) and (15.25) to give

$$\hat{M}_d = \hat{V}_d + \hat{B}_d^2. \quad (15.26)$$

Note that $E_\mathbf{u}(\hat{B}_d^2) = \{E_\mathbf{u}(\hat{B}_d)\}^2 + Var_\mathbf{u}(\hat{B}_d)$, so (15.26) is a conservative estimator of this conditional MSE.

There is nothing to stop the conditional MSE estimator (15.26) being used with any small area estimator that can be written in *pseudo-linear* form, that is as a linear combination of the sample values of Y, where the weights defining the linear combination are asymptotically fixed. To illustrate this, we develop the form of (15.26) for the EBLUP (15.12). To start, we note that this predictor is pseudo-linear since it can equivalently be written

$$\hat{\bar{y}}_d^{EBLUP} = \sum_s w_i^{EBLUPd} y_i = (\mathbf{w}_s^{EBLUPd})' \mathbf{y}_s$$

where

$$\mathbf{w}_s^{EBLUPd} = (w_i^{EBLUPd}) = N_d^{-1}\left[\Delta_{sd} + \left\{\hat{\mathbf{H}}_s' \mathbf{Z}_r' + \left(\mathbf{I}_n - \hat{\mathbf{H}}_s' \mathbf{Z}_s'\right)\hat{\mathbf{V}}_{ss}^{-1} \hat{\mathbf{V}}_{sr}\right\} \Delta_{rd}\right]. \tag{15.27}$$

Here Δ_{sd} and Δ_{rd} are the vectors of zeros and ones of size n and $N-n$ that 'pick out' the sample and non-sampled units in area d respectively, and $\hat{\mathbf{H}}_s = \left(\mathbf{Z}_s'\hat{\mathbf{V}}_{ss}^{-1}\mathbf{Z}_s\right)^{-1}\mathbf{Z}_s'\hat{\mathbf{V}}_{ss}^{-1}$. Note that the weight vector (15.27) is a function of the estimated variance components of (15.6), and for large samples can be assumed to be fixed. Redefining $\theta_i = w_i^{EBLUPd} - 1$, we see that (15.24) can still be used to estimate the conditional prediction variance of (15.12). Similarly, we can estimate the conditional prediction bias of (15.12) by first noting that

$$E_\mathbf{u}(\hat{\bar{y}}_d^{EBLUP} - \bar{y}_d) \cong (\bar{\mathbf{z}}_{wd}^{EBLUPd} - \bar{\mathbf{z}}_d)'\beta + \sum_h \left(\sum_{s_h} w_i^{EBLUPd} \mathbf{g}_i'\right)\mathbf{u}_h - \bar{\mathbf{g}}_d' \mathbf{u}_d$$
$$= \sum_h \left(\sum_{s_h} w_i^{EBLUPd} \mathbf{g}_i'\right)\mathbf{u}_h - \bar{\mathbf{g}}_d' \mathbf{u}_d$$

since the EBLUP weights are 'locally calibrated' on Z, that is $\bar{\mathbf{z}}_{wd}^{EBLUPd} = \sum_s w_i^{EBLUPd} \mathbf{z}_i = \bar{\mathbf{z}}_d$. The same argument that leads to the use of (15.22) then leads to an estimator of this conditional bias of the form

$$\hat{B}_d = \sum_h \left(\sum_{s_h} w_i^{EBLUPd} \mathbf{g}_i'\right)\tilde{\mathbf{u}}_h - \bar{\mathbf{g}}_d' \tilde{\mathbf{u}}_d \tag{15.28}$$

where

$$\tilde{\mathbf{u}}_d = \mathbf{g}_i'\left(\mathbf{G}_{sd}' \mathbf{G}_{sd}\right)^{-1} \mathbf{G}_{sd}' \left(\mathbf{y}_{sd} - \mathbf{Z}_{sd}\hat{\beta}^{EBLUE}\right).$$

The value of $\hat{V}_d$ generated by setting $\theta_i = w_i^{EBLUPd} - 1$ and the square of the value of $\hat{B}_d$ defined by (15.28) can be combined, as in (15.26), to then generate an estimate of the conditional MSE of the EBLUP (15.12).

For a more detailed discussion of this approach to conditional MSE estimation, see Chambers, Chandra and Tzavidis (2009). See Salvati *et al.* (2010) for an application of this approach.

15.6 Simulation-Based Comparison of EBLUP and MBD Prediction

How well does the MBD predictor (15.20) compare with the EBLUP (15.12)? Chambers, Chandra and Tzavidis (2009) report results from a large number of simulation studies that investigate the comparative performances of a number of small area predictors as well as estimators of their MSE for a range of sample sizes, ranging down to as low as $n_d = 5$ in each area. Both model-based (varying population) and design-based (fixed population) simulations were carried out, with the model-based simulations based on repeatedly generating population data under the linear mixed model (15.6), so that the area effect for the same small area varied from simulation to simulation. These model-based simulations showed that for population data generated under the linear mixed model, the EBLUP (15.12) and the unconditional MSE estimator (15.13) worked very well. In contrast, the MBD predictor was unbiased but more variable, as were the conditional MSE estimators for both the MBD predictor and the EBLUP based on (15.26).

Of more practical interest, however, are the simulation results reported by these authors when the fit of the linear mixed model could only be considered approximate. These are based on two design-based (i.e. repeated sampling from a fixed finite population) studies. The first used data from a sample of 3591 households spread across $D = 36$ districts of Albania that participated in the 2002 Albanian Living Standards Measurement Study. These sample data were bootstrapped to create data for a realistic population of $N = 724,782$ households by re-sampling from the sample households with replacement with probability proportional to a household's sample weight. A total of 1000 independent stratified random samples were then drawn from this bootstrap population, with total sample size equal to that of the original sample and with districts defining the strata. Sample sizes within districts were the same as in the original sample, and varied between 8 and 688 (with median district sample size equal to 56). The Y variable of interest was household per capita consumption expenditure (HCE) and Z was defined by three zero-one variables (ownership of television, parabolic antenna and land). The aim was to estimate the average value of HCE for each district. In the original 2002 survey, the linear relationship between HCE and the three variables making up Z was rather weak, with very low predictive power. In particular, only ownership of land was significantly related to HCE at the 5% level. This fit was considerably improved by extending the linear model to include random intercepts, defined by independent district effects. These explained approximately 10% of the residual variation in this model.

The second design-based simulation study was based on sampling a population of Australian broadacre farms. This population was defined by bootstrapping a sample of 1652 farms that participated in the Australian Agricultural and Grazing Industries Survey (AAGIS) to create a population of $N = 81,982$

Table 15.1 Design-based simulation performances of EBLUP and MBD predictors.

Predictor	Albanian Population		AAGIS Population	
	RB	RRMSE	RB	RRMSE
EBLUP (15.12)	0.42	5.90	1.60	15.90
MBD (15.20)	0.03	6.14	−0.82	14.45

farms by re-sampling from the original AAGIS sample with probability proportional to a farm's sample weight. The small areas of interest in this case were the $D = 29$ broadacre farming regions represented in this sample. The design-based simulation was carried out by 1000 independent stratified random samples from this bootstrap population, with strata defined by the regions and with stratum sample sizes defined by those in the original AAGIS sample. These sample sizes vary from 6 to 117, with a median region sample size of 55. Here Y is Total Cash Costs (TCC) associated with operation of the farm, and Z is a vector that includes farm area (Area), effects for six post-strata defined by three climatic zones and two farm size bands as well as the interactions of these variables. In the original AAGIS sample the relationship between TCC and Area varies significantly between the six post-strata, with an overall R-squared value of approximately 0.48 after the deletion of two outliers. The fixed effects in the prediction model were therefore specified as corresponding to a separate linear fit of TCC in terms of Area in each post-stratum. Random effects were defined as independent regional effects (i.e. a random intercepts specification) on the basis that in the original AAGIS sample the between-region variance component is highly significant, explaining just over 10% of the total residual variability with the two outliers removed. The aim was to estimate the regional averages of TCC.

Table 15.1 shows the median relative biases (RB) and the median relative RMSEs (RRMSE) of the EBLUP and the MBD predictors based on the 1000 independent stratified samples taken from the Albanian and AAGIS populations respectively. Note that these medians are over 36 areas in the case of the Albanian population and over 29 areas in the case of the AAGIS population. Similarly, Table 15.2 shows the median relative biases and median relative RMSEs of estimators of the MSEs of these estimators calculated from the same samples. It is noteworthy that the EBLUP, although the best performer in terms of MSE for the Albanian population, also records the highest biases for both.

Reflecting its model-dependent basis, the unconditional MSE estimator (15.13) for the EBLUP displays a substantial upward bias in both sets of design-based simulations as well as the largest instability. In contrast, for the Albanian population both versions of the conditional MSE estimator (15.26) are essentially unbiased, while for the AAGIS population it remains unbiased for the

Table 15.2 Design-based simulation performances of MSE estimators for EBLUP and MBD predictors.

MSE Estimator	Albanian Population		AAGIS Population	
	RB	RRMSE	RB	RRMSE
EBLUP/Unconditional (15.13)	14.6	44	23.7	209
EBLUP/Conditional (15.26)	0.1	24	11.5	157
MBD/Conditional (15.26)	−0.8	25	−0.8	190

MBD predictor and displays an upward bias for the EBLUP, though not to the same extent as the unconditional MSE estimator (15.13) of the EBLUP.

An insight into the reasons for this difference in behaviour can be obtained by examining the area specific RMSE values displayed in Fig. 15.1 for the Albanian population and in Fig. 15.2 for the AAGIS population. Thus, in Fig. 15.1 we see that conditional MSE estimator (15.26) tracks the district-specific repeated sampling RMSEs of both the EBLUP and the MBD predictor exceptionally well. In contrast, the unconditional MSE estimator (15.13) does not seem to be able to capture between-district differences in the repeated sampling RMSE of the EBLUP. In Fig. 15.2 we see that the estimator (15.26) of the conditional MSE of the MBD predictor tracks the RMSE of this predictor well in all regions except one (region 6) where it substantially overestimates the RMSE. This region is noteworthy because samples that are unbalanced with respect to Area within the region lead to negative weights under the assumed linear mixed model. The picture becomes more complex when one considers the region-specific RMSE estimation performance of the EBLUP in Fig. 15.2. Here we see that the estimator (15.26) of the conditional MSE of the EBLUP clearly tracks the region-specific RMSE of this predictor better than the estimator (15.23) of its unconditional MSE, with the noteworthy exception of region 21, where (15.26) shows significant overestimation. This region contains a number of massive outliers (all replicated from a single outlier in the original AAGIS sample) and these lead to a 'blow out' in the value of (15.26) when they appear in sample. With the exception of region 6 (where sample balance is a problem), there seems to be little regional variation in the value of the estimator (15.13) of the unconditional MSE of the EBLUP, indicating a serious bias problem with this approach.

It is important to note that these results should not be interpreted as indicating superiority of the MBD predictor over the EBLUP. Their main purpose is to demonstrate that when model specification is uncertain, one should not rely on efficiency results that are model dependent. In this context, using a direct estimator like the MBD predictor, though theoretically inefficient, may actually be justified from model robustness considerations. In fact, in the above simulations, results reported for the simple sample mean in each area were virtually

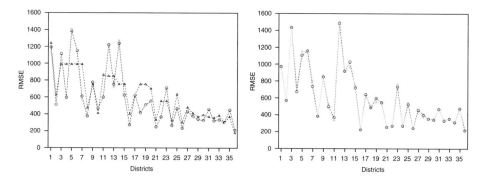

Fig. 15.1 District level values of true repeated sampling RMSE (solid grey line) and average estimated RMSE (dashed line) obtained in the design-based simulations using the Albanian household population. Districts are ordered in terms of increasing population size. Values for the unconditional MSE estimator (15.13) are indicated by Δ while those for the conditional MSE estimator (15.26) are indicated by o. Plots show results for the EBLUP (left) and MBDE (right) predictors.

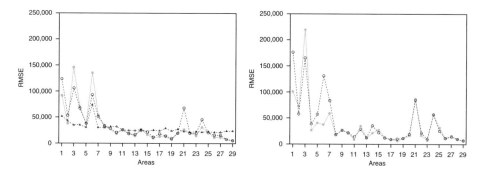

Fig. 15.2 Region-level values of true repeated sampling RMSE (solid grey line) and average estimated RMSE (dashed line) obtained in the design-based simulations using the AAGIS population. Regions are ordered in terms of increasing population size. Values for the unconditional MSE estimator (15.13) are indicated by Δ while those for the conditional MSE estimator (15.26) are indicated by o. Plots show results for the EBLUP (left) and MBDE (right) predictors.

identical to those of the MBD predictor, so the main advantage is from use of a direct estimator, rather than the use of a particular type of weighting scheme. An important lesson is that if one wants to use an efficient model-dependent predictor like the EBLUP, then choosing the 'right' model to underpin this predictor is an important consideration in small area estimation.

15.7 Generalised Linear Mixed Models in Small Area Prediction

Survey data are often categorical, and generalised linear models (GLMs) are widely used for modelling such data (Nelder and Wedderburn, 1972). Their application in the small area context leads to synthetic GLM-based small area predictors. Similarly, the generalised linear mixed model (GLMM) extension of GLM (Breslow and Clayton, 1993) can be used if a model with random area effects is required for small area prediction. Such models are typically written $E(y_i|\mathbf{u}_d) = h(\eta_i) = h\left(\mathbf{z}'_i\boldsymbol{\beta} + \mathbf{g}'_i\mathbf{u}_d\right)$ for a specified function h, so that the population values of the 'linear predictor' can be written $\boldsymbol{\eta}_U = \mathbf{Z}_U\boldsymbol{\beta} + \mathbf{G}_U\mathbf{u}$. As in (15.6), $\mathbf{u}' = (\mathbf{u}'_1\mathbf{u}'_2\cdots\mathbf{u}'_D)$, where $\mathbf{u}_d$ is a vector of q-dimensional random area d effects. These area effects are typically assumed to be normally distributed, uncorrelated between areas, and with zero mean vector and variance-covariance matrix $Var(\mathbf{u}_d) = \sigma_u^2\Gamma(\varphi)$. Note that for a GLM specification, $\sigma_u^2 = 0$.

To illustrate, suppose Y is a Bernoulli random variable, that is Y takes the values 1 and 0, with distinct population values of Y independently distributed. Suppose further that the logit of the probability π_d that a population unit i in area d is a 'success' (i.e. has $Y = 1$) follows a random mean (RM) model. That is,

$$\log\left\{\pi_d(1-\pi_d)^{-1}\right\} = \mu + u_d \tag{15.29}$$

or equivalently

$$\pi_d = \frac{exp(\mu + u_d)}{1 + exp(\mu + u_d)}$$

where u_d is a random area effect assumed to be normally distributed with zero mean and variance σ_u^2. Predicted values of the proportion $\bar{y}_d$ of 'successes' in each small area are required. For simplicity we assume there is sample in each area. The small area sample counts $t_{s1}, t_{s2}, \ldots, t_{sD}$ of successes are then realisations of independent binomial random variables with different area specific success probabilities π_d. The model (15.29) can be fitted using these data, typically via ML or penalized quasi-likelihood combined with REML, leading to an estimate $\hat{\mu}$ of μ and predictions $\hat{u}_d$ of the area effects u_d. The predicted proportion of successes in area d is then the plug-in EB predictor

$$\hat{\bar{y}}_d^{EBP} = N_d^{-1}[n_d\bar{y}_{sd} + (N_d - n_d)\hat{\pi}_d]$$

where

$$\hat{\pi}_d = \frac{exp(\hat{\mu} + \hat{u}_d)}{1 + exp(\hat{\mu} + \hat{u}_d)}.$$

In general, the EB predictor of the area d mean of Y under a GLMM is

$$\hat{\bar{y}}_d^{EBP} = N_d^{-1}\left[n_d\bar{y}_{sd} + \sum\nolimits_{r_d} h\left(\mathbf{z}'_i\hat{\boldsymbol{\beta}} + \mathbf{g}'_i\hat{\mathbf{u}}_d\right)\right] \tag{15.30}$$

where $\hat{\boldsymbol{\beta}}$ and $\hat{\mathbf{u}}_d$ are the estimated value of the regression coefficient $\boldsymbol{\beta}$ and the predicted value of the area effect $\mathbf{u}_d$ respectively that are obtained when we fit the GLMM to the sample data. An important point to note here is that since h is generally non-linear, we need to know the individual values of $\mathbf{z}_i$ and $\mathbf{g}_i$ for the non-sampled population units in area d before we can compute (15.30).

A more detailed description of the methodology for fitting GLMMs is outside the scope of this book. Saei and Chambers (2003) review the use of these models in small area prediction and also discuss estimation of the unconditional MSEs of predictors like (15.30) using the approach described in Section 15.3.

15.8 Prediction of Small Area Unemployment

This section presents empirical results on the application of generalised linear models (GLMs) and generalised linear mixed models (GLMMs) in small area prediction. In particular it describes how such models performed when used to predict the annual numbers of unemployed people within each of the 406 local authority districts (LADs) of Great Britain for a single year between 2000 and 2005. The definition of unemployment used in this case is the one provided by the International Labour Organisation, and the unemployment estimates that are based on it are the official measures of UK unemployment. At national and regional levels, these estimates are obtained from the UK Labour Force Survey (UKLFS). Direct estimates of numbers of unemployed, by age group by gender within LADs, are also available from the UKLFS, and correspond to sample weighted sums of the numbers of unemployed survey respondents at these levels. Since the UKLFS sample sizes at LAD level were typically small over the period of interest, many of these direct estimates were not precise enough to be released by the UK Office for National Statistics.

However, there is another source of information about UK unemployment. This is the *claimant count*, that is the number of people who register in order to claim unemployment benefits. The official measure of unemployment and the claimant count are not the same, although they are closely related. However, because the claimant count is based on a register, it is available at much lower levels of aggregation than the UKLFS-based unemployment estimates. Small area prediction methods were therefore used to produce estimates of numbers of unemployed people at low levels of aggregation. In particular, LAD-level predictions of numbers of unemployed people were obtained by combining data from the UKLFS with LAD-level claimant count data.

Since unemployment levels vary by age and gender, the models used for this purpose treated claimant count data, broken down by age group and gender within an LAD, as an auxiliary variable. Also, since unemployment levels vary both by socio-economic level as well as by geography within the UK, the regional affiliation of the LAD and the socio-economic grouping of the LAD were also included as factors in these models.

Three model specifications were investigated as part of a larger study of LAD-level unemployment estimation using the UKLFS data. These are set out in Table 15.3. They define the matrix $\mathbf{Z}_s$ when fitting a logistic regression model

Table 15.3 Model specifications for prediction of LAD-level unemployment.

Model	Specification	Number of Parameters
A	Age-gender effect	5
	+ logit(claimant count proportion) effect	+1
	+ age-gender × logit(claimant count proportion) interaction	+5
	+ region effect	+11
	+ socio-economic group effect	+6
	+ logit(LAD-level claimant count proportion) effect	+1 = 29
L	Age-gender effect	5
	+ region effect	+11
	+ socio-economic group effect	+6
	+ logit(LAD-level claimant count proportion) effect	+1 = 23
M	Age-gender effect	5
	+ region effect	+11
	+ socio-economic group effect	+6 = 22

to the sample values of the one-zero variable defined by whether a UKLFS respondent was unemployed or not.

From Table 15.3 we see that model A allows the logit of the probability of unemployment to be a different linear function of the logit of the claimant count proportion in each age-gender group. It also includes an effect based on the total claimant count proportion in the LAD, reflecting the potential additional impact of overall employment conditions in the area. Conversely, model L assumes that the relationship between probability of unemployment and claimant count is the same for all age-gender groups in an LAD. Finally, model M has no claimant count effects at all, assuming that the probability of unemployment just varies by age-gender, region and socio-economic groupings in the population.

All three models were used to obtain GLM and GLMM-based predictions for each age-gender group within an LAD. These were then summed over these groups to obtain a prediction of the total LAD unemployment level. Note that the GLMM predictions included an LAD-specific random effect. GLMM model parameters were estimated using a penalised quasi-likelihood/REML algorithm and the estimated prediction variances of the resulting small area predictors were estimated using an algorithm based on the MSE decomposition described in Section 15.3. See Saei and Chambers (2003) for a description of these algorithms. In all cases, the predictions generated by the different models were calibrated (via iterative re-scaling) to agree with UKLFS estimates of numbers of unemployed by age-gender, region and socio-economic group for Great Britain.

Table 15.4 shows the values of two diagnostics for the goodness of fit of the different small area predictions generated by models A, L and M. Both diagnostics work by comparing the LAD-level predictions generated under these

Table 15.4 Small area diagnostics for LAD-level unemployment estimates.

Model	W	Non-Overlap
A	421	18/406
L	426	17/406
M	584	38/406

models with corresponding UKLFS direct estimates, which are assumed to be unbiased. The W statistic is defined as

$$W = \sum_d \frac{\left(\hat{\bar{y}}_d^{EBP/Syn} - \hat{\bar{y}}_d^{UKLFS}\right)^2}{\hat{V}\left(\hat{\bar{y}}_d^{EBP/Syn}\right) + \hat{V}\left(\hat{\bar{y}}_d^{UKLFS}\right)} \tag{15.31}$$

where $\hat{\bar{y}}_d^{EBP/Syn} = N_d^{-1}\left[n_d\bar{y}_{sd} + \sum_{r_d} h\left(\mathbf{z}_i'\hat{\beta}\right)\right]$ is the synthetic version of (15.30) generated under a GLM and $\hat{\bar{y}}_d^{UKLFS}$ is the corresponding direct estimate obtained from the UKLFS. Here $\hat{V}\left(\hat{\bar{y}}_d^{EBP/Syn}\right)$ denotes an estimate of the prediction variance of $\hat{\bar{y}}_d^{EBP/Syn}$ and $\hat{V}\left(\hat{\bar{y}}_d^{UKLFS}\right)$ denotes an estimate of the prediction variance of $\hat{\bar{y}}_d^{UKLFS}$. Note that the use of a synthetic EB predictor in (15.31) ensures that the correlation between it and the direct estimator is negligible (since the model fit used in the synthetic EB predictor is based on data for the entire sample), and so this statistic has an asymptotic chi-square distribution with D degrees of freedom when the synthetic EB predictor has expectation equal to the direct estimator. This statistic therefore measures closeness of model-based predictions to expected values of the direct estimates. The more its value exceeds $D = 406$ (i.e. the number of LADs) the worse the fit.

Similarly, the Non-Overlap statistic is based on the fact that if X and Y are two independent normal random variables, with the same mean but with different standard deviations, σ_x and σ_y respectively, and if $z(\alpha)$ is such that the probability that a standard normal variable takes values greater than $z(\alpha)$ is $\alpha/2$, then a sufficient condition for there to be probability of α that the two intervals $X \pm z(\beta)\sigma_x$ and $Y \pm z(\beta)\sigma_y$ do not overlap is when

$$z(\beta) = z(\alpha)\left(1 + \frac{\sigma_x}{\sigma_y}\right)^{-1}\sqrt{1 + \frac{\sigma_x^2}{\sigma_y^2}}.$$

This diagnostic takes $z(\alpha) = 1.96$, calculates $z(\beta)$ using the above formula, with σ_x replaced by the estimated standard error of the GLM-based synthetic EB predictor and σ_y replaced by the estimated standard error of the direct estimate and then computes the overlap proportion between the corresponding $z(\beta)$-based confidence intervals generated by the two estimation methodologies. Nominally, for $z(\alpha) = 1.96$, this overlap proportion should be 95%. Consequently there should be 5% of 406 (i.e. approximately 20) non-overlapping confidence intervals if the model-based prediction variance estimates are unbiased.

Both the W and Non-Overlap statistics are discussed in Brown et al. (2001). Inspection of the results set out in Table 15.4 indicates that the GLM versions of models A and L seem reasonable for the unemployment data. With model M however, it is clear that the GLM specification is inadequate.

We illustrate the improvements from using the small area predictions based on models A, L and M by showing their estimated gain plots in Fig. 15.3. The estimated gain value for a particular LAD is defined as the estimated standard error of the UKLFS direct estimate for that LAD divided by the square root of the estimated prediction MSE of the corresponding model-based prediction. Estimated gain values greater than one indicate improvement by the model-based method. The plots show how these values for different model/methods change when the LADs are ordered by their claimant counts. We observe that the estimated gain decreases as the claimant count increases and that the estimated gains associated with GLM-based methods are generally larger than those based on the GLMM-based methods. Both of these features are as one would expect – the introduction of a random effect tends to decrease potential bias at the expense of decreasing precision, while increasing claimant count is highly correlated with increasing LAD population size and hence increasing UKLFS sample size for the LAD. Given that the GLM versions of model A and model L based predictions appear to be essentially unbiased in this case (see Table 15.4), the argument for including random effects in these models seems weak. What is rather worrying, however, is that the estimated gains under the GLM version of model M are shown as being rather substantial, even though this model/method has already been observed to generate poor diagnostics (again, see Table 15.4). In contrast, the price for the lack of fit of model M is evident in the estimated gains associated with the GLMM version of this model. These are comparatively small.

Unfortunately, the superior estimated gains under the GLM specification so evident in Fig. 15.3 are not what they seem. Their validity depends on the GLM version of each model (A, L, M) adequately representing the population distribution of unemployment. That is, there is no 'extra binomial' variation in the unemployment data due to remaining differences in the probability of unemployment between LADs after accounting for differences due to variation of model covariates. What happens if this is not true? As the following simulation study demonstrates, the estimated gains shown in Fig. 15.3 for the GLM versions of models A, L and M then evaporate.

In this simulation study, independent Bernoulli random variables $\{y_i; i = 1, \ldots, N = 390,192\}$ were drawn from $D = 30$ areas. For an observation from area d, the probability $\pi_i = \Pr(y_i = 1)$ satisfied

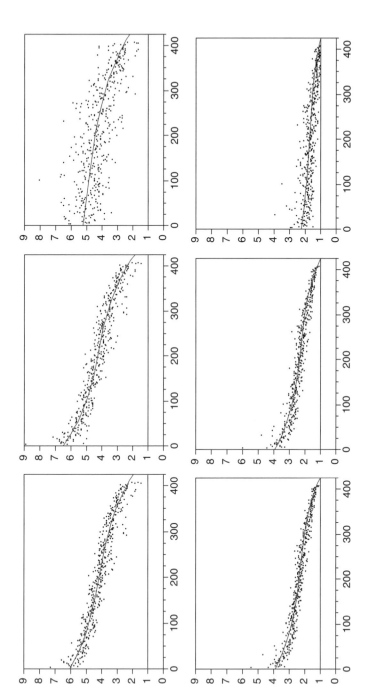

Fig. 15.3 Estimated gain plots for LAD-level unemployment estimates. First line is GLM-based results, models A, L and M from left to right. Second line is corresponding GLMM results. Smooth curve shown on each plot is cubic fit to values. Horizontal axis is rank of LAD with respect to claimant count.

$$\log\left(\frac{\pi_i}{1-\pi_i}\right) = -3 + 0.05 z_i + u_d$$

where z_i was randomly assigned to zero or one with equal probability and the u_d were random area effects that were independently drawn from a normal distribution with zero mean and variance σ_u^2. The Z-values were assumed known for all population units. A stratified random sample of size $n = 20,683$ was then drawn from this population, with stratification on Z, and predicted values for the area Y-means calculated. Three methods of prediction were investigated. These were (i) direct estimates corresponding to stratum weighted sample proportions in each area; (ii) predicted values based on fitting a fixed effects linear logistic model based on Z to the sample values of Y (i.e. a GLM); and (iii) predicted

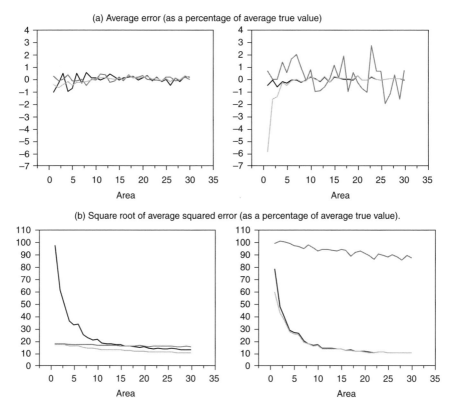

Fig. 15.4 Simulation performances by area for direct estimates (black), fixed effects logistic model-based predictions (grey) and random effects logistic model-based predictions (light grey). Plot on the left is for $\sigma_u^2 = 0.03$ (similar individual, i.e. binomial, variability and between-area variability) while plot on the right is for $\sigma_u^2 = 1.0$ (between-area variability much larger than individual variability). Areas ordered by increasing population size on horizontal axis.

values based on fitting the same linear logistic model, but now with random area effects (i.e. a GLMM), to these sample values.

(a) Coverage rate (%) by area for nominal 95 % 'two-sigma' confidence intervals. Solid black line shows coverage rate of intervals based on direct estimates, solid grey line is for intervals based on fixed effects logistic model-based predictions and dotted line is for intervals based on random effects logistic model-based predictions.

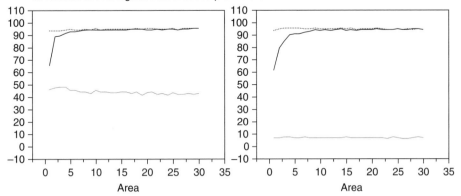

(b) True gain vs. estimated gain. True gain for fixed effects logistic model-based predictions are black, while estimated gain is dotted line. True gain for random effects logistic model-based predictions is grey, while estimated gain is a dashed line.

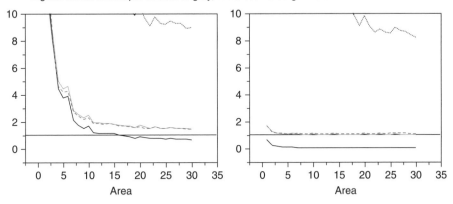

Fig. 15.5 Simulation performances by area for direct estimates, fixed effects logistic model-based predictions and random effects logistic model-based predictions. Plot on the left is for $\sigma_u^2 = 0.03$ (similar individual, i.e. binomial, variability, and between-area variability) while plot on the right is for $\sigma_u^2 = 1.0$ (between area variability much larger than individual variability). Areas ordered by increasing population size on horizontal axis.

The population generation and sample selection process described in the previous paragraph was then independently repeated 5000 times for two different values of σ_u^2. Each time, the prediction errors associated with the different area specific predictions were calculated, as well as the estimated values of the variance of these prediction errors. Figure 15.4(a) shows the averages of these errors over the 5000 repetitions, while Fig. 15.4(b) shows the square roots of the averages of the squares of these errors, again over the 5000 repetitions. The superiority of the predictions generated by the logistic mixed model is clear. Also, we see that the predictions generated by the fixed effects version of this model are inefficient, especially for large values of σ_u^2.

Figure 15.5(a) shows the 'two-sigma' coverages recorded by the nominal 95% confidence intervals generated by the three prediction methods investigated in the simulation study. Again, we note the good performance of the estimated prediction variances based on the logistic mixed model. In contrast, the estimated prediction variances based on the fixed effects version of this model appear to have a downward bias. This bias increases with increasing σ_u^2, leading to undercoverage. This bias in estimated prediction variance under the fixed effects logistic model stands out in Fig. 15.5(b). Here we see that the average estimated gain under this model is very large (too large in fact to be completely shown on the plot) while the average actual gain is in fact negligible. This is in contrast to the results for the mixed model, where estimated and actual gains are very close. There is a clear lesson that can be drawn from these results. Reliance on estimated standard errors generated under a small area model that is incorrectly specified can lead to quite inflated expectations of efficiency gains from use of this model to predict individual area characteristics. Consequently, when we carry out small area estimation it is almost always a good strategy to use a model that allows for potential between area variation that is not captured by the covariates in the model. Including area random effects in the model is one way – though not the only way, see Chambers and Tzavidis (2006) – of achieving this aim. The price one pays for this 'insurance' is a higher estimate of MSE for the resulting small area estimate. In most cases, this is a price worth paying. For example, in Fig. 15.3 we see that for the GLMM versions of both 'acceptable' models (A and L) there is still a substantial efficiency gain relative to direct estimation, indicating that LAD estimates of unemployment based on either model should provide real improvement over direct estimation using the UKLFS data. After careful consideration of data from many different sources, the UK Office for National Statistics decided to publish LAD level estimates of unemployment based on a version of model A. See ONS (2006).

15.9 Concluding Remarks

The focus of this chapter has been on so-called *unit level* models for small areas, that is where unit level sample data are available for modelling purposes. This reflects the overall focus of this book. However, there are situations where such data are not available, but direct estimates for the areas are available, together with area level characteristics for use as covariates. In such cases, *area level models* can be specified, linking the values of the direct estimates to the values of the area level characteristics, and the fitted values generated by these models used for small area estimation.

To illustrate, we consider a very popular model of this type, originally proposed by Fay and Herriot (1979). This assumes that for each area d, a direct estimate $\bar{y}_{wd}$ of the unknown area mean $\bar{y}_d$ is available, where the w subscript indicates that $\bar{y}_{wd}$ is a weighted average of the sample data from area d. Furthermore, it is assumed that

$$\bar{y}_{wd} = \bar{y}_d + e_d \tag{15.32}$$

and

$$\bar{y}_d = \mathbf{x}'_d \beta + u_d \tag{15.33}$$

where $\mathbf{x}_d$ is a vector of known area characteristics, and e_d and u_d are independent random errors, with zero expectations and with variances σ_{ed}^2 and σ_u^2 respectively. That is, $\bar{y}_{wd}$ is an unbiased predictor of $\bar{y}_d$ with prediction error e_d and the regression of $\bar{y}_d$ on $\mathbf{x}_d$ is linear with regression error u_d. In passing, we observe that (15.32) allows the prediction variance of $\bar{y}_{wd}$ to vary from area to area, which is necessary since it is extremely unlikely that the underlying sample sizes in the different areas will be the same. It immediately follows from (15.32) and (15.33) that

$$\bar{y}_{wd} = \mathbf{x}'_d \beta + u_d + e_d \tag{15.34}$$

and, if we further assume that e_d and u_d are jointly normally distributed, then (15.33) and (15.34) imply

$$E(\bar{y}_d | \bar{y}_{wd}, \mathbf{x}_d) = \mathbf{x}'_d \beta + \frac{\sigma_u^2}{\sigma_u^2 + \sigma_{ed}^2} (\bar{y}_{wd} - \mathbf{x}'_d \beta)$$

$$= \left(\frac{\sigma_{ed}^2}{\sigma_u^2 + \sigma_{ed}^2} \right) \mathbf{x}'_d \beta + \left(\frac{\sigma_u^2}{\sigma_u^2 + \sigma_{ed}^2} \right) \bar{y}_{wd}.$$

An EB predictor for $\bar{y}_d$ given $\bar{y}_{wd}$ and $\mathbf{x}_d$ follows directly, and is of the form

$$\left(\frac{\hat{\sigma}_{ed}^2}{\hat{\sigma}_u^2 + \hat{\sigma}_{ed}^2} \right) \mathbf{x}'_d \hat{\beta} + \left(\frac{\hat{\sigma}_u^2}{\hat{\sigma}_u^2 + \hat{\sigma}_{ed}^2} \right) \bar{y}_{wd} \tag{15.35}$$

where, as usual, a 'hat' denotes an estimate. It is clear from (15.34) that an unbiased estimator $\hat{\beta}$ can be calculated by regressing the area level estimates $\bar{y}_{wd}$ on the area level covariates $\mathbf{x}_d$. Unfortunately, calculation of $\hat{\sigma}_{ed}^2$ and $\hat{\sigma}_u^2$ is not quite so straightforward. If an unbiased estimate $\hat{v}(\bar{y}_{wd})$ of the prediction variance of $\bar{y}_{wd}$ under (15.34) is available, then this can replace $\hat{\sigma}_{ed}^2$ in (15.35).

Furthermore, if $\hat{\sigma}^2_{u+ed}$ denotes the estimated variance of the regression error in area d under (15.34), then an estimate of σ^2_u is $\hat{\sigma}^2_u = \hat{\sigma}^2_{u+ed} - \hat{v}(\bar{y}_{wd})$. However, it should be noted that $\hat{v}(\bar{y}_{wd})$ can be rather unstable (since it is based on a small sample), and so $\hat{\sigma}^2_u$ can become unstable as well and may take on negative values. If the variation in sample sizes in the areas is large enough, one practical approach to overcoming this problem is to 'smooth' both the values $\hat{v}(\bar{y}_{wd})$ and the differences $\hat{\sigma}^2_{u+ed} - \hat{v}(\bar{y}_{wd})$ against these sample sizes before using them in (15.35). In many applications, the values of $\hat{\sigma}^2_{ed}$ and $\hat{\sigma}^2_u$ used in (15.35) are chosen somewhat arbitrarily, and the optimality properties of this EB predictor therefore need to be treated with a certain amount of caution.

Even when unit level sample data are available for small areas, there are many different ways that small area estimates can be computed, depending on the particular characteristics of the variable of interest. For example, models for small area estimation that allow area effects to be spatially correlated have been investigated, see Singh *et al.* (2005) and Pratesi and Salvati (2008), as have models that allow a non-linear relationship between this variable and the available covariates, see Opsomer *et al.* (2008) and Ugarte *et al.* (2009). Other developments have focused on the use of loglinear models for small area estimation of tabulated data, see Zhang and Chambers (2004), and generalised quantile regression-based alternatives to mixed models for small area estimation, see Chambers and Tzavidis (2006), Tzavidis *et al.* (2008, 2010) and Salvati *et al.* (2010). Outlier-robust small area estimation methods are explored in Sinha and Rao (2009), while Pereira and Coelho (2010) consider models for small area estimation that borrow strength across time and space.

16 Model-Based Inference for Distributions and Quantiles

The population mean $\bar{y}_U$ of a survey variable Y is the standard measure of the location of the population distribution of this variable. In many surveys, however, an important secondary aim is to gain a better understanding of the actual distribution of the N values of Y across this population. This is typically achieved by predicting selected quantiles of this distribution, where the α quantile is the smallest value $Q_{Ny}(\alpha)$ such that a proportion $1-\alpha$ of the Y-values in population are greater than $Q_{Ny}(\alpha)$. A convenient alternative way of defining $Q_{Ny}(\alpha)$ is via the estimating equation

$$N^{-1}\sum_{U} I\{y_i \leq Q_{Ny}(\alpha)\} = F_{Ny}\{Q_{Ny}(\alpha)\} = \alpha. \tag{16.1}$$

Here $F_{Ny}(t) = N^{-1}\sum_{U} I(y_i \leq t)$ is the *finite population distribution function* defined by the population Y-values.

The median corresponds to the $\alpha = 0.5$ quantile, and we have already looked at issues associated with prediction of the finite population median in Chapter 11. There we assumed that this variable is predicted by solving a sample-weighted analogue of the population estimating equation (16.1). In this chapter we take a more indepth look at efficient prediction of $F_{Ny}(t)$ for given t, and, as a consequence, better prediction of finite population quantiles (including the median). Additional methods for estimating distribution functions and quantiles can be found in Dorfman (2009a).

16.1 Distribution Inference for a Homogeneous Population

To start, consider prediction of $F_{Ny}(t)$ (and hence $Q_{Ny}(\alpha)$) when the working model for the population values of Y is the homogeneous population model (3.1). Since $F_{Ny}(t)$ is not a linear combination of the population Y-values, an unbiased predictor of this population parameter under this model that is also linear in the sample Y-values cannot be defined. Instead, we consider the indicator variable $D_t = I(Y \leq t)$ with population values $d_{it} = I(y_i \leq t); i = 1,\ldots,N$. That is, d_{it} takes the value 1 if y_i is less than or equal to t and is zero otherwise. By definition $F_{Ny}(t)$ is the population mean of D_t. Let $F_y(t) = \Pr(Y \leq t)$ denote the distribution of Y under the working homogeneous population model. Then

$$E(d_{it}) = F_y(t) \tag{16.2a}$$
$$Var(d_{it}) = F_y(t)\{1 - F_y(t)\} \tag{16.2b}$$
$$Cov(d_{it}, d_{jt}) = 0. \tag{16.2c}$$

That is, the assumption of a homogeneous population model for Y induces a similar model for D_t. It follows that the BLUP for $F_{Ny}(t)$ in this case is the sample mean of D_t,

$$\hat{F}_{Ny}(t) = n^{-1} \sum_s d_{it} \tag{16.3}$$

with prediction variance

$$Var\left\{\hat{F}_{Ny}(t) - F_{Ny}(t)\right\} = n^{-1}(1 - n/N)F_y(t)\{1 - F_y(t)\}. \tag{16.4}$$

An unbiased estimator of this prediction variance is

$$\hat{V}\left(\hat{F}_{Ny}(t)\right) = (n-1)^{-1}(1 - n/N)\hat{F}_{Ny}(t)\left\{1 - \hat{F}_{Ny}(t)\right\}. \tag{16.5}$$

See exercise E.35 for a proof of the unbiasedness of (16.5).

Given $\hat{F}_{Ny}(t)$, prediction of the finite population quantile function $Q_{Ny}(\alpha)$ proceeds by appropriately inverting $\hat{F}_{Ny}(t)$. That is, a predictor $\hat{Q}_{Ny}(\alpha)$ of $Q_{Ny}(\alpha)$ is defined via the equation

$$\hat{F}_{Ny}\left(\hat{Q}_{Ny}(\alpha)\right) = \alpha. \tag{16.6}$$

Since $\hat{F}_{Ny}(t)$ is a step function, (16.6) does not have a unique solution when t is not equal to one of the sample Y-values. In practice a smoothing operator is usually applied to $\hat{F}_{Ny}(t)$ so as to make it continuous and monotonically increasing over the interval $c < y_{(1)} \leq t \leq y_{(n)} < d$. Here $y_{(k)}$ is the kth sample order statistic for Y. Let $\hat{F}^*_{Ny}(t)$ denote this smoothed version of $\hat{F}_{Ny}(t)$. For $n^{-1} \leq \alpha \leq n^{-1}(n-1)$ the corresponding smoothed quantile function predictor $\hat{Q}^*_{Ny}(\alpha)$ is then defined as the solution of the equation

$$\hat{F}^*_{Ny}\left(\hat{Q}^*_{Ny}(\alpha)\right) = \alpha. \tag{16.7}$$

The smoothed quantile predictor $\hat{Q}^*_{Ny}(\alpha)$ need not exist for values of α outside these bounds.

Linear interpolation is an easy way of smoothing $\hat{F}_{Ny}(t)$. Let m_t be the smallest positive integer such that $y_{(k)} < t < y_{(k+m_t)}$. Our smoothed predictor of $F_{Ny}(t)$ is

$$\hat{F}^*_{Ny}(t) = \frac{y_{(k+m_t)} - t}{y_{(k+m_t)} - y_{(k)}} \hat{F}_{Ny}(y_{(k)}) + \frac{t - y_{(k)}}{y_{(k+m_t)} - y_{(k)}} \hat{F}_{Ny}(y_{(k+m_t)}) \tag{16.8a}$$

with estimated prediction variance

$$\hat{V}\left(\hat{F}^*_{Ny}(t)\right) = (n-1)^{-1}(1-n/N)\hat{F}^*_{Ny}(t)\left\{1-\hat{F}^*_{Ny}(t)\right\}. \qquad (16.8b)$$

A simple method for deriving a prediction interval for $Q_{Ny}(\alpha)$ was suggested by Woodruff (1952). Let z denote an appropriately chosen percentile of a standard normal distribution and put

$$f_1 = \hat{F}_{Ny}\left(\hat{Q}_{Ny}(\alpha)\right) - z\sqrt{\hat{V}\left(\hat{F}_{Ny}\left(\hat{Q}_{Ny}(\alpha)\right)\right)}$$

$$f_2 = \hat{F}_{Ny}\left(\hat{Q}_{Ny}(\alpha)\right) + z\sqrt{\hat{V}\left(\hat{F}_{Ny}\left(\hat{Q}_{Ny}(\alpha)\right)\right)}.$$

The interval $\left[\hat{Q}_{Ny}(f_1), \hat{Q}_{Ny}(f_2)\right]$ is then a prediction interval for $Q_{Ny}(\alpha)$ with confidence coefficient $2\{1 - \Phi(z)\}$, where Φ is the (cumulative) distribution function of the standard normal distribution.

Obviously, one can substitute $\hat{F}^*_{Ny}(t)$ for $\hat{F}_{Ny}(t)$ in Woodruff's method, and the result is still approximately valid. Additional methods of constructing confidence intervals for distribution and quantile function values can be found in Dorfman (2009a).

16.2 Extension to a Stratified Population

It is straightforward to extend the results of the previous section to where the stratified population model (4.1) is appropriate. In this case the strata correspond to independent homogeneous populations. Within stratum h it follows

$$E(d_{it}) = F_{hy}(t) \qquad (16.9a)$$
$$Var(d_{it}) = F_{hy}(t)\{1 - F_{hy}(t)\} \qquad (16.9b)$$
$$Cov(d_{it}, d_{jt}) = 0. \qquad (16.9c)$$

where $F_{hy}(t)$ is the stratum h distribution function for the random variable Y.

Since (16.9) is just a special case of (4.1), the BLUP of the finite population distribution function $F_{Ny}(t)$ is

$$\hat{F}^S_{Ny}(t) = N^{-1}\sum_h N_h n_h^{-1}\sum_{s_h} d_{it} = N^{-1}\sum_h N_h \hat{F}_{hy}(t). \qquad (16.10)$$

Development of the prediction variance $\hat{F}^S_{Ny}(t)$ is left as an exercise, as is development of an unbiased estimator of this variance. See exercise E.36.

Observe that $\hat{F}^S_{Ny}(t)$ is a weighted sum of step functions and so is also a step function. A smoothed version of this predictor is easily obtained, however, by taking the same weighted sum of appropriately smoothed stratum level distribution function predictors. For example, if the linear interpolation based smoother (16.8a) is used in each stratum, then the corresponding smoothed version of (16.10) is

$$\hat{F}_{Ny}^{*S}(t) = N^{-1} \sum_h N_h \hat{F}_{hy}^*(t)$$

where

$$\hat{F}_{hy}^*(t) = \frac{y_{(k_h+m_{th})} - t}{y_{(k_h+m_{th})} - y_{(k_h)}} \hat{F}_{hy}(y_{(k_h)}) + \frac{t - y_{(k_h)}}{y_{(k_h+m_{th})} - y_{(k_h)}} \hat{F}_{hy}(y_{(k_h+m_{th})}).$$

Here k_h and m_{th} are stratum h versions of k and m_t in (16.8a).

Again, prediction of the population quantile $Q_{Ny}(\alpha)$ is carried out by inverting either $\hat{F}_{Ny}^S(t)$ or $\hat{F}_{Ny}^{S*}(t)$, and the Woodruff method for calculating a confidence interval for this quantile is still applicable, with appropriate modifications.

16.3 Distribution Function Estimation under a Linear Regression Model

Knowledge of an auxiliary size variable Z that is correlated with the survey variable of interest can be used to substantially improve prediction of the finite population distribution function $F_{Ny}(t)$ and, as a consequence, the finite population quantile function $Q_{Ny}(\alpha)$. In particular, we now show how a linear regression model for Y in terms of Z can be used for this purpose.

To illustrate, assume that the regression model is the one considered in exercise E.14. That is, where the regression of Y on Z is a straight line through the origin, with slope β but with residual variance not necessarily proportional to Z. This is the model

$$E(y_i|z_i) = \beta z_i \qquad (16.11a)$$

$$Var(y_i|z_i) = \sigma^2 v(z_i) \qquad (16.11b)$$

$$Cov(y_i, y_j|z_i, z_j) = 0 \text{ for all } i \neq j \qquad (16.11c)$$

where $v(z)$ is a known non-decreasing function of z. Obviously $v(z) = z$ corresponds to the model under which the ratio estimator is the BLUP for the population total of Y. We extend this model by assuming that the standardised errors $e_i = v_i^{-1/2}(y_i - \beta z_i)$ are independent and identically distributed with common distribution function G. Here $v_i = v(z_i)$. It follows that the Y-values of distinct population units are independent given their respective values of Z.

To motivate our approach, recollect that $F_{Ny}(t)$ is the mean of the population values of the variable D_t. The real problem, given the sample data, is therefore prediction of the non-sample total of these values. An EB predictor of this total, given distinct population units are independent under (16.11), is defined by an estimator of its expected value under this model, which is

$$\sum_{j \in r} \Pr(y_j \leq t) = \sum_{j \in r} \Pr\left\{e_j \leq v_j^{-1/2}(t - \beta z_j)\right\}$$
$$= \sum_{j \in r} G\left(v_j^{-1/2}(t - \beta z_j)\right). \qquad (16.12)$$

That is, an EB predictor of $F_{Ny}(t)$ is

$$\hat{F}_{Ny}^R(t) = N^{-1}\left\{\sum_{i \in s} d_{it} + \sum_{j \in r} \hat{G}\left(v_j^{-1/2}(t - b_v z_j)\right)\right\} \qquad (16.13)$$

where $\hat{G}$ and b_v are efficient estimators of G and β. In particular (see exercise E.14)

$$b_v = \sum_s v_i^{-1} y_i z_i \Big/ \sum_s v_i^{-1} z_i^2$$

is the BLUE of the regression parameter β.

Given the sample data, a consistent and asymptotically unbiased estimator of $G(u)$ is the empirical distribution function defined by the standardised sample residuals $r_i = v_i^{-1/2}(y_i - b_v z_i)$. That is,

$$\hat{G}(u) = n^{-1} \sum_s I(r_i \leq u). \qquad (16.14)$$

Substituting this expression in (16.13) we can write down the EB predictor as

$$\hat{F}_{Ny}^R(t) = N^{-1}\left\{\sum_{i \in s} d_{it} + n^{-1} \sum_{j \in r} \sum_{i \in s} I\left(r_i \leq v_j^{-1/2}(t - b_v z_j)\right)\right\}. \qquad (16.15)$$

The predictor (16.15) was suggested by Chambers and Dunstan (1986). It can be very efficient, provided the model (16.11) holds. However, it can also be biased if this working model assumption fails. In particular, it is sensitive to misspecification of the variance function (16.11b), as we now show.

In Chapter 8 we saw that use of an appropriately balanced sample ensures that a linear model-based BLUP for a population total remains unbiased under a wide variety of alternative non-linear regression models for the survey variable Y. An obvious question to ask therefore is whether balanced sampling imparts a similar robustness to the linear model-based EB predictor $\hat{F}_{Ny}^R(t)$ defined in (16.15). Unfortunately, the following analysis indicates that as far as misspecification of the variance function $v(z)$ is concerned, balanced sampling has little effect on the bias robustness of $\hat{F}_{Ny}^R(t)$.

For ease of analysis, instead of (16.15), we look at the behaviour of the pseudo statistic

$$\tilde{F}_{Ny}^{R}(t) = N^{-1}\left\{\sum_{i\in s} d_{it} + n^{-1}\sum_{j\in r}\sum_{i\in s} I\left(v_i^{-1/2}(y_i - \beta z_i)\right.\right.$$
$$\left.\left. \leq v_j^{-1/2}(t - \beta z_j)\right)\right\}. \tag{16.16}$$

Suppose now that (16.11b) is misspecified and in reality $Var(y_i|z_i) = \sigma^2 w(z_i) = \sigma^2 w_i$. The asymptotic bias of (16.16) is then

$$N^{-1}E\left\{n^{-1}\sum_{j\in r}\sum_{i\in s} I\left(v_i^{-1/2}(y_i - \beta z_i) \leq v_j^{-1/2}(t-\beta z_j)\right) - \sum_{j\in r} I(y_j \leq t)\right\}$$

Put $\theta_{ij} = \sqrt{(w_j v_i)/(w_i v_j)}$. This bias can then be written

$$N^{-1}\sum_{j\in r}\left\{n^{-1}\sum_{i\in s} G(\theta_{ij}e_{tj}) - G(e_{tj})\right\} \tag{16.17}$$

where $e_{tj} = w_j^{-1/2}(t - \beta z_j)$. Expanding each of the $N - n$ terms in the square brackets in (16.17) around $e_{tj} = 0$ via a Taylor series leads to the first order approximation (where g is the first derivative of G)

$$g(0)N^{-1}\sum_{j\in r}\left\{\left(n^{-1}\sum_{i\in s}\theta_{ij} - 1\right)e_{tj}\right\}.$$

In particular, when $v(z)/w(z) = z^{2\delta}$, where $-1 \leq \delta \leq 1$, this first order approximation is

$$g(0)N^{-1}\sum_{j\in r}\left\{\left(n^{-1}\sum_{i\in s} z_i^\delta - z_j^\delta\right)v_j^{-1/2}(t - \beta z_j)\right\}. \tag{16.18}$$

Suppose now a balanced sample is selected, and, in particular, one such that

$$n^{-1}\sum_s z_i^\delta = (N-n)^{-1}\sum_r z_j^\delta.$$

For such a sample the first order approximation (16.18) to the bias at $t = \beta \bar{z}_r$ is

$$g(0)\beta N^{-1}\sum_{j\in r} v_j^{-1/2}(z_j^\delta - \bar{z}_r^{(\delta)})(z_j - \bar{z}_r). \tag{16.19}$$

where $\bar{z}_r^{(\delta)} = (N-n)^{-1}\sum_{k\in r} z_k^\delta$. Typically, β and the auxiliary variable Z will be positive. In such situations (16.19) will be strictly positive when $\delta > 0$ and strictly negative when $\delta < 0$. That is, in balanced samples and for values of t in the middle of the range of Y-values in the population, $\tilde{F}_{Ny}^R(t)$, and therefore $\hat{F}_{Ny}^R(t)$, will tend to be biased upwards when $v(z)$ overstates the true variance, and will tend to be biased downwards when it understates this variance.

16.4 Use of Non-parametric Regression Methods for Distribution Function Estimation

In order to get around the bias robustness problem with (16.15), Chambers, Dorfman and Wehrly (1993), see also Kuk and Welsh (2001), have suggested weakening the assumption that the errors $y_i - \beta z_i$ have variances specified by (16.11b) to one where these variances are locally constant. That is, $Var(y_i|z_i) \cong Var(y_j|z_j)$ when z_i and z_j are close. In particular, we assume that the two errors $y_i - \beta z_i$ and $y_j - \beta z_j$ are identically distributed only when z_i and z_j are close. In this situation, the estimator (16.14) underpinning (16.15) is inappropriate, since it treats all standardised sample residuals as identically distributed, irrespective of the differences in their values of Z. In order to develop an alternative to (16.15) we therefore replace (16.12) by

$$\sum_{j \in r} \Pr(y_j \leq t) = \sum_{j \in r} \Pr(y_j - \beta z_j \leq t - \beta z_j) = \sum_{j \in r} G_j(t - \beta z_j) \tag{16.20}$$

where G_j denotes the distribution of $y_j - \beta z_j$. The resulting EB predictor of $F_{Ny}(t)$ is

$$\hat{F}_{Ny}^{rob}(t) = N^{-1} \left\{ \sum_{i \in s} d_{it} + \sum_{j \in r} \hat{G}_j(t - b_v z_j) \right\} \tag{16.21}$$

where $\hat{G}_j$ denotes an efficient estimate of the distribution function G_j.

In order to define $\hat{G}_j$, Chambers, Dorfman and Wehrly (1993) recommend local mean smoothing. That is, rather than attempting to estimate the distribution function G_j by averaging over the entire set of sample residuals, we estimate it by averaging only over those sample residuals with Z-values that are close enough to z_j so that it is reasonable to assume that they have the same distribution as $y_j - \beta z_j$. In effect, we replace (16.14) by

$$\hat{G}_j(u) = \sum_{i \in s} w_{ij} I(y_i - b_v z_i \leq u) / \sum_{i \in s} w_{ij} \tag{16.22}$$

where the weights w_{ij} only take non-zero values for those sample units i with z_j close to z_i. Put $w_{ij}^* = w_{ij} / \sum_{k \in s} w_{kj}$. Then (16.21) can be written

$$\hat{F}_{Ny}^{RNP}(t) = N^{-1} \left\{ \sum_{i \in s} d_{it} + \sum_{j \in r} \sum_{i \in s} w_{ij}^* I(y_i - b_v z_i \leq t - b_v z_j) \right\}. \tag{16.23}$$

Clearly (16.23) reduces to (16.15) when $v_i = 1$ and $w_{ij} = 1$ for all sample units i and non-sample units j.

We now develop an approximation to the prediction variance of (16.23). This is based on the general model

$$y_i = \mu(z_i) + \sigma(z_i) e_i \tag{16.24}$$

where the errors e_i are assumed to be independent and identically distributed, and both $\mu(z)$ and $\sigma(z)$ are smooth functions of z. To save notation, we write $\mu_i = \mu(z_i)$ and $\sigma_i = \sigma(z_i)$ from now on. Note that (16.24) includes (16.11) as a special case and is essentially the model underlying the development in Kuk and Welsh (2001). Let $\hat{\mu}_i = \hat{\mu}(z_i)$ denote a consistent estimator of μ_i based on the sample data. The obvious generalisation of (16.23) under (16.24) is

$$\hat{F}_{Ny}^{NP}(t) = N^{-1}\left\{\sum_{i \in s} d_{it} + \sum_{j \in r}\sum_{i \in s} w_{ij}^* I(y_i - \hat{\mu}_i \leq t - \hat{\mu}_j)\right\} \quad (16.25)$$

and so

$$\mathrm{Var}\left(\hat{F}_{Ny}^{NP}(t) - F_{Ny}(t)\right) = N^{-2}(V_s + V_r) \quad (16.26)$$

where

$$V_s = \mathrm{Var}\left(\sum_{j \in r} \hat{P}_j(t)\right) = \sum_{j \in r}\sum_{k \in r} \mathrm{Cov}\left(\hat{P}_j(t), \hat{P}_k(t)\right)$$
$$V_r = \sum_{j \in r} \mathrm{Var}\left(I(y_j \leq t)\right)$$

and

$$\hat{P}_j(t) = \sum_{i \in s} w_{ij}^* I(y_i - \hat{\mu}_i \leq t - \hat{\mu}_j).$$

In order to estimate V_s in (16.26) we note that in large samples $\hat{P}_j(t)$ can be replaced by $P_j(t) = \sum_{i \in s} w_{ij}^* I(e_i \leq u_{ij}(t))$ where $u_{ij}(t) = \sigma_i^{-1}(t - \mu_j)$. Furthermore

$$\mathrm{Cov}(P_j(t), P_k(t)) = \sum_{i \in s} w_{ij}^* w_{ik}^* \mathrm{Cov}\left(I(e_i \leq u_{ij}(t)), I(e_i \leq u_{ik}(t))\right)$$
$$= \Psi_{jk}^{(1)} - \Psi_{jk}^{(2)}$$

with

$$\Psi_{jk}^{(1)} = \sum_{i \in s} w_{ij}^* w_{ik}^* \Pr\left(e_i \leq \min(u_{ij}(t), u_{ik}(t))\right)$$
$$= \sum_{i \in s} w_{ij}^* w_{ik}^* \Pr\left(y_i - \mu_i \leq t - \max(\mu_j, \mu_k)\right)$$
$$= \sum_{i \in s} w_{ij}^* w_{ik}^* G_i\left(t - \max(\mu_j, \mu_k)\right)$$

and

$$\Psi_{jk}^{(2)} = \sum_{i \in s} w_{ij}^* w_{ik}^* \Pr\left(e_i \leq u_{ij}(t)\right)\Pr\left(e_i \leq u_{ik}(t)\right)$$
$$= \sum_{i \in s} w_{ij}^* w_{ik}^* \{\Pr(y_i - \mu_i \leq t - \mu_j)\Pr(y_i - \mu_i \leq t - \mu_k)\}$$
$$= \sum_{i \in s} w_{ij}^* w_{ik}^* G_i(t - \mu_j) G_i(t - \mu_k)$$

where $G_i(u) = \Pr(y_i - \mu_i \leq u)$. Without loss of generality, assume that the non-sampled units are labelled from 1 to $N - n$ in the same order as their mean values under (16.24). That is, $\mu_j < \mu_k$ when $j < k$. Then

$$V_s = \sum_{i \in s} \left(\sum_{j=1}^{N-n} \sum_{k=1}^{N-n} w_{ij}^* w_{ik}^* \{G_i(t - \max(\mu_j, \mu_k)) - G_i(t - \mu_j)G_i(t - \mu_k)\} \right)$$

$$= \sum_{i \in s} \left(\sum_{j=1}^{N-n} \sum_{k=1}^{N-n} w_{ij}^* w_{ik}^* \{G_i(t - \max(\mu_j, \mu_k))(1 - G_i(t - \min(\mu_j, \mu_k)))\} \right)$$

$$= \sum_{i \in s} \left[\begin{array}{l} \left\{ \sum_{j=1}^{N-n} w_{ij}^* (1 - G_i(t - \mu_j)) \right\} \left\{ \sum_{j=1}^{N-n} w_{ij}^* G_i(t - \mu_j) \right\} \\ + \sum_{j=1}^{N-n} \sum_{k=1}^{j} w_{ij}^* w_{ik}^* \{G_i(t - \mu_j) - G_i(t - \mu_k)\}. \end{array} \right]$$

An estimator of this expression is

$$\hat{V}_s = \sum_{i \in s} \left[\begin{array}{l} \left\{ \sum_{j=1}^{N-n} w_{ij}^* \left(1 - \hat{G}_i(t - \mu_j)\right) \right\} \left\{ \sum_{j=1}^{N-n} w_{ij}^* \hat{G}_i(t - \mu_j) \right\} \\ + \sum_{j=1}^{N-n} \sum_{k=1}^{j} w_{ij}^* w_{ik}^* \left\{ \hat{G}_i(t - \mu_j) - \hat{G}_i(t - \mu_k) \right\} \end{array} \right] \quad (16.27)$$

where

$$\hat{G}_i(u) = \sum_{k \in s} w_{ki}^* I(y_k - \hat{\mu}_k \leq u).$$

A corresponding estimator of V_r is more straightforward to derive. Since

$$V_r = \sum_{j \in r} \Pr(y_j \leq t)\{1 - \Pr(y_j \leq t)\}$$

all we need to do is to replace $\Pr(y_j \leq t)$ by its obvious estimator $\hat{P}_j(t)$ to give

$$\hat{V}_r = \sum_{j \in r} \hat{P}_j(t) \left\{1 - \hat{P}_j(t)\right\}. \quad (16.28)$$

Combining (16.26) with (16.27) and (16.28) leads to an estimator of the prediction variance of (16.25) of the form

$$\hat{V}\left(\hat{F}_{Ny}^{NP}(t)\right) = N^{-2}\left(\hat{V}_s + \hat{V}_r\right). \quad (16.29)$$

Note that in using (16.29) we are assuming that there is a negligible contribution to the prediction variance due to estimation of μ_i by $\hat{\mu}_i$. This assumption is reasonable for large sample sizes. For smaller sample sizes we need to include this extra component of variance. Under the assumption $\mu_i = \beta z_i$, with β estimated via b_v this was done for (16.15) by Chambers and Dunstan (1986) and for (16.23) by Kuk and Welsh (2001).

The expression (16.27) for $\hat{V}_s$ above can be extremely time consuming to calculate, especially if n and N are large. However, smoothness of the mean function $\mu(z_i)$ with respect to variation in z implies that we can speed up computation of (16.27) by replacing individual non-sample Z-values by grouped data (Dunstan and Chambers, 1989). In particular, let $\{z_h; h = 1, \cdots, H\}$ denote the mid-points of a partition of the non-sample Z-values into H groups, with sizes $\{m_h; h = 1, \cdots, H\}$ and such that $\hat{\mu}(z_g) < \hat{\mu}(z_h)$ when $g < h$. We can replace (16.27) by

$$\hat{V}_s^{grp} = \sum_{i \in s} \left[\begin{array}{l} \left\{ \sum_{h=1}^H m_h w_{ih}^* \left(1 - \hat{G}_i(t - \hat{\mu}_h)\right) \right\} \left\{ \sum_{h=1}^H m_h w_{ih}^* \hat{G}_i(t - \hat{\mu}_h) \right\} \\ + \sum_{h=1}^H \sum_{g=1}^h m_h m_g w_{ih}^* w_{ig}^* \left\{ \hat{G}_i(t - \hat{\mu}_h) - \hat{G}_i(t - \hat{\mu}_g) \right\} \end{array} \right]$$

where $\hat{\mu}_h = \hat{\mu}(z_h)$ and (16.28) by

$$\hat{V}_r^{grp} = \sum_{h=1}^H m_h \hat{P}_h(t) \left\{ 1 - \hat{P}_h(t) \right\}.$$

The final (grouped) estimator of the prediction variance is

$$\hat{V}^{grp} = N^{-2} \left(\hat{V}_s^{grp} + \hat{V}_r^{grp} \right). \tag{16.30}$$

For small values of H (16.30) will be a conservative approximation to (16.29). This is because the absolute value of the second term on the right hand side of $\hat{V}_s^{grp}$ above decreases to zero as the number of groups, H, decreases. That is, the value of (16.30) increases as the number of groups decreases. This is consistent with the fact that (16.30) actually corresponds to a large sample variance estimator for a grouped version of (16.25) and that a decreased number of groups implies increased aggregation of the non-sample Z-values and hence an increased loss of efficiency for this grouped estimator.

Finally, we observe that prediction of the population quantile function $Q_{Ny}(\alpha)$ proceeds by inverting (16.25), with Woodruff's method providing a quick and convenient way of calculating a confidence interval for any specified value of α.

16.5 Imputation vs. Prediction for a Wages Distribution

In order to illustrate application of the methods developed in the previous section, we consider data obtained in a large-scale UK business survey that collected data on employee salaries, hours worked and hourly rates of pay. A key objective of this survey was measurement of the distribution of hourly pay rates (Y, measured in pence) in the target population. These hourly rates could not be

Table 16.1 Distribution of survey data classified by whether values of Y provided or not.

Y available?		Quantiles of distribution		
		25%	50%	75%
Yes	Y	482	597	843
	X	492	634	892
No	X	717	1014	1491

Table 16.2 Distributions of Y and X for subset of employees providing both values.

Quantile	Y	X
100.0%	1995	2955
90.0%	1120	1190
75.0%	820	870
50.0%	600	635
25.0%	495	500
10.0%	435	440
0.0%	300	300

obtained from all responding employees, since many were not paid by the hour. However, given that data on total earnings and hours worked were also obtained in the survey, an implicit hourly rate (X, also measured in pence) could be calculated for all respondents. As might be expected, Y and X were not the same, even when both were available. This is illustrated in Table 16.1, which shows the distributions of these variables for the survey respondents that provided values for both, as well as for respondents that provide values for X alone. In particular, we see that values of X in the latter group tend to be considerably larger than values of X in the former group. In Table 16.2 we focus on the marginal distributions of Y and X for a subset of the survey respondents that provided data for both these variables. This subset is defined by excluding all respondents that provided these data but had implausibly small values for either Y or X, or where either of these values was very large. Figure 16.1 is the scatterplot of Y and X values underpinning the data contributing to Table 16.2. Here we see that although there are clearly many employees where Y and X are very similar, there is also a large amount of variability.

Let s_1 denote the n_1 employees that provided data for Y and X and let s_2 denote the n_2 employees that provided data for X alone. If data on Y were available for all $n = n_1 + n_2$ employees, the predictor of the proportion of wage earners with hourly pay rates less than or equal to t would be

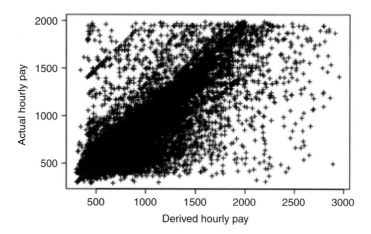

Fig. 16.1 Observed hourly pay rate (y-axis) versus derived hourly pay rate (x-axis) for employees in Table 16.2.

$$\hat{F}_s(t) = n^{-1}\left\{\sum_{s_1} I(y_i \le t) + \sum_{s_2} I(y_j \le t)\right\}. \tag{16.31}$$

Since the second term on the right is unknown, (16.31) cannot be calculated. One option is to replace it by the *available data* predictor,

$$\hat{F}_{s1}(t) = n_1^{-1}\sum_{s_1} I(y_i \le t). \tag{16.32}$$

Another option is to impute each missing value of Y, leading to the *imputation* predictor

$$\hat{F}_{imp}(t) = n^{-1}\left\{\sum_{s_1} I(y_i \le t) + \sum_{s_2} I(Y^*(x_j) \le t)\right\}. \tag{16.33}$$

Here $Y^*(x_j)$ denotes the imputed value for the s_2 unit with $X = x_j$. In the simulation study described below, we use a fuzzy type of *nearest neighbour imputation*. This is defined by first identifying the unit $i \in s_1$ that minimises $|x_i - x_j|$. We then identify the two largest Y values smaller than or equal to y_i, say $y_{2L(i)} \le y_{1L(i)}$ and the two smallest Y values larger than or equal to y_i, say $y_{1U(i)} \le y_{2U(i)}$. Finally, we select $Y^*(x_j)$ from the ordered set $(y_{2L(i)}, y_{1L(i)}, y_i, y_{1U(i)}, y_{2U(i)})$ with probabilities $(0.14, 0.22, 0.28, 0.22, 0.14)$.

Clearly, it is possible to make multiple independent draws (with replacement) from the above distribution. We can average the resulting single imputation-based predictions over these draws in order to reduce the Monte Carlo variability associated with the single draw predictor (16.33). The limiting form of this averaged predictor can be shown to be of the form

$$\hat{F}_{imp}^{\infty}(t) = n^{-1}\sum_{s_1}\left(1 + \sum_{j \in s_2} p_{ij}\right) I(y_i \le t). \tag{16.34}$$

Here $\{p_{ij}; i \in s_1\}$ are the (known) probabilities defining the distribution of nearest neighbour values used to generate the imputed value $Y^*(x_j)$ in s_2.

An alternative to prediction using imputed values is to predict the distribution function (16.31) using the sample X values as auxiliary information. That is, the original sample is considered to be the population of interest, the employees in this sample who provided data on both Y and X are treated as the sampled units and the remainder who only provided data on X are treated as the non-sampled units. Here we consider the locally weighted distribution function predictor (16.25) with $\mu(x) = \alpha + \beta x$. There are many ways the local weights w_{ij} in this predictor can be defined. We use the simple specification

$$w_{ij} = I\left(\|x_i - x_j\| \leq \frac{range(x)}{f}\right)$$

provided $\sum_{i \in s} w_{ij} > 5$, otherwise $w_{ij} = 1$. Note that $1/f$ plays the role of a bandwidth here and the parametric predictor (16.15) corresponds to $f = 1$. A straightforward way of choosing a value f to use in (16.25) is via an ordered half-sample *cross-validation* procedure, defined as follows:

- Order the sample x-values: $x_{(1)}, x_{(2)}, x_{(3)}, \cdots, x_{(n-1)}, x_{(n)}$.
- Create two sets $E = \{x_{(1)}, x_{(3)}, \cdots\}$ and $V = \{x_{(2)}, x_{(4)}, \cdots\}$.
- For given f and t compute (16.25) treating E as the sample and V as the non-sample.
- Choose the value of f that minimises $\sum_{g=1}^{G} \left\{\hat{F}_{Ny}^{NP}(t_g) - n^{-1}\sum_s I(y_i \leq t_g)\right\}^2$ over a pre-specified grid $\{t_g\}$ of t-values.

Note that this procedure will tend to oversmooth, since it is actually optimising predictive performance for a sample of size $n/2$. However, since n is large in this application, this is not an important issue.

A simulation study was carried out in order to evaluate these different options, based on the data underpinning Table 16.2. In particular, the respondents who provided these data were randomly split into 2 groups, U_1 and U_2 of approximately the same size so that Pr(inclusion in U_2) was proportional to X^2. Five hundred random samples s_1 and s_2 each of size 500 were then independently drawn from U_1 and U_2 respectively. Values of Y and X were assumed to be available on s_1, while values of X only were assumed to be available on s_2. Table 16.3 shows the quantiles of the two subpopulations (Y available/Y not available) defined as a consequence.

Twenty five 'target' values of t (400, 425, 450, ..., 950, 975, 1000) were used in the simulation study, and Table 16.4 shows the resulting biases and root mean squared errors of the available data predictor (16.32), denoted Fn, the limiting form (16.34) of the 'fuzzy' nearest neighbour predictor, denoted NNI, the parametric predictor (16.15) under (16.24) with $\mu(z) = \alpha + \beta z$ and $\sigma(z) = 1$, denoted L1, and the non-parametric form (16.23) of this predictor, denoted LCV, with local smoothing weights and bandwidth chosen as described above. Results

Table 16.3 Quantiles of populations used in simulations.

Population		25%	50%	75%
U_1	Y	465	530	650
	X	465	545	670
U_2	Y	550	725	1000
	X	589	790	1065

Table 16.4 Values of prediction bias and root mean squared error (both $\times 10^4$) for simulations.

t	Bias				Root Mean Squared Error			
	Fn	NNI	L1	LCV	Fn	NNI	L1	LCV
400	34	0	135	41	41	54	137	49
425	434	3	63	−11	443	139	72	39
450	726	2	7	−41	735	181	43	60
475	950	−1	4	−13	958	196	44	49
500	1259	−2	−111	−82	1266	219	122	101
525	1437	9	−134	−65	1444	233	146	98
550	1577	−3	−205	−96	1584	242	213	128
575	1665	3	−209	−70	1672	245	217	111
600	1724	−13	−242	−81	1730	247	250	119
625	1753	−24	−262	−84	1759	253	269	123
650	1768	−32	−275	−85	1774	253	283	123
675	1756	−34	−307	−109	1762	261	315	144
700	1747	−29	−318	−118	1753	263	325	151
725	1721	−35	−297	−97	1726	273	305	135
750	1695	−23	−282	−91	1700	265	290	132
775	1650	−27	−248	−69	1655	255	256	116
800	1588	−30	−255	−96	1593	250	263	134
825	1538	−32	−233	−87	1543	256	242	125
850	1474	−36	−233	−101	1479	258	241	133
875	1421	−35	−217	−96	1426	253	226	130
900	1362	−16	−210	−99	1367	259	219	133
925	1293	−2	−215	−110	1298	259	225	141
950	1228	0	−207	−107	1233	261	216	138
975	1166	11	−198	−106	1171	262	207	138
1000	1087	26	−209	−127	1092	267	217	154

were also obtained for the single draw version (16.33) of the imputation predictor. However, since these were always marginally worse than those of (16.34), we do not show them here.

Inspection of Table 16.4 allows one to reach some clear conclusions. Ignoring the information in s_2 and predicting (16.31) via the available data predictor Fn is generally a very poor choice, since this predictor is heavily biased. In contrast, the imputation predictor NNI is unbiased, but with a relatively large variance. The parametric predictor L1 is also biased, though not as badly as (16.32). This appears to be principally due to the misspecification of $\sigma(z)$. Finally, the non-parametric predictor LCV seems to the best performer. Its bias is typically less than half of that of L1, and its root mean squared error is smaller than the alternative estimators for all except three values of t.

16.6 Distribution Inference for Clustered Populations

We finally consider prediction of the finite population distribution function of Y when the underlying population (and hence the sample) has a clustered structure. In doing so, we use the notation developed in Chapter 6. That is, we write this distribution function in the form

$$F_{Ny}(t) = N^{-1} \sum_g \sum_{i \in g} I(y_i \leq t)$$

where N is the total number of population elements and the random variable Y follows (6.1), that is the Y-values all have the same mean μ and variance σ^2, with observations in different clusters uncorrelated, and observations in the same cluster having the same intra-cluster correlation, ρ. As usual, we note that $F_{Ny}(t)$ is the population mean of the derived variable $D_t = I(Y \leq t)$.

A clustered pattern of observations for Y should lead to a corresponding clustering of the values of D_t. Furthermore, since the Y-values are exchangeable within a cluster, (16.2) holds. That is

$$E(d_{it}) = F_{1y}(t) \tag{16.35a}$$

$$Var(d_{it}) = F_{1y}(t)\left[1 - F_{1y}(t)\right] \tag{16.35b}$$

$$Cov(d_{it}, d_{jt}) = \begin{cases} F_{2y}(t,t) - F_{1y}^2(t) & i, j \in g \\ 0 & \text{otherwise} \end{cases} \tag{16.35c}$$

where $F_{1y}(t) = \Pr(Y \leq t)$ and $F_{2y}(t, u) = \Pr(Y_a \leq t, Y_b \leq u)$. Here Y_a and Y_b are values of Y corresponding to distinct population units in the same cluster.

Clearly (16.35) is just a special case of the model (6.1), so the EB predictor of $F_{Ny}(t)$ is given by (7.7), which in this case is

$$\hat{F}_{Ny}^{EB}(t) = N^{-1} \left\{ \begin{array}{l} \sum_s m_g \bar{d}_{sgt}^{(1)} + \sum_s (M_g - m_g) \left[(1 - \hat{\alpha}_{gt}^{(21)}) \hat{F}_{1y}(t) + \hat{\alpha}_{gt}^{(21)} \bar{d}_{sgt}^{(1)} \right] \\ + \hat{F}_{1y}(t) \left(N - \sum_s M_g \right) \end{array} \right\}$$

(16.36)

where

$$\bar{d}_{sgt}^{(1)} = m_g^{-1} \sum_{i \in s_g} I(y_i \leq t) \tag{16.37a}$$

$$\hat{\alpha}_{gt}^{(21)} = \hat{\rho}_t^{(21)} m_g \left(1 - \hat{\rho}_t^{(21)} + \hat{\rho}_t^{(21)} m_g \right)^{-1} \tag{16.37b}$$

$$\hat{F}_{1y}(t) = \sum_s \hat{\alpha}_{gt}^{(21)} \bar{d}_{sgt}^{(1)} / \sum_s \hat{\alpha}_{gt}^{(21)} \tag{16.37c}$$

$$\hat{\rho}_t^{(21)} = \frac{\hat{F}_{2y}(t) - \hat{F}_{1y}^2(t)}{\hat{F}_{1y}(t) \left\{ 1 - \hat{F}_{1y}(t) \right\}} \tag{16.37d}$$

and $\hat{F}_{2y}(t)$ is the BLUE of $F_{2y}(t) = \Pr(Y_a \leq t, Y_b \leq t | a, b \in \text{same cluster})$.

In order to define $\hat{F}_{2y}(t)$, we note that the cluster level second order means

$$\bar{d}_{sgt}^{(2)} = m_g^{-1}(m_g - 1)^{-1} \sum_{i \in s_g} \sum_{\substack{j \in s_g \\ j \neq i}} I(y_i \leq t) I(y_j \leq t)$$

are independently distributed, with common expected value $F_{2y}(t)$. Consequently, the BLUE of this expected value is

$$\hat{F}_{2y}(t) = \sum_s \text{Var}^{-1}\left(\bar{d}_{sgt}^{(2)}\right) \bar{d}_{sgt}^{(2)} / \sum_s \text{Var}^{-1}\left(\bar{d}_{sgt}^{(2)}\right)$$

where

$$\text{Var}(\bar{d}_{sgt}^{(2)}) = m_g^{-2}(m_g - 1)^{-2} \left[\begin{array}{l} m_g(m_g - 1) F_{2y}(t) \left\{ 1 - F_{2y}(t) \right\} \\ + m_g(m_g - 1)(4m_g - 7) \left\{ F_{3y}(t) - F_{2y}^2(t) \right\} \\ + m_g(m_g - 1)(m_g - 2)(m_g - 3) \left\{ F_{4y}(t) - F_{2y}^2(t) \right\} \end{array} \right]$$

$$= m_g^{-1}(m_g - 1)^{-1} F_{2y}(t) [1 - F_{2y}(t)] \left[1 + (4m_g - 7)\rho_t^{32} + (m_g - 2)(m_g - 3)\rho_t^{42} \right]$$

with

$$\rho_t^{32} = \frac{F_{3y}(t) - F_{2y}^2(t)}{F_{2y}(t) [1 - F_{2y}(t)]}$$

$$\rho_t^{42} = \frac{F_{4y}(t) - F_{2y}^2(t)}{F_{2y}(t) [1 - F_{2y}(t)]}.$$

Here
$$F_{3y}(t) = \Pr(Y_a \leq t, Y_b \leq t, Y_c \leq t | a, b, c \in \text{samecluster})$$
and
$$F_{4y}(t) = \Pr(Y_a \leq t, Y_b \leq t, Y_c \leq t, Y_d \leq t | a, b, c, d \in \text{samecluster}).$$

The above development illustrates what is sometimes referred to as the 'infinite staircase of inference', where in order to optimally estimate a lower order parameter of a model for a distribution, we need to know one or more higher order parameters of this distribution. Thus, our optimal estimator of $F_{1y}(t)$ depends on $F_{2y}(t)$, while our optimal estimator of $F_{2y}(t)$ depends on $F_{3y}(t)$ and $F_{4y}(t)$, and so on ad infinitum.

One way of getting off this staircase is to replace $\hat{\rho}_t^{(21)}$ defined by (16.37d) by

$$\hat{\rho}_t^{(21)} = \frac{\tilde{F}_{2y}(t) - \tilde{F}_{1y}^2(t)}{\tilde{F}_{1y}(t)\left\{1 - \tilde{F}_{1y}(t)\right\}} \tag{16.38}$$

where

$$\tilde{F}_{1y}(t) = \sum_s m_g \bar{d}_{sgt}^{(1)} / \sum_s m_g$$
$$\tilde{F}_{2y}(t) = \sum_s m_g(m_g - 1)\bar{d}_{sgt}^{(2)} / \sum_s m_g(m_g - 1).$$

Unfortunately, as the following small-scale simulation study demonstrates, the gains from incorporating the cluster structure of the population into prediction of $F_{Ny}(t)$ appear minimal. This study used a population made up of $Q = 100$ clusters, with the fixed distribution of cluster sizes shown in Table 16.5. The value of Y for population unit i in cluster g was generated from the linear mixed model $y_i = 100 + u_g + e_i$, where the $\{u_g; g = 1, \ldots, Q\}$ were random cluster effects distributed independently as $N\left(0, \sigma_u^2\right)$ and the $\{e_i; i = 1, \ldots, N\}$ were random individual 'effects' distributed independently as $N\left(0, \sigma_e^2\right)$. All effects were generated independently of one another, so $\mu = 100$, $\sigma^2 = \sigma_u^2 + \sigma_e^2$ and $\rho = \sigma_u^2/\sigma^2$. A total of 1000 populations, each consisting of $N = 4927$ individuals, were independently simulated for each of two values of ρ and two methods of sampling. In all cases $\sigma^2 = 2400$, with $\rho = 0.2$ or $\rho = 0.5$. The first method of sampling

Table 16.5 Quantiles of distribution of cluster sizes ($Q = 100$, average cluster size = 49.27).

Min	25%	50%	75%	Max
21.00	30.75	51.00	63.00	79.00

Table 16.6 Average error (AE) and RMSE under random sampling of clusters.

q	AE			RMSE		
	Fn	FBLUP	FEBLUP	Fn	FBLUP	FEBLUP
			$\rho = 0.2$			
0.1	−0.0005	−0.0006	−0.0006	0.0470	0.0476	0.0480
0.2	−0.0005	−0.0006	−0.0002	0.0678	0.0683	0.0691
0.3	−0.0016	−0.0018	−0.0015	0.0817	0.0819	0.0828
0.4	−0.0012	−0.0012	−0.0009	0.0893	0.0896	0.0903
0.5	0.0000	−0.0001	0.0001	0.0917	0.0923	0.0930
0.6	−0.0015	−0.0016	−0.0013	0.0903	0.0910	0.0915
0.7	−0.0014	−0.0014	−0.0010	0.0828	0.0834	0.0838
0.8	−0.0007	−0.0006	−0.0002	0.0692	0.0698	0.0701
0.9	0.0004	0.0005	0.0006	0.0473	0.0479	0.0484
			$\rho = 0.5$			
0.1	−0.0025	−0.0026	−0.0024	0.0658	0.0659	0.0661
0.2	−0.0010	−0.0012	−0.0011	0.0955	0.0956	0.0958
0.3	−0.0003	−0.0005	−0.0005	0.1160	0.1162	0.1164
0.4	0.0004	0.0003	0.0005	0.1272	0.1275	0.1276
0.5	0.0008	0.0006	0.0007	0.1315	0.1318	0.1320
0.6	−0.0006	−0.0007	−0.0007	0.1282	0.1285	0.1287
0.7	−0.0023	−0.0024	−0.0024	0.1193	0.1196	0.1199
0.8	−0.0019	−0.0021	−0.0020	0.0996	0.0999	0.1001
0.9	−0.0023	−0.0025	−0.0025	0.0713	0.0716	0.0717

corresponded to taking a random sample of five clusters, while the second was an unequal probability method that sampled five clusters with probabilities that increased with the size M_g of a cluster. Under both sampling designs, a fixed number $m = 20$ of individuals were sampled at random in each sampled cluster. The total sample size was therefore $n = mq = 100$.

The target parameters for prediction were the values $F_{Ny}(t_{Nyq})$, with t_{Nyq} defined as the qth quantile of the distribution of Y-values across the population. A total of nine values of q were investigated, $q = 0.1, \ldots, 0.9$, using three predictors. The first (Fn) ignores the cluster structure and corresponds to the simple sample mean of D_t, which is equivalent to setting $\rho_t^{(21)}$ to zero in (16.36). The second (FEBLUP) corresponded to (16.36) combined with the moment estimator (16.38) of the intra-cluster correlation $\rho_t^{(21)}$, while the third (FBLUP) also corresponded to (16.36), but with this intra-cluster correlation now replaced by a close approximation to its true value. Tables 16.6 and 16.7 show the average error (AE) and the square root of the mean squared error (RMSE) over the 1000 simulated population/sample combinations for each value of q for each of these predictors.

Table 16.7 Average error (AE) and RMSE under unequal probability sampling of clusters.

q	AE			RMSE		
	Fn	FBLUP	FEBLUP	Fn	FBLUP	FEBLUP
			$\rho = 0.2$			
0.1	0.0028	0.0027	0.0027	0.0480	0.0481	0.0484
0.2	0.0042	0.0042	0.0043	0.0710	0.0712	0.0714
0.3	0.0060	0.0063	0.0063	0.0839	0.0842	0.0845
0.4	0.0065	0.0067	0.0067	0.0888	0.0893	0.0895
0.5	0.0039	0.0041	0.0042	0.0901	0.0904	0.0908
0.6	0.0034	0.0037	0.0038	0.0867	0.0867	0.0868
0.7	0.0026	0.0029	0.0028	0.0800	0.0800	0.0804
0.8	0.0005	0.0007	0.0006	0.0676	0.0676	0.0680
0.9	−0.0001	0.0000	−0.0001	0.0468	0.0470	0.0474
			$\rho = 0.5$			
0.1	−0.0012	−0.0013	−0.0013	0.0691	0.0690	0.0690
0.2	−0.0026	−0.0027	−0.0029	0.1006	0.1006	0.1008
0.3	−0.0036	−0.0037	−0.0038	0.1193	0.1193	0.1193
0.4	−0.0029	−0.0031	−0.0033	0.1315	0.1314	0.1315
0.5	−0.0032	−0.0034	−0.0036	0.1349	0.1348	0.1347
0.6	−0.0039	−0.0040	−0.0041	0.1310	0.1309	0.1310
0.7	−0.0027	−0.0027	−0.0027	0.1195	0.1194	0.1196
0.8	−0.0019	−0.0019	−0.0020	0.1002	0.1000	0.1002
0.9	0.0013	0.0012	0.0012	0.0680	0.0679	0.0680

Examination of Tables 16.6 and 16.7 shows that in virtually every case the simple sample average Fn, that is

$$F_n(t) = n^{-1} \sum_s m_g \bar{d}_{sgt}^{(1)}$$

which ignores the cluster structure of the population, is the either the best, or almost the best, predictor. Consequently, there seems little point in allowing for the presence of intra-cluster correlation when predicting the value of $F_{Ny}(t)$ if the sample size in each sampled cluster is the same. The situation is completely analogous to that discussed at the end of Section 6.2, where we noted that in this situation there was no point in going beyond the sample mean when predicting the overall population mean of Y.

Development of an expression for the variance of $F_n(t)$ above is left as an exercise, as is that of an estimator of this prediction variance (see exercise E.38).

17 Using Transformations in Sample Survey Inference

This book is about the use of models in sample survey inference. In this context, we have emphasised linear models because of their wide applicability. However, data structures in the real world are rarely exactly linear, and so the use of non-linear models in finite population prediction is an important topic. In Section 15.8 we recognise this in our treatment of small area estimation of unemployment, where a logistic mixed model is used. In this last chapter of this book we focus on another important aspect of non-linear modelling, which is the use of transformations to linearity.

17.1 Back Transformation Prediction

In exercise E.31 we described a situation that is quite common in business surveys. This is where both the survey variable Y and the auxiliary variable Z are strictly positive and where the relationship between these variables is not obviously linear, but where $\log(Y)$ and $\log(Z)$ are clearly linearly related. In particular, a good working model for these variables is

$$\log(y_i) = \mathbf{x}'_i \beta + e_i \tag{17.1}$$

where $\mathbf{x}'_i = (1 \; \log(z_i))$. Following standard practice, the errors e_i in (17.1) are assumed to be identically distributed realisations from a normal distribution with zero mean and variance σ^2. As a consequence, the distribution of Y given Z is lognormal with $E(y_i|z_i) = e^{\mathbf{x}'_i \beta + \sigma^2/2}$. If we let $\hat\beta$ denote the ordinary least squares estimates of β based on the sample values of $\log(y_i)$ and $\mathbf{x}_i$, then an initial predictor of the population total t_y of Y under (17.1) is the simple *back transformed* predictor

$$\hat t_y^{LL} = \sum_s y_i + \sum_r e^{\mathbf{x}'_i \hat\beta}. \tag{17.2}$$

This predictor is negatively biased, since $Var(\hat\beta)$ is $O(n^{-1})$ and

$$E(e^{\mathbf{x}'_i \hat\beta} - y_i) = e^{\mathbf{x}'_i \beta + \mathbf{x}'_i Var(\hat\beta) \mathbf{x}_i/2} - e^{\mathbf{x}'_i \beta + \sigma^2/2} = e^{\mathbf{x}'_i \beta}\left(e^{\mathbf{x}'_i Var(\hat\beta)\mathbf{x}_i/2} - e^{\sigma^2/2}\right).$$

The usual correction for the bias of (17.2) is based on the fact that $e^{\mathbf{x}'_i \hat\beta + s^2/2}$ is a consistent estimator of $E(y_i|z_i)$ under (17.1), where $s^2 = (n-2)^{-1} \sum_s \left(\log(y_i) - \mathbf{x}'_i \hat\beta\right)^2$. Consequently, the predictor

$$\hat{t}_y^{LL1} = \sum_s y_i + \sum_r e^{\mathbf{x}_i'\hat{\beta}+s^2/2} \tag{17.3}$$

should be asymptotically consistent. However, (17.3) still has a bias that is $O(Nn^{-1})$ since

$$\begin{aligned}
E(e^{\mathbf{x}_i'\hat{\beta}+s^2/2} - y_i) &= \frac{e^{\mathbf{x}_i'\beta+\sigma^2/2}}{2}\left(\mathbf{x}_i' Var(\hat{\beta})\mathbf{x}_i + \tfrac{1}{4}Var(s^2)\right) + o(n^{-1}) \\
&= \frac{e^{\mathbf{x}_i'\beta+\sigma^2/2}}{2}\left(\sigma^2\mathbf{x}_i'(\mathbf{X}_s'\mathbf{X}_s)^{-1}\mathbf{x}_i + \frac{\sigma^4}{2(n-2)}\right) + o(n^{-1})
\end{aligned}$$

where $\mathbf{X}_s$ is the $n \times 2$ matrix with i^{th} row $\mathbf{x}_i'$. This bias of (17.3) can therefore be corrected further. If we define

$$k_i = 1 + \frac{1}{2}\left(\sigma^2\mathbf{x}_i'(\mathbf{X}_s'\mathbf{X}_s)^{-1}\mathbf{x}_i + \frac{\sigma^4}{2(n-2)}\right)$$

then $E\left(k_i^{-1}e^{\mathbf{x}_i'\hat{\beta}+s^2/2} - y_i\right) = o(n^{-1})$. Hence, setting

$$\hat{k}_i = 1 + \frac{1}{2}\left(s^2\mathbf{x}_i'(\mathbf{X}_s'\mathbf{X}_s)^{-1}\mathbf{x}_i + \frac{s^4}{2n}\right) = 1 + \frac{s^2 a_i}{2} + \frac{s^4}{4n}$$

where $a_i = \mathbf{x}_i'\left(\mathbf{X}_s'\mathbf{X}_s\right)^{-1}\mathbf{x}_i$, we can define a first order bias corrected predictor

$$\hat{t}_y^{LL2} = \sum_s y_i + \sum_r \hat{k}_i^{-1} e^{\mathbf{x}_i'\hat{\beta}+s^2/2}. \tag{17.4}$$

The predictor (17.4) was first suggested by Karlberg (2000a; 2000b), though in a slightly different form. Her development is based on exact results under the lognormal distribution, and leads to a predictor that is very close to (17.4),

$$\hat{t}_y^{LLK} = \sum_s y_i + \sum_r e^{\mathbf{x}_i'\hat{\beta}+s^2(1-a_i)/2-\frac{s^4}{4n}} = \sum_s y_i + \sum_r \hat{l}_i^{-1} e^{\mathbf{x}_i'\hat{\beta}+s^2/2} \tag{17.5}$$

where $\hat{l}_i = e^{\frac{s^2 a_i}{2} + \frac{s^4}{4n}} \cong \hat{k}_i$.

17.2 Model Calibration Prediction

The bias correction methods discussed in the previous section are intimately linked to an assumption that the error term in the linear model (17.1) for $\log(Y)$ is normally distributed. Is it possible to develop a bias correction for the simple back transformed predictor (17.2) that does not require this assumption? Chambers and Dorfman (2003) have suggested two approaches that one can use for bias correction following a log transformation that do not require one to assume that the original data are lognormally distributed. In the rest of this chapter we summarise their key results.

First, we consider a concept that is sometimes referred to as *model calibration* (Wu and Sitter, 2001). Suppose that our working model specifies

$$E(y_i|z_i) = h(z_i; \theta) \tag{17.6}$$

for some known non-linear function h and unknown vector of parameters θ. A naive plug-in predictor of t_y under this working model is then

$$\hat{t}_y^{h1} = \sum_s y_i + \sum_r h(z_i; \hat{\theta}) \tag{17.7}$$

where $\hat{\theta}$ is an unbiased (or at least consistent) estimate of θ based on the sample values of Y and Z. Clearly $\hat{t}_y^{h1}(\hat{\theta})$ is biased since $E\left[h(z_i; \hat{\theta})|z_i\right] \neq E(y_i|z_i)$ in general. However, it is also true that if (17.6) is at least approximately valid then a plot of the sample values $\{y_i; i \in s\}$ of Y against their corresponding fitted values $\{h(z_i; \hat{\theta}); i \in s\}$ under this model will appear more linear than the corresponding plot of Y against Z. That is, we either have

$$E(y_i|h(z_i; \hat{\theta})) = \gamma h(z_i; \hat{\theta}) \tag{17.8a}$$

or

$$E(y_i|h(z_i; \hat{\theta})) = \gamma_0 + \gamma_1 h(z_i; \hat{\theta}). \tag{17.8b}$$

Depending on which of these two 'fitted value' models seems more appropriate, we can then replace the biased predictor (17.7) by the ratio-adjusted predictor

$$\hat{t}_y^{hR} = \sum_s y_i + \frac{\sum_s y_i}{\sum_s h(z_i; \hat{\theta})} \sum_r h(z_i; \hat{\theta}) \tag{17.9a}$$

if (17.8a) is the model of choice, or by the regression-adjusted predictor

$$\hat{t}_y^{hL} = \sum_s y_i + \sum_r \left\{\hat{\gamma}_0 + \hat{\gamma}_1 h(z_i; \hat{\theta})\right\} \tag{17.9b}$$

if (17.8b) seems more appropriate. Here $\hat{\gamma}_0$ and $\hat{\gamma}_1$ are the ordinary least squares estimates of γ_0 and γ_1 in (17.8b). In either case we expect the predictors (17.9a) and (17.9b) to have smaller bias than the naive plug-in predictor (17.7).

Note that there is nothing special about the choice of the ratio form (17.9a) or the regression form (17.9b). Both are simple ways of using the (hopefully, close to) linear relationship between the observed values y_i and the fitted values $h(z_i; \hat{\theta})$ to further reduce bias in prediction. Both choices also include hidden assumptions about the nature of this relationship. For example, the ratio-adjusted predictor (17.9a) implicitly assumes that $Var(y_i|h(z_i; \hat{\theta})) \propto h(z_i; \hat{\theta})$. If the sample data indicate a variance model of the form $Var(y_i|h(z_i; \hat{\theta})) \propto h^2(z_i; \hat{\theta})$ is more appropriate, then a more efficient version of (17.9a) is the mean of the ratios adjusted predictor

$$\hat{t}_y^{hR2} = \sum_s y_i + n^{-1} \sum_{i \in s} \frac{y_i}{h(z_i; \hat{\theta})} \sum_{j \in r} h(z_j; \hat{\theta}). \tag{17.9c}$$

Similarly the regression-adjusted predictor (17.9b) implicitly assumes that $Var(y_i|h(z_i;\hat{\theta}))$ is constant.

In the case of the working model (17.1), $\theta = (\beta, \sigma^2)$, $h(z_i; \theta) = e^{x'_i\beta + \sigma^2/2}$ and $h(z_i; \hat{\theta}) = e^{x'_i\hat{\beta} + s^2/2}$. Here it makes sense to use the homogeneous linear specification (17.8a) since both y_i and $h(z_i; \hat{\theta})$ are intrinsically positive. If one further decides to use a ratio-type predictor under this model then one is lead to (17.9a), which in this case reduces to

$$\hat{t}_y^{\exp R} = \frac{\sum_s y_i}{\sum_s e^{x'_i\hat{\beta}+s^2/2}} \sum_U e^{x'_i\hat{\beta}+s^2/2} = \sum_s y_i + \frac{\sum_s y_i}{\sum_s z_i^{\hat{\beta}_1}} \sum_r z_i^{\hat{\beta}_1}. \quad (17.10)$$

Here $\hat{\beta}_1$ is the OLS estimator of the coefficient of $\log(z_i)$ in (17.1).

Before moving on, it is worthwhile to comment a bit further on model calibration as a method of bias correction. In general, one can see that this approach requires two modelling steps:

- An initial modelling step that leads to the specification (17.6).
- A second modelling step that specifies the nature of the relationship between the fitted values generated by (17.6) and the actual values. Ideally this relationship will be homogeneous and linear, but it does not have to be.

A little thought should lead to the realisation that this process is in much the same spirit as the bias correction ideas (parametric and non-parametric) introduced in Sections 10.2 and 10.3. Under model calibration however, rather than adding a residual-based bias correction term to the working model-based predictor, one uses the fitted values generated under the working model as auxiliary variables in a new predictor. Obviously, there is no requirement that this new predictor be linear. Note also that implementation of this approach does not require any distributional assumptions.

To more clearly show the relationship between bias correction and model calibration, suppose that we can write the true conditional expectation of y_i given z_i as

$$E(y_i|z_i) = h(z_i) + b_i$$

where b_i represents bias in the specification (17.6). Note that we have dropped explicit reference to parameters here for the sake of exposition. Then bias correction (or twicing) is equivalent to predicting y_i given z_i using the efficient predictor

$$y_i^{bc} = h(z_i) + E(y_i - h(z_i)|z_i) = E(y_i|z_i) = h(z_i) + b_i.$$

On the other hand, model calibration predicts y_i given z_i using

$$y_i^{mc} = E(y_i|h(z_i)) = E(E(y_i|z_i)|h(z_i)) = h(z_i) + E(b_i|z_i).$$

Observe that $E\left(y_i^{bc}|z_i\right) = E\left(y_i^{mc}|z_i\right)$, so both approaches are predicting the same thing. However, all other things being equal, we expect that model calibration will be less efficient.

17.3 Smearing Prediction

Our second approach to bias correction under (17.1) is based on the concept of smearing (Duan, 1983). This starts from the additive model in the transformed scale

$$g(y_i) = \mu(z_i; \beta) + e_i \qquad (17.11)$$

where $g(y)$ is a monotone (and hence invertible) function, $\mu(z_i; \beta)$ is the conditional expectation of y_i in the transformed scale (i.e. $E(g(y_i)|z_i)$) and the errors e_i are assumed to be independently and identically distributed. Let j index a non-sample unit in the population. Then

$$y_j = g^{-1}(g(y_j)) = g^{-1}(\mu(z_j; \beta) + e_j)$$

so

$$E(y_j|z_j) = \int g^{-1}(\mu(z_j; \beta) + e) f(e) d\varepsilon. \qquad (17.12)$$

Here $f(e)$ denotes the density function of the error term e_i in (17.11). An estimator of (17.12) follows directly when this unknown density is substituted by its sample empirical, based on the transform scale residuals $\left\{r_i = g(y_i) - \mu(z_i; \hat{\beta}); i \in s\right\}$. This is the smearing estimator of $E(y_j|z_j)$,

$$\begin{aligned}\hat{E}(y_j|z_j) &= n^{-1} \sum_{i \in s} g^{-1}\left(\mu(z_j; \hat{\beta}) + r_i\right) \\ &= n^{-1} \sum_{i \in s} g^{-1}\left(g(y_i) + \left\{\mu(z_j; \hat{\beta}) - \mu(z_i; \hat{\beta})\right\}\right).\end{aligned}$$

The EB predictor of t_y based on this smearing estimator (which we refer to from now on as the *smearing predictor*) is then

$$\begin{aligned}\hat{t}_y^{g^{-1}S} &= \sum_s y_i + \sum_{j \in r} \hat{E}(y_j|z_j) \\ &= \sum_s y_i + n^{-1} \sum_{j \in r} \sum_{i \in s} g^{-1}\left(g(y_i) + \left\{\mu(z_j; \hat{\beta}) - \mu(z_i; \hat{\beta})\right\}\right).\end{aligned} \qquad (17.13)$$

We apply this approach to the log-linear model (17.1). Here $g(y) = \log(y)$, so $g^{-1}(x) = \exp(x)$, and $\mu(z_i; \hat{\beta}) = \mathbf{x}_i'\beta$, and the smearing predictor (17.13) takes the form

$$\hat{t}_y^{\exp S} = \sum_s y_i + n^{-1} \sum_{j \in r} \sum_{i \in s} e^{x_j' \hat{\beta} + r_i} = \sum_s y_i + \left(n^{-1} \sum_{i \in s} \frac{y_i}{z_i^{\hat{\beta}_1}} \right) \sum_{j \in r} z_j^{\hat{\beta}_1}$$
(17.14)

where, as before, $\hat{\beta}_1$ is the OLS estimator of the coefficient of $\log(z_i)$ in (17.1).

Although based on rather different modelling approaches, it is interesting that the ratio-adjusted predictor (17.10) and the smearing predictor (17.14) are so similar under the loglinear model (17.1). Both essentially correspond to regression model-based predictors under the non-linear regression model $E(y_i|z_i) \propto z_i^{\beta_1}$. In particular, we see that smearing via (17.14) is equivalent to the mean of the ratios adjustment (17.9c) under (17.1). Furthermore, since under the lognormal version of (17.1) we have $\text{Var}\left(y_i|z_i^{\beta_1}\right) \propto z_i^{2\beta_1}$, we expect that smearing will be more efficient than ratio adjustment if this lognormal assumption is actually true.

17.4 Outlier Robust Model Calibration and Smearing

Both the model calibration and the smearing approaches to bias correction implicitly assume that the working model fits the observed sample data. In reality, however, even if these data are generally linear in the log scale, they can contain values (e.g. zero) that are impossible under a log scale linear model, as well as values that are extreme on the raw scale. The log transformation effectively controls the influence of extreme raw scale values, but is susceptible to 'small' outliers (i.e. values near zero). This has implications for the impact of such outliers on inference, since unlike the situation considered in Chapter 10, where the concern was prediction bias due to the presence of 'large' representative sample outliers, here the concern is more about the increased variability of estimates of parameters of log scale linear models due to the presence of these 'small' sample outliers.

In the following development we shall ignore the issue of zero values in the sample data. See Karlberg (2000b) for a model-based treatment of this problem. However, we control the impact of outliers in the log scale sample data by using outlier robust estimates of the parameters β and σ^2 of (17.1) when applying the lognormal-based predictors described in Section 17.1. That is, we estimate these parameters using a method (e.g. M-estimation) that effectively downweights outliers.

Even though β and σ^2 may be estimated in a robust fashion, it is wise to also downweight outliers in the bias correction process. In order to do so, we assume that the outlier robust parameter estimation method used (e.g. the *rlm* function in R, see Venables and Ripley, 2002, Section 8.3) generates non-negative *outlier weights* $(\alpha_i; i \in s)$ that are close to one for non-outliers in the sample data and close to zero for outliers in these data, where outlier/non-outlier status is relative to the log scale linear model (17.1). Without loss of generality we assume that these outlier weights sum to n over the sample.

Consider first the general model calibration predictors (17.9a)–(17.9c). Introduction of the outlier weights $(\alpha_i; i \in s)$ into these predictors is straightforward. We replace (17.9a) by

$$\hat{t}_y^{hR\alpha} = \sum_s y_i + \frac{\sum_s \alpha_i y_i}{\sum_s \alpha_i h(z_i; \hat{\theta}_{rob})} \sum_r h(z_i; \hat{\theta}_{rob}) \qquad (17.15a)$$

and (17.9c) by

$$\hat{t}_y^{hR2\alpha} = \sum_s y_i + n^{-1} \sum_{i \in s} \alpha_i \frac{y_i}{h(z_i; \hat{\theta}_{rob})} \sum_{j \in r} h(z_j; \hat{\theta}_{rob}). \qquad (17.15c)$$

Here $\hat{\theta}_{rob}$ denotes the robust estimator of the parameters $\theta = (\beta, \sigma^2)$ referred to in the previous paragraph. In the case of (17.9b), we replace this regression predictor by its robust equivalent

$$\hat{t}_y^{hL\alpha} = \sum_s y_i + \sum_r \left\{ \hat{\gamma}_{0\alpha} + \hat{\gamma}_{1\alpha} h(z_i; \hat{\theta}_{rob}) \right\} \qquad (17.15b)$$

where $\hat{\gamma}_{0\alpha}$ and $\hat{\gamma}_{1\alpha}$ are the weighted least squares estimates of γ_0 and γ_1 in (17.8b), based on the outlier weights $(\alpha_i; i \in s)$ and with $h(z_i; \hat{\theta})$ replaced by $h(z_i; \hat{\theta}_{rob})$. In all cases, these outlier weights are being used to ensure that log scale outliers have little or no effect on estimation of the parameters in (17.8a) and (17.8b).

The general smearing predictor (17.13) is 'robustified' in exactly the same way, leading to

$$\hat{t}_y^{g^{-1}S\alpha} = \sum_s y_i + n^{-1} \sum_{j \in r} \sum_{i \in s} \alpha_i g^{-1} \left[g(y_i) + \left\{ \mu(z_j; \hat{\beta}_{rob}) - \mu(z_i; \hat{\beta}_{rob}) \right\} \right] \qquad (17.16)$$

where $\hat{\beta}_{rob}$ is an outlier robust estimator of the vector β of parameters defining the transformation scale mean function in (17.11).

Application of these robustified ratio-adjusted and smearing predictors under the log-linear model (17.1) follows directly. In particular, (17.10) is replaced by

$$\hat{t}_y^{\exp R\alpha} = \sum_s y_i + \frac{\sum_s \alpha_i y_i}{\sum_s \alpha_i z_i^{\hat{\beta}_{1rob}}} \sum_r z_i^{\hat{\beta}_{1rob}}. \qquad (17.17)$$

while (17.14) becomes

$$\hat{t}_y^{\exp S\alpha} = \sum_s y_i + \left(n^{-1} \sum_{i \in s} \alpha_i \frac{y_i}{z_i^{\hat{\beta}_{1rob}}} \right) \sum_{j \in r} z_j^{\hat{\beta}_{1rob}}. \qquad (17.18)$$

Both (17.17) and (17.18) are pseudo-linear predictors, in the sense that they can be written in the weighted form

$$\hat{t}_{wy} = \sum_s w_i y_i \qquad (17.19)$$

where the weights w_i are functions of the population values $\mathbf{Z}_U$ of Z, the robust estimate of the slope coefficient β_1 in (17.1) and the outlier weights $\alpha_s = (\alpha_i; i \in s)$. In particular, under (17.17)

$$w_i = 1 + \frac{\alpha_i \sum_{j \in r} z_j^{\hat{\beta}_{1rob}}}{\sum_{k \in s} \alpha_k z_k^{\hat{\beta}_{1rob}}}$$

while under (17.18)

$$w_i = 1 + \frac{\alpha_i \sum_{j \in r} z_j^{\hat{\beta}_{1rob}}}{n z_i^{\hat{\beta}_{1rob}}}.$$

First order estimators of the prediction variances of the ratio-adjusted and smearing predictors under (17.1) can be obtained by treating them as linear weighted predictors, with weights w_i as defined above, and using the robust prediction variance estimation methods discussed in Section 9.2. See, for example, the development of such a robust prediction variance estimator in the context of a small area estimator in Section 15.5. Alternatively, we can use jackknife or bootstrap methods of prediction variance estimation. See Section 12.2 and 12.3. In particular, the simulation results described in the next section are based on the use of the jackknife.

17.5 Empirical Results I

Figures 17.1 and 17.2 show the values of two survey variables, $Y =$ WAGES and $Y =$ EMP plotted against an auxiliary $Z =$ Register EMP on the raw scale as well as on the log scale for two sectors (A and B) of the economy. These data are extracted from a business survey carried out in the late 1990s. Here WAGES corresponds to the total wages paid to all employees of a business over a specific period while EMP is the number of employees of the business who were paid these wages. The auxiliary Register EMP is a past value of EMP, and is known for all businesses on the sampling frame for the survey. Note the clear linearity in the log scale for these variables.

These two data sets were used as study populations in a simulation study of the efficiency of the different transformation-based prediction methods described in this chapter. The sampling design used in the study was one that is commonly employed in business surveys. That is, stratified random sampling based on 4 size strata, defined by equal aggregate stratification (Section 4.10) on Z, with the stratum coresponding to the largest businesses completely enumerated, and

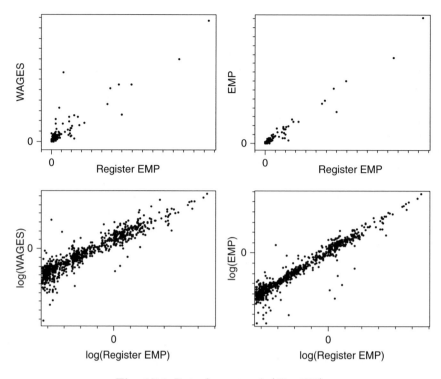

Fig. 17.1 Data for sector A ($N = 768$).

the remaining strata allocated approximately equal sample sizes. For Sector B this resulted in stratum sample allocations of 15, 15, 15, 5, while for sector A the corresponding allocations were 13, 13, 12, 12. The total sample size was therefore $n = 50$ in both sectors. A total of 1000 independent samples were selected via this design.

Table 17.1 lists ten of the predictors that were investigated in the simulation study. Note that all are stratified predictors, in the sense that they are calculated within the strata defined by the sample design, and then summed over these strata. Both EE and RE are raw scale predictors. The remaining predictors all use the loglinear model (17.1) in prediction. Predictors that use an outlier robust method of parameter estimation are denoted by the modifier/R. Standard linearisation-based methods of prediction variance estimation were used for EE and RE. Jackknife variance estimation (Section 12.2) was used for the loglinear model-based predictors.

Results from the simulation study are set out in Tables 17.2 to 17.5. These show the average error (AE), the root mean squared error (RMSE) and the average estimated standard error (AvSE), all expressed as a percentage of the true population value, as well as the coverage proportion of 'two sigma' prediction

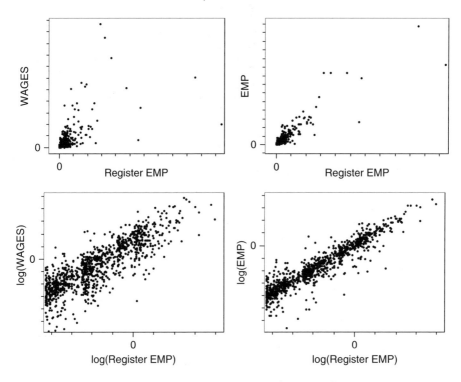

Fig. 17.2 Data for sector B ($N = 1005$).

Table 17.1 Predictors used in the simulation study. Note that/R indicates the outlier robust version of a predictor.

Name	Description
EE	Expansion estimator
RE	Ratio estimator
TA, TA/R	Naive back-transform predictor (17.2)
TK, TK/R	Lognormal-based Karlberg predictor (17.5)
RA, RA/R	Ratio-adjusted predictor (17.10)
SM, SM/R	Smearing predictor (17.14)

intervals (CR) generated by the different methods. Nominally, CR should be 0.95. The lowest value of RMSE, as well as any other RMSE value within 5% of this minimum value, is highlighted in each table.

Consideration of the results set out in Tables 17.2 to 17.5 leads to some clear conclusions. If we use values of AE to measure bias and values of RMSE to measure efficiency, then the simple separate ratio estimator (RE) does rather well.

Table 17.2 Simulation results for sector A and $Y = $ WAGES.

Predictor	AE	RMSE	AvSE	CR
EE	0.51	14.05	12.74	0.87
RE	1.03	12.23	9.46	0.89
TA	−6.32	11.38	11.03	0.85
TA/R	−6.75	11.08	11.08	0.82
TK	2.47	12.71	12.87	0.96
TK/R	−2.00	**10.11**	12.47	0.93
RA	1.68	12.25	13.98	0.94
RA/R	−2.52	**10.54**	12.84	0.91
SM	7.75	32.90	19.21	0.96
SM/R	−1.74	**10.37**	12.98	0.93

Table 17.3 Simulation results for sector A and $Y = $ EMP.

Predictor	AE	RMSE	AvSE	CR
EE	0.15	9.23	8.66	0.89
RE	0.11	4.85	3.76	0.89
TA	−2.52	5.58	5.29	0.94
TA/R	−1.35	**4.19**	5.06	0.94
TK	0.73	5.19	5.47	0.98
TK/R	−0.13	**4.24**	5.26	0.96
RA	0.24	5.21	5.49	0.97
RA/R	0.06	4.76	5.58	0.95
SM	0.82	6.84	6.01	0.97
SM/R	−0.12	**4.27**	5.28	0.96

Table 17.4 Simulation results for sector B and $Y = $ WAGES.

Predictor	AE	RMSE	AvSE	CR
EE	0.14	16.26	16.35	0.91
RE	0.27	**13.05**	12.62	0.93
TA	−21.77	24.37	12.90	0.57
TA/R	−18.16	21.68	15.68	0.71
TK	2.15	15.02	16.97	0.96
TK/R	2.23	15.79	21.93	0.97
RA	−0.53	**13.67**	16.13	0.95
RA/R	−1.78	**13.60**	17.61	0.95
SM	0.97	15.39	17.51	0.96
SM/R	−1.18	14.16	18.39	0.95

Table 17.5 Simulation results for sector B and $Y = \text{EMP}$.

Predictor	AE	RMSE	AvSE	CR
EE	−0.31	11.86	11.18	0.90
RE	−0.16	7.41	6.43	0.93
TA	−7.92	10.55	7.89	0.84
TA/R	−3.46	7.13	8.27	0.95
TK	1.86	7.86	8.59	0.98
TK/R	0.29	**6.33**	8.44	0.98
RA	−0.44	7.76	8.35	0.97
RA/R	0.24	**6.62**	8.54	0.98
SM	0.87	8.16	8.81	0.98
SM/R	−0.31	11.86	11.18	0.90

In contrast, the non-robust transformation-based predictors need to be treated with considerable care. The naive back-transformed predictor (TA) can be very biased, the Karlberg predictor (TK) is less efficient than RE, the smearing predictor (SM) can be very unstable and the ratio-adjusted predictor (RA), although much less biased, is often inefficient. In contrast, the outlier robust versions of transformation-based predictors generally work quite well. In particular, the variability of the ratio-adjusted predictor (RA/R) is reduced without an increase in bias, and there is a substantial reduction in the bias and the variability of the Karlberg (TK/R) and the smearing (SM/R) predictors.

The linearisation-based prediction variance estimators used with the stratified expansion estimator (EE) and the separate ratio estimator (RE) are generally biased low, with corresponding undercoverage. In contrast, the jackknife prediction variance estimators used with the loglinear model-based predictors seem to be much better. They are slightly biased upwards, but their prediction interval coverage is often very good.

Overall, these empirical results indicate that taking proper account of non-linearity in the prediction process can bring gains, provided modelling is carried out robustly. In particular, log scale outliers can have a much more dramatic effect on non-linear estimators than raw scale outliers have on linear estimators.

17.6 Robustness to Model Misspecification

The transformation-based predictors investigated in the simulation study reported in the previous section all assumed a linear model (17.1) in the log scale applied to most, if not all, of the population units in a stratum. What if this is not true? For example, suppose we apply these prediction methods to the entire sample, rather than within strata. This would help reduce the variability of parameter estimates, but would also mean that (17.1) would need to hold across the entire population. This may not be the case. For example, the WAGES population shown in Fig. 17.1 has a significant non-linear component in the log scale. Are there ways to modify the ratio-adjusted and smearing predictors to allow them to retain their unbiasedness properties in such situations?

Consider first the ratio-adjusted predictor (17.9a). An obvious generalisation of this predictor is to the general weighted ratio form

$$\hat{t}_{wy}^{hR} = \sum_s y_i + \frac{\sum_s w_i y_i}{\sum_s w_i h(z_i; \hat{\theta})} \sum_r h(z_i; \hat{\theta}). \tag{17.20}$$

Note that the w_i above are generic $O(Nn^{-1})$ expansion weights, and should not be confused with the outlier weights α_i used in (17.17). In general, we can think of a weighting function $w(z)$ that generates these weights. Let $f_s(z)$ and $f_r(z)$ denote the densities of the sample and non-sample values of Z generated by our sample design. Chambers and Dorfman (2003) define a weighting function as *balanced* if it satisfies $n\,w(z)f_s(z) = (N-n)f_r(z)$. They also prove the following result.

Result 17.1 *Suppose that $g_{true}(y_i) = \mu(z_i) + \sqrt{\nu(z_i)}\eta_i$ in the population, where the η_i are independently and identically distributed with zero mean and variance τ^2, while our working model is $g_{work}(y_i) = \mathbf{x}_i'\beta + e_i$, where the e_i are independently and identically distributed with zero mean and variance σ^2. Then, under balanced weighting and with $h(z_i; \hat{\theta}) \equiv g_{work}^{-1}(\mathbf{x}_i'\hat{\beta})$,*

$$E\left(\hat{t}_{wy}^{hR} - t_y\right) \propto \left(\frac{\int \Psi(z) f_r(z) dz}{\int w(z)\Psi(z) f_s(z) dz} - \frac{\int \Omega(z) f_r(z) dz}{\int w(z)\Omega(z) f_s(z) dz}\right) \cong 0$$

where $\Omega(z_i) = E\left\{g_{true}^{-1}\left(\mu(z_i) + \sqrt{\nu(z_i)}\eta_i\right)\right\}$ and $\Psi(z_i) = E\left\{g_{work}^{-1}(\mathbf{x}_i'\beta)\right\}$.

Next consider the smearing predictor (17.14). Again, a straightforward generalisation of this predictor is the weighted form

$$\hat{t}_{\varphi y}^{g^{-1}S} = \sum_s y_i + \sum_{j \in r} \sum_{i \in s} \varphi_{ij} g^{-1}\left(g(y_i) + \{\mu(z_j; \hat{\beta}) - \mu(z_i; \hat{\beta})\}\right). \tag{17.21}$$

The weights φ_{ij} here can vary between different $j \in r$ but are all $O(n^{-1})$ and satisfy $\sum_{i \in s} \varphi_{ij} = 1$. Furthermore, a bias corrected (i.e. 'twiced') version of (17.21) based on the same expansion weights as in (17.20) is then

$$\hat{t}_{w\varphi y}^{g^{-1}S} = \hat{t}_{\varphi y}^{g^{-1}S} + \sum_{j \in s} w_j \left\{y_j - \sum_{i \in s} \varphi_{ij} g^{-1}\left[g(y_i) + \{\mu(z_j; \hat{\beta}) - \mu(z_i; \hat{\beta})\}\right]\right\}. \tag{17.22}$$

Given this set up, Chambers and Dorfman (2003) prove a second result.

Result 17.2 *Under the same conditions as in Result 17.1, including balanced weighting, the twiced weighted smearing predictor (17.22) remains approximately unbiased.*

Both Result 17.1 and Result 17.2 point to the need to use balanced weighting in the ratio-adjusted predictor and the (twiced) smearing predictor if we want these predictors to be robust to misspecification of the transformation scale regression function. A simple way of approximating balanced weights is via non-parametric weighting (see Section 8.4), and a simple type of non-parametric weighting is *histogram weighting*, that is where we put

$$\varphi_{ij} = \frac{I\left(\left|g^{-1}(\mathbf{x}'_j \hat{\beta}) - g^{-1}(\mathbf{x}'_i \hat{\beta})\right| \leq c\right)}{\sum_{k \in s} I\left(\left|g^{-1}(\mathbf{x}'_j \hat{\beta}) - g^{-1}(\mathbf{x}'_k \hat{\beta})\right| \leq c\right)} \qquad (17.23a)$$

and

$$w_i = \sum_{j \in r} \varphi_{ij}. \qquad (17.23b)$$

The simulation results reported in Section 17.7 are based on $c = n^{-1} \left\{ \max_{1 \leq i \leq N} \left(\mathbf{x}'_i \hat{\beta}\right) - \min_{1 \leq i \leq N} \left(\mathbf{x}'_i \hat{\beta}\right) \right\}$. These weights can then be multiplied by the outlier weights $\{\alpha_i; i \in s\}$ used in (17.15) and (17.16).

17.7 Empirical Results II

We return to the simulation study described in Section 17.5, but now investigate what happens if we ignore the stratification in prediction. We calculate the combined ratio estimator (see exercise E.16) as well as a number of new unstratified weighted transformation-based predictors that employ histogram weighting (17.23). See Table 17.6. By construction, these aim to control possible misspecification bias caused by ignoring the stratification. That is, they implicitly assume (17.1) holds across the entire population. Again, RMSE values within 5% of the lowest in each table are highlighted.

As we expect, the bias of the ratio-adjusted predictor is substantially reduced by histogram weighting. Furthermore, the efficiency of this predictor is improved by the use of outlier robust parameter estimation, but at the cost of an increase in bias. In contrast, outlier robust histogram weighting is insufficient to guarantee unbiased smearing prediction. This bias is reduced but not eliminated by twicing. Rather surprisingly, the Karlberg predictor (TK) seems to perform reasonably well in this case, indicating that this approach really requires a larger sample size for both bias correction and stability.

Table 17.6 Additional predictors used in the simulation study. As in Table 17.1, /R indicates the outlier robust version of a predictor.

Name	Description
RE/A	Combined ratio estimator
RAH, RAH/R	Histogram weighted ratio-adjusted predictor (17.20)
SMH, SMH/R	Histogram weighted smearing predictor (17.21)
SMH2/R	Twiced histogram weighted smearing predictor (17.22)

Table 17.7 Simulation results for sector A and $Y =$ WAGES when stratification is ignored.

Predictor	AE	RMSE	AvSE	CR
RE/A	0.31	10.80	9.83	0.91
TA	−12.75	13.65	7.37	0.63
TA/R	−12.15	12.83	6.47	0.58
TK	−3.74	**8.17**	9.18	0.88
TK/R	−7.97	9.05	6.49	0.79
RA	−7.94	10.15	8.28	0.81
RA/R	−9.38	10.41	6.66	0.75
RAH	−0.84	9.37	10.65	0.95
RAH/R	−5.26	**7.81**	8.04	0.91
SM	−1.66	17.19	11.88	0.87
SM/R	−8.47	9.54	6.54	0.77
SMH	4.54	51.58	17.52	0.91
SMH/R	−7.02	8.41	7.01	0.85
SMH2/R	−5.25	**7.81**	8.05	0.92

The other major effect that comes through clearly in Tables 17.7 to 17.10 is that the jackknife variance estimator is extremely conservative in this case, leading to coverages close to 100% in many cases. Why this happens is unclear, and requires further research.

Finally, we can compare the results in Tables 17.2 to 17.5 with those in Tables 17.7 to 17.10. Recollect that the samples and data underpinning the results in these two sets of tables are identical. Here we see substantial reductions in root mean squared errors when the stratification is ignored in prediction. However, this is at the cost of increases in biases in most cases. This indicates that the simple histogram weighting used to control bias in the transformation scale may need further development, perhaps in terms of choice of bandwidth c in (17.23a).

Table 17.8 Simulation results for sector A and $Y = \text{EMP}$ when stratification is ignored.

Predictor	AE	RMSE	AvSE	CR
RE/A	0.03	4.79	4.17	0.93
TA	−4.54	5.17	4.99	0.98
TA/R	−2.15	2.79	3.82	1.00
TK	−0.53	2.59	4.47	1.00
TK/R	−1.29	**2.25**	3.81	1.00
RA	−1.89	3.47	4.93	1.00
RA/R	−1.80	2.62	4.10	1.00
RAH	−0.07	3.40	4.81	1.00
RAH/R	−0.59	2.48	4.17	1.00
SM	−1.42	3.53	4.65	1.00
SM/R	−1.29	**2.25**	3.82	1.00
SMH	0.19	6.72	5.62	1.00
SMH/R	−0.93	**2.18**	3.94	1.00
SMH2/R	−0.48	2.65	4.27	1.00

Table 17.9 Simulation results for sector B and $Y = \text{WAGES}$ when stratification is ignored.

Predictor	AE	RMSE	AvSE	CR
RE/A	0.14	12.96	13.01	0.94
TA	−27.71	28.64	10.11	0.23
TA/R	−24.87	26.08	11.00	0.39
TK	0.82	**11.04**	14.48	0.98
TK/R	0.69	13.55	20.69	0.96
RA	−13.97	16.85	14.60	0.89
RA/R	−13.69	16.46	14.13	0.88
RAH	−2.12	14.05	15.76	0.93
RAH/R	−4.04	12.90	14.80	0.94
SM	−3.98	**10.95**	13.41	0.96
SM/R	−5.64	**11.29**	13.21	0.95
SMH	−8.27	16.50	16.89	0.89
SMH/R	−10.41	14.94	15.23	0.90
SMH2/R	−6.00	12.83	14.44	0.92

17.8 Efficient Sampling under Transformation and Balanced Weighting

Under lognormality, the loglinear model (17.1) for Y implies that its conditional variance increases as the square of Z, so an efficient sampling method is one that

Table 17.10 Simulation results for sector B and $Y = \text{EMP}$ when stratification is ignored.

Predictor	AE	RMSE	AvSE	CR
RE/A	−0.26	7.30	6.57	0.94
TA	−9.34	10.66	7.10	0.80
TA/R	−3.12	5.28	6.83	0.99
TK	2.42	5.97	7.21	1.00
TK/R	0.88	**4.25**	6.58	1.00
RA	−3.69	6.64	8.62	0.99
RA/R	−2.50	4.83	8.23	1.00
RAH	0.01	7.54	8.39	0.97
RAH/R	0.90	5.48	7.17	0.99
SM	−0.32	5.63	7.17	0.99
SM/R	0.62	**4.21**	6.50	1.00
SMH	0.17	8.89	9.71	0.99
SMH/R	−0.08	4.54	7.50	1.00
SMH2/R	0.56	4.95	7.05	0.99

favours units with large values of Z compared to units with small values of this variable. A sampling method that achieves this is one where units are selected with probability proportional to their values of Z. Under this method of sampling, which we denote PPZ sampling, we then have $f_s(z) = z\, f_U(z)/\mu_Z$, where $f_U(z)$ denotes the density of Z over the population and μ_Z is the population mean of Z. A balanced weighting function for this case satisfies $w(z) \approx N\mu_Z/nz$, in which case weighted sample moments of Z are equal to their population moments, that is

$$\frac{\int z^k w(z) f_s(z)\, dz}{\int w(z) f_s(z)\, dz} = \int z^k f_U(z)\, dz.$$

Valliant et al. (2000) show that linear predictors based on balanced weights have the lowest prediction variances when samples satisfy the empirical version of this identity

$$\frac{\sum_s z_i^{-1} z_i^k}{\sum_s z_i^{-1}} = \frac{\sum_U z_i^k}{N}. \tag{17.24}$$

Such samples are said to have *weighted balance of order k*, or equivalently, to be k^{th} order *overbalanced* on Z.

To illustrate the impact of this type of balanced sampling, we present one last set of simulation results from Chambers and Dorfman (2003). These are based on independently selecting 1000 samples via Z-ordered systematic PPZ sampling from the sector A population. The 100 of these 1000 samples closest

Efficient Sampling under Transformation and Balanced Weighting 231

Table 17.11 Values of AE and RMSE for sector A and Y = WAGES under PPZ sampling. Smallest RMSE value is highlighted for both PPZ and Overbalanced sampling.

Predictor	AE		RMSE	
	Systematic PPZ	Overbalanced	Systematic PPZ	Overbalanced
RE/wtd	−0.38	2.25	13.67	5.28
TA	−10.90	−7.14	11.97	7.51
TA/R	−10.56	−7.78	11.30	7.96
TK	−2.75	−0.43	**7.26**	3.17
TK/R	−6.75	−4.70	8.05	5.09
RA	−4.11	−1.62	8.07	3.82
RA/R	−6.32	−5.23	8.10	5.78
RAH	−2.59	1.54	8.20	5.56
RAH/R	−5.08	−2.76	7.70	3.71
SM	−2.56	−0.40	7.86	3.52
SM/R	−6.69	−4.74	8.06	5.13
SMH	−3.44	0.42	13.35	8.72
SMH/R	−5.61	−4.55	7.95	5.32
SMH2/R	−4.68	−1.66	7.81	**3.01**

Table 17.12 Values of AvSE and Coverage for sector A and Y = WAGES under PPZ sampling.

Predictor	AvSE		Coverage	
	Systematic PPZ	Overbalanced	Systematic PPZ	Overbalanced
RE/wtd	7.91	7.74	0.86	1.00
TA	7.64	6.90	0.73	0.98
TA/R	6.64	5.88	0.71	0.94
TK	8.67	7.59	0.86	1.00
TK/R	6.85	5.93	0.84	0.98
RA	8.74	7.83	0.90	1.00
RA/R	7.09	6.38	0.85	0.98
RAH	9.39	9.51	0.89	1.00
RAH/R	7.98	7.17	0.89	1.00
SM	9.30	7.95	0.86	1.00
SM/R	6.99	5.96	0.84	0.98
SMH	9.68	11.18	0.88	1.00
SMH/R	8.00	7.33	0.88	0.98
SMH2/R	8.37	7.32	0.90	1.00

to both zero-order and second order overbalance (17.24) were then identified. Results specific to these samples are denoted as Overbalanced in Tables 17.11 and 17.12. Note that the ratio estimator RE/wtd in these tables is the combined ratio estimator defined by weighting by the inverses of the sample inclusion probabilities. The substantial gains from overbalanced sampling for all the predictors considered in the simulation study are clear.

Bibliography

Anderson, T.W. (1984). *An Introduction to Multivariate Statistical Analysis*. Second Edition. New York: Wiley.

Bankier, M.D., Rathwell, S. and Majkowski, M. (1992). Two step generalised least squares estimation in the 1991 Canadian census. *Proceedings of the Workshop on Uses of Auxiliary Information in Surveys*, Statistics Sweden, Örebro, October 5–7.

Battese, G., Harter, R. and Fuller, W. (1988). An error-components model for prediction of county crop areas using survey and satellite data. *Journal of the American Statistical Association*, **83**, 28–36.

Bardsley, P. and Chambers, R.L. (1984). Multipurpose estimation from unbalanced samples. *Applied Statistics*, **33**, 290–299.

Beaton, A.E. and Tukey, J.W. (1974). The fitting of power series, meaning polynomials, illustrated on band-spectroscopic data. *Technometrics*, **16**, 147–185.

Box, G.E.P. and Draper, N.R. (1987). *Empirical Model-Building and Response Surfaces*. New York: Wiley.

Breidt, F.J. and Opsomer, J.D. (2009). Nonparametric and semiparametric estimation in complex surveys. Chapter 27 of *Handbook of Statistics 29: Sample Surveys: Design, Methods and Applications* (Editors Pfeffermann, D. and Rao, C.R.). Amsterdam: North Holland.

Breiman, L., Friedman, J.H., Olshen, R.A. and Stone, C.J. (1984). *Classification and Regression Trees*. Belmont, California: Wadsworth.

Breslow, N.E. and Clayton, D.G. (1993). Approximate inference in generalized linear mixed models. *Journal of the American Statistical Association*, **88**, 9–25.

Brewer, K. (1963). Ratio estimation and finite populations: some results deducible from the assumption of an underlying stochastic process. *Australian Journal of Statistics*, **5**, 93–105.

Brewer, K. and Gregoire, T.G. (2009). Introduction to survey sampling. Chapter 1 of *Handbook of Statistics 29: Sample Surveys: Design, Methods and Applications* (Editors Pfeffermann, D. and Rao, C.R.). Amsterdam: North Holland.

Brown, G., Chambers, R., Heady, P. and Heasman, R. (2001). Evaluation of small area estimation methods – An application to the unemployment estimates from the UK LFS. *Proceedings of the XVIIIth International Symposium on Methodological Issues*, Statistics Canada, Ottawa, October 2001.

Casady, R.J., Dorfman, A.H. and Wang, S. (1998). Confidence intervals for domain parameters when the sample sizes are random. *Survey Methodology*, **24**, 57–67.

Chambers, R.L. (1982). Robust Finite Population Estimation. Ph.D thesis, Department of Biostatistics, The Johns Hopkins University, Baltimore.

Chambers, R.L. (1986). Outlier robust finite population estimation. *Journal of the American Statistical Association*, **81**, 1063–1069.

Chambers, R.L. (1996). Robust case-weighting for multipurpose establishment surveys. *Journal of Official Statistics*, **12**, 3–32.

Chambers, R.L. (1997). Weighting and calibration in sample survey estimation. *Conference on Statistical Science Honouring the Bicentennial of Stefano Franscini's Birth* (Editors C. Malaguerra, S. Morgenthaler and E. Ronchetti). Basel: Birkhäuser Verlag.

Chambers, R., Chandra, H. and Tzavidis, N. (2009). On bias-robust mean squared error estimation for linear predictors for domains. *Working Paper 09-08*, Centre for Statistical and Survey Methodology, The University of Wollongong, Australia. (http://ro.uow.edu.au/cssmwp/31 [accessed 15 July 2011]).

Chambers, R.L. and Dunstan, R. (1986). Estimating distribution functions from survey data. *Biometrika*, **73**, 597–604.

Chambers, R.L., Dorfman, A.H. and Wehrly, T.E. (1993). Bias robust estimation in finite populations using nonparametric calibration. *Journal of the American Statistical Association*, **88**, 260–269.

Chambers, R.L. and Dorfman, A.H. (1994). Robust sample survey inference via bootstrapping and bias correction: The case of the ratio estimator. *Proceedings of the Joint Statistical Meetings*, Toronto, August 13–18, 1994, Alexandria: American Statistical Association.

Chambers, R.L. and Dorfman, A.H. (2003). Transformed variables in survey sampling. *S3RI Methodology Working Papers M03/21*, Southampton Statistical Sciences Research Institute (http://eprints.soton.ac.uk/8171/ [accessed 15 July 2011]).

Chambers, R.L. and Kokic, P.N. (1993). Outlier robust sample survey inference. *Bulletin of the International Statistical Institute 55, Proceedings of the 49th Session of the International Statistical Institute*, 55–72, The Hague: International Statistical Institute.

Chambers, R.L. and Skinner, C.J. (Eds.) (2003). *Analysis of Survey Data*. New York: Wiley.

Chambers, R. and Tzavidis, N. (2006). M-quantile models for small area estimation. *Biometrika*, **93**, 255–268.

Chandra, H. and Chambers, R. (2009). Multipurpose weighting for small area estimation. *Journal of Official Statistics*, **25**, 1–18.

Chandra, H. and Chambers, R. (2011). Small area estimation under transformation to linearity. *Survey Methodology*, **37**, 39–51.

Clark, R.G. and Chambers, R.L. (2008). Adaptive calibration for prediction of finite population totals. *Survey Methodology*, **34**, 2, 163–172.

Cressie, N. (1982). Playing safe with misweighted means. *Journal of the American Statistical Association*, **77**, 754–759.

Cochran, W.G. (1977). *Sampling Techniques*. Third edition. New York: Wiley.

Cox, D.R. and Hinkley, D.V. (1974). *Theoretical Statistics*. London: Chapman and Hall.

Dalenius, T. and Hodges, J.L. (1959). Minimum variance stratification. *Journal of the American Statistical Association*, **54**, 88–101.

Datta, G.S. (2009). Model-based approaches to small area estimation. Chapter 32 of *Handbook of Statistics 29: Sample Surveys: Design, Methods and Applications* (Editors Pfeffermann, D. and Rao, C.R.). Amsterdam: North Holland.

Datta, G.S. and Lahiri, P. (2000). A unified measure of uncertainty of estimates for best linear unbiased predictors in small area estimation problem. *Statistica Sinica*, **10**, 613–627.

Deming, W.E. (1960). *Sample Design in Business Research*. New York: Wiley.

Deville, J.C. and Särndal, C.E. (1992). Calibration estimators in survey sampling. *Journal of the American Statistical Association*, **87**, 376–382.

Dorfman, A.H. and Hall, P. (1992). Estimators of the finite population distribution function using nonparametric regression. *Annals of Statistics*, **21**, 1452–1475.

Dorfman, A.H. (1994). Open questions in the application of smoothing methods to finite population inference. *Computationally Intensive Statistical Methods: Proceedings of the 26th Symposium on the Interface of Computing Science and Statistics*, 201–205, Fairfax Station: Interface Foundation of North America.

Dorfman, A.H. (2009a). Inference on distribution functions and quantiles. Chapter 36 of *Handbook of Statistics 29: Sample Surveys: Design, Methods and Applications* (Editors Pfeffermann, D. and Rao, C.R.). Amsterdam: North Holland.

Dorfman, A.H. (2009b). Nonparametric regression and the two sample problem. *Proceedings of the Joint Statistical Meetings*, 277–290. American Statistical Association: Alexandria, VA.

Duan, N. (1983). Smearing estimate: A nonparametric retransformation method. *Journal of the American Statistical Association*, **78**, 605–610.

Dunstan, R. and Chambers, R.L. (1989). Estimating distribution functions from survey data with limited benchmark information. *Australian Journal of Statistics*, **31**, 1–11.

Efron, B. (1982). *The Jacknife, the Bootstrap and Other Resampling Plans*. SIAM: Philadelphia.

Fay, R.E. and Herriot, R.A. (1979). Estimation of income from small places: an application of James-Stein procedures to census data. *Journal of the American Statistical Association*, **74**, 269–277.

Fisher, R.A. and Yates, F. (1963). *Statistical Tables for Biological, Agricultural and Medical Research.* Edinburgh: Oliver and Boyd.

Foreman, E.K. (1991). *Survey Sampling Principles.* New York: Marcel Dekker.

Gasser, T., Müller, H.G. and Mammitzsch, V. (1985). Kernels for nonparametric curve estimation. *Journal of the Royal Statistical Society Series B*, **47**, 238–252.

Gershunskaya, J., Jiang, J. and Lahiri, P. (2009). Resampling methods in surveys. Chapter 28 of *Handbook of Statistics 29: Sample Surveys: Design, Methods and Applications* (Editors Pfeffermann, D. and Rao, C.R.). Amsterdam: North Holland.

Ghosh, M. (2009). Bayesian developments in survey sampling. Chapter 29 of *Handbook of Statistics 29: Sample Surveys: Design, Methods and Applications* (Editors Pfeffermann, D. and Rao, C.R.). Amsterdam: North Holland.

Ghosh, M. and Meeden, G. (1997). *Bayesian Methods for Finite Population Sampling.* London: Chapman and Hall.

Goldstein, H. (2003). *Multilevel Statistical Models.* Third edition. London: Arnold.

Gonzalez, M.E. and Hoza, C. (1978). Small-area estimation with application to unemployment and housing estimates. *Journal of the American Statistical Association*, **73**, 7–15.

Groves, R.M., Fowler, F.F., Couper, M.P., Lepkowski, J.M., Singer, E. and Tourangeau, R. (2004). *Survey Methodology.* New York: Wiley.

Gwet, J.-P. and Rivest, L.-P. (1992). Outlier resistant alternatives to the ratio estimator. *Journal of the American Statistical Association*, **87**, 1174–1182.

Hall, P.G. (1992). *The Bootstrap and Edgeworth Expansion.* New York: Springer.

Härdle, W. (1990). *Applied Nonparametric Regression Analysis.* Cambridge: Cambridge University Press.

Henderson, C.R. (1953). Estimation of variance and covariance components. *Biometrics*, **9**, 226–252.

Huang, E.T. and Fuller, W.A. (1978). Nonnegative regression estimation for survey data. *Proceedings of the Social Statistics Section*, 300–303, Alexandria: American Statistical Association.

Huber, P.J. (1964). Robust estimation of a location parameter. *Annals of Mathematical Statistics*, **35**, 73–101.

Kacker, R.N. and Harville, D.A. (1984). Approximations for standard errors of estimations of fixed and random effects in mixed linear models. *Journal of the American Statistical Association*, **79**, 853–862.

Karlberg, F. (2000a). Population total prediction under a lognormal superpopulation model, *Metron*, **LVIII**, 53–80.

Karlberg, F. (2000b). Survey estimation for highly skewed populations in the presence of zeroes, *Journal of Official Statistics*, **16**, 229–241.

Kish, L. (1965). *Survey Sampling.* New York: Wiley.

Kuo, L. (1988). Classical and prediction approaches to estimating distribution functions from survey data. *Proceedings of the Section on Survey Research Methods*, 280–285, Alexandra: American Statistical Association.

Kuk, A. (1993). A kernel method for estimating finite population distribution functions using auxiliary information, *Biometrika* **80**, 385–392.

Kuk, A.Y.C. and Welsh, A.H. (2001). Robust estimation for finite populations based on a working model. *Journal of the Royal Statistical Society Series B*, **63**, 277–292.

Liu, R.Y. (1988). Bootstrap procedures under some non-i.i.d. models. *Annals of Statistics*, **16**, 1696–1708.

Lohr, S.L. (1999). *Sampling: Design and Analysis*. Pacific Grove: Duxbury Press.

Mahalonobis, P.C. (1946). Recent experiments in statistical sampling in the Indian Statistical Institute. *Journal of the Royal Statistical Society*, **109**, 325–370.

McCulloch, C.E. and Searle, S.R. (2001). *Generalized, Linear, and Mixed Models*. London: Chapman and Hall.

Nelder, J.A. and Wedderburn, R.W. (1972). Generalized linear models. *Journal of the Royal Statistical Society Series A*, **135**, 370–384.

Neyman, J. (1934). On the two different aspects of the representative method: the method of stratified sampling and the method of purposive selection. *Journal of the Royal Statistical Society*, **97**, 558–606.

ONS (2006). *Model-Based Estimates of ILO Unemployment for LAD/UAs in Great Britain: Guide for Users*. London: United Kingdom Office for National Statistics. (http://www.statistics.gov.uk/downloads/theme_labour/User_Guide.pdf [accessed 15 July 2011]).

Opsomer, J.D. and Miller, C.P. (2003). Selecting the amount of smoothing in local polynomial regression estimation for complex surveys. *Proceedings of the Joint Statistical Meetings*, 3115–3122. American Statistical Association: Alexandria, VA.

Opsomer, J.D., Claeskens, G., Ranalli, M.G., Kauermann, G. and Breidt, F.J. (2008). Nonparametric small area estimation using penalized spline regression. *Journal of the Royal Statistical Society, Series B*, **70**, 265–286.

Pereira, L.N. and Coelho, P.S. (2010). Small area estimation of mean price of habitation transaction using time series and cross-sectional area-level models. *Journal of Applied Statistics*, **37**, 651–666.

Pinheiro, J., Bates, D., DebRoy, S. and Sarkar, D. (2005). nlme: Linear and nonlinear mixed effects models. *R package version 3.1-60*.

Prasad, N.G.N and Rao, J.N.K. (1990). The estimation of the mean squared error of small-area estimators. *Journal of the American Statistical Association*, **85**, 163–171.

Pratesi, M. and Salvati, N. (2008). Small area estimation: the EBLUP estimator based on spatially correlated random area effects. *Statistical Methods and Applications*, **17**, 113–141.

Quenouille, M. (1956). Notes on bias in estimation. *Biometrika*, **43**, 353–360.

R Development Core Team (2005). R: A language and environment for statistical computing. *R Foundation for Statistical Computing*, Vienna, Austria. ISBN 3-900051-07-0, (http://www.R-project.org [accessed 15 July 2011]).

Rao, J.N.K. (2003). *Small Area Estimation*. New York: Wiley.

Rivest, L.P. and Rouillard, E. (1991). M-estimators and outlier resistant alternatives to the ratio estimator, In *Proceedings of the 1990 Symposium of Statistics Canada*, 271–285.

Royall, R.M. (1970). On finite population sampling theory under certain linear regression models. *Biometrika*, **57**(2), 377–387.

Royall, R.M. (1976). The linear least squares prediction approach to two-stage sampling. *Journal of the American Statistical Association*, **71**, 657–664.

Royall, R.M. and Herson, J. (1973a). Robust estimation in finite populations I. *Journal of the American Statistical Association*, **68**, 880–889.

Royall, R.M. and Herson, J. (1973b). Robust estimation in finite populations II. *Journal of the American Statistical Association*, **68**, 890–893.

Royall, R.M. and Cumberland, W.G. (1978). Variance estimation in finite population sampling. *Journal of the American Statistical Association*, **73**, 351–358.

Royall, R.M. and Cumberland, W.G. (1981). An empirical study of the ratio estimator and estimators of its variance. *Journal of the American Statistical Association*, **76**, 66–88.

Royall, R.M. and Eberhardt, K.A. (1975). Variance estimates for the ratio estimator. *Sankhya Series C*, **37**, 43–52.

Sadooghi-Alvandi, S.M. (1988). A note on the efficiency of proportional stratification. *Australian Journal of Statistics*, **30**, 196–199.

Saei, A. and Chambers, R. (2003). Small area estimation under linear and generalized linear mixed models with time and area effects. *S3RI Methodology Working Papers M03/15*, Southampton Statistical Sciences Research Institute (http://eprints.soton.ac.uk/8165/ [accessed 15 July 2011]).

Salant, P. and Dillman, D.A. (1994). *How to Conduct Your Own Survey*. New York: Wiley.

Salvati, N., Chandra, H., Ranali, M.G. and Chambers, R. (2010). Small area estimation using a nonparametric model-based direct estimator. *Computational Statistics and Data Analysis*, **54**, 2159–2171.

Salvati, N., Pratesi, M., Tzavidis, N. and Chambers, R. (2010). Small area estimation via M-quantile geographically weighted regression. *Test*, doi: 10.1007/s11749-010-0231-1.

Särndal, C.E., Swensson, B. and Wretman, J. (1992). *Model-Assisted Survey Sampling*. New York: Springer-Verlag.

Scott, A.J., Brewer, K.R.W. and Ho, E.W.H. (1978). Finite population sampling and robust estimation. *Journal of the American Statistical Association*, **73**, 359–361.

Searle, S.R., Casella, G. and McCulloch, C.E. (2006). *Variance Components*. New York: Wiley.

Serfling, R.J. (1980). *Approximation Theorems of Mathematical Statistics*. New York: Wiley.

Shao, J. and Tu, D. (1995). *The Jackknife and Bootstrap*. New York: Springer-Verlag.
Silva, P.L.N. and Skinner, C.J. (1997). Variable selection for regression estimation in finite populations. *Survey Methodology*, **23**, 23–32.
Silverman, B.W. (1986). *Density Estimation for Statistics and Data Analysis*. London: Chapman and Hall.
Singh, B.B., Shukla, G.K. and Kundu, D. (2005). Spatio-temporal models in small area estimation. *Survey Methodology*, **31**, 183–195.
Sinha, S.K. and Rao, J.N.K. (2009). Robust small area estimation. *Canadian Journal of Statistics*, **37**, 381–399.
Skinner, C.J., Holt, D. and Smith, T.M.F. (Eds.) (1989). *Analysis of Complex Surveys*. Chichester, Wiley.
Sonquist, J.N., Baker, E.L. and Morgan, J.A. (1971). *Searching for Structure*. Institute for Social Research, University of Michigan.
Sverchkov, M. and Pfeffermann, D. (2004). Prediction of finite population totals based on the sample distribution. *Survey Methodology*, **30**, 79–92.
Tallis, G.M. (1978). Note on robust estimation in finite populations. *Sankhya Series C*, **40**, 136–138.
Tam, S.M. (1986). Characteristics of best model-based predictors in survey sampling. *Biometrika*, **73**, 232–235.
Thompson, S.K. (1992). *Sampling*. New York: Wiley.
Tillé, Y. (2001). *Théorie des Sondages*. Paris: Dunod.
Tukey, J.W. (1958). Bias and confidence in not quite large samples. *Annals of Mathematical Statistics*, 29, 614.
Tukey, J.W. (1977). *Exploratory Data Analysis*. Reading: Addison-Wesley.
Tzavidis, N., Marchetti, S. and Chambers, R. (2010). Robust estimation of small area means and quantiles. *Australian and New Zealand Journal of Statistics*, **52**, 167–186.
Tzavidis, N., Salvati, N., Pratesi, M. and Chambers, R. (2008). M-quantile models with application to poverty mapping. *Statistical Methods and Applications*, **17**, 393–411.
Ugarte, M.D., Goicoa, T.A., Militino, A.F. and Durban, M. (2009). Spline smoothing in small area trend estimation and forecasting. *Computational Statistics and Data Analysis*, 53, 3616–3629.
Valliant, R. (2009). Model-based prediction of finite population totals. Chapter 23 of *Handbook of Statistics 29: Sample Surveys: Design, Methods and Applications* (Editors Pfeffermann, D. and Rao, C.R.). Amsterdam: North Holland.
Valliant, R., Dorfman, A.H. and Royall, R.M. (2000). *Finite Population Sampling and Inference*. New York: John Wiley.
Venables, W.N. and Ripley, B.D. (2002). *Modern Applied Statistics with S*. New York: Springer.
Vitter, J.S. (1984). Faster methods for random sampling. *Communications of the ACM* 27, 703–718, New York: Association for Computing Machinery.

Wand, M.P. and Jones, M.C. (1995). *Kernel Smoothing*. London: Chapman and Hall.

Wolter, K.M. (1985). *Introduction to Variance Estimation*. New York: Springer-Verlag.

Woodruff, K.M. (1971). A simple method for approximating the variance of a complicated estimate. *Journal of the American Statistical Association*, **66**, 411–414.

Woodruff, R.S. (1952). Confidence intervals for medians and other position measures. *Journal of the American Statistical Association*, **47**, 635–646.

Wu, C. and Sitter, R.R. (2001). A model calibration approach to using complete auxiliary information from survey data. *Journal of the American Statistical Association*, **96**, 185–193.

Zhang, Li-C. and Chambers, R.L. (2004). Small area estimates for cross-classifications. *Journal of the Royal Statistical Society, Series B*, **66**, 479–496.

Exercises

E.1 Let W and V be two random variables, and let $g(V)$ be any function of V. Show that we then always have $E(W - g(V))^2 = E(W - E(W|V))^2 + E(g(V) - E(W|V))^2$ and hence deduce that $E(W|V)$ is the minimum mean squared error predictor of W given V.

E.2 Consider the urn model described in Section 3.1 with $w =$ the number of white balls observed in a random sample of n balls selected from the N balls in the urn.

(a) Show that $\hat{W} = N(w/n)$ is an unbiased estimator of W, the total number of white balls in the urn.

(b) Show that the variance of $\hat{W}$ is $\frac{(N-n)W(N-W)}{n(N-1)}$.

E.3 Assume the population Y-values follow model (3.10). This is a generalisation of the homogenous model which allows every pair of units to have a common correlation ρ. Model (3.1) is the special case when $\rho = 0$. Show that when model (3.10) applies:

(a) The expansion estimator (3.2) is unbiased for the population total t_y.

(b) The expansion estimator (3.2) is the BLUP of this population total.

(c) $E(y_i - \bar{y}_s)^2 = n^{-1}(n-1)\sigma^2(1-\rho)$.

(d) The estimator (3.6) is an unbiased estimator of the prediction variance of the expansion estimator (3.2).

E.4 A household survey is designed to estimate the proportion of families possessing a certain attribute. For the principal item of interest, this proportion is expected to lie between 0.3 and 0.7. Assuming that the distribution of this attribute within the population can be modelled as equivalent to independent and identically distributed realisations of a Bernoulli random variable (i.e. a random variable that takes the value one when the attribute is present and is zero otherwise), how large are the sample sizes required to estimate the following population parameters with relative standard errors not exceeding 3%:

(a) The overall population proportion (P).

(b) The individual proportions (P_1, P_2, P_3) for each of the household income classes; under $25000, $25000 to $40000, over $40000. It is known that these three classes make up 35, 50 and 15% of the population respectively.

E.5 A consumer survey was carried out on a simple random sample of 14 households obtained from a community of 150 households. Data were obtained in the sample on household size, annual household income and expenditure on food (in £100 units) as follows:

Sample Household	Household Size (Z)	Household Income (X)	Food Expenditure (Y)	Y/X
1	1	50	9	0.1800
2	3	122	25	0.2049
3	2	96	19	0.1979
4	3	156	31	0.1987
5	4	144	27	0.1875
6	1	65	13	0.2000
7	2	87	17	0.1954
8	2	82	21	0.2561
9	5	146	35	0.2397
10	4	127	22	0.1732
11	5	115	28	0.2435
12	1	106	22	0.2075
13	2	77	14	0.1818
14	2	85	19	0.2235

The distribution of household sizes for the community is known, and is given by

Household Size	1	2	3	4	5	6	7
Number	15	25	45	35	20	5	5

(a) Making any necessary assumptions, use these data to predict R_1, the ratio of total food expenditure to total income in the community of 150 households, and provide a 95% prediction interval for this parameter. Justify the estimator you use.

(b) An alternative population parameter measuring the relationship between food expenditure and income in this community is the average of the individual household ratios, ($N = 150$)

$$R_2 = N^{-1} \sum_{i=1}^{N} y_i/x_i.$$

Given the above survey data, predict R_2 and provide a 95% prediction interval for its value.

(c) Which quantity, R_1 or R_2, would you choose as a measure of the average relationship between food expenditure and income? Why?

E.6 Let $Q_1, Q_2, \ldots, Q_H$ denote a sequence of H strictly positive numbers. Show that in order to minimise $\sum_h Q_h^2/n_h$ subject to $\sum_h n_h = n$, where n is fixed, we need to make n_h proportional to Q_h.

E.7 Using the Cities' population (Table 4.1) as the target population, and the 1930 population count as the variable of interest, estimate the prediction variance of the stratified expansion estimator under the homogeneous strata model corresponding to each of the following stratified sampling strategies. In all cases the total number of strata is 4 and the total sample size is 16.

(a) Strata defined by splitting the population into equal sized groups after ordering on the size variable (1920 count) and proportional allocation.

(b) Same as (a), except allocation is optimal based on the 1920 population counts.

(c) Stratum boundaries defined using the Dalenius and Hodges technique (see Section 4.8) based on the 1920 population counts and optimal allocation based on the same 1920 counts.

(d) Strata defined to have equal totals for the 1920 population counts, with equal allocation.

Which strategy is the most efficient for this population? In the case of strategies (b) and (c), what loss in efficiency occurs compared to optimal allocation based on the 1930 population counts?

E.8 Let C denote the total funds allocated to carry out a stratified sample survey. Further, assume that the unit sampling cost in stratum h is equal to c_h, with the administrative cost associated with sampling in stratum h equal to a_h. Show that the optimal sample allocation required to minimise the prediction variance of the stratified expansion estimator (assuming a stratified model for the target population) is then

$$n_h = \frac{N_h \sigma_h}{\sqrt{c_h}} \left(C - \sum_g a_g \right) \left(\sum_g N_g \sigma_g \sqrt{c_g} \right)^{-1}.$$

E.9 Show that the total sample size n required for the stratified expansion estimator to achieve a relative standard error of α per cent under proportional allocation is

$$n = (10^4/\alpha^2) \left[\sum_h F_h \sigma_h^2 \bigg/ \left(\sum_h F_h \bar{y}_h \right)^2 \right],$$

where $F_h = N^{-1}N_h$. Show also that the corresponding sample size under equal allocation to H strata is

$$n = (10^4/\alpha^2)\left[H\sum_h N_h^2\sigma_h^2 \Big/ \left(\sum_h N_h \bar{y}_h\right)^2\right].$$

E.10 Consider a population of businesses with Z = employment size (number of employees), as set out in the table below.

Employment size	Number of businesses
0–1	632
2–5	674
6–10	391
11–20	156
21–30	168
31–50	93
51–100	62
101–300	21

(a) Apply the Dalenius–Hodges stratification procedure (see Section 4.8) to determine the boundaries of three strata defined on this population.

(b) Given a total sample size of 380, allocate this sample to the three strata defined in (a) above using Neyman allocation based on Z. (Hint: calculate the stratum variances by assuming the element Z-values in each employment size group above are all equal to the mid-point $\bar{z}_g$ of the group. That is, there are 632 businesses with employment 0.5, 674 with employment 3.5, and so on.)

(c) Suppose the survey has a total budget of £64,000, with a start-up cost of £15,000 and per unit sampling costs of £14 for stratum 1 (small Z), £98 for stratum 2 (medium Z) and £150 for stratum 3 (large Z). Using the result in (a), and calculating the stratum variances as in (b) above, determine the optimal sample sizes for the three strata given this budget.

(d) Contrast the design in (b) with a design defined by an equal aggregate Z stratification based on the grouped Z values above, together with equal allocation.

E.11 Using the Cities' population (Table 4.1), draw the sample of size $n = 16$ defined by those cities with the largest 1920 counts. Compute the ratio estimate for the total 1930 count based on this sample and a corresponding estimate of its prediction variance under the ratio population model

(5.2). Compare these estimates with those obtained from taking a first-order balanced sample of the same size and using the regression estimator. Is there any loss in efficiency in using the balanced strategy? Compare both sets of estimates with those obtained from a stratified balanced sample and the separate ratio estimator, based on 4 equal-sized strata and equal allocation (see **E.7(d)** above).

E.12 The following table below sets out the data for the Boys population. This corresponds to measurements on a group of 150 boys, aged between 10 and 12. The measurements are height (H) in inches, chest circumference (C) in centimetres and weight (W) in pounds.

i	C	H	W	i	C	H	W	i	C	H	W
1	72	66.0	127.0	51	61	58.5	88.0	101	72	55.7	86.3
2	71	64.5	106.0	52	62	58.3	89.8	102	67	55.7	77.3
3	72	64.0	115.4	53	67	58.3	77.3	103	68	55.5	69.3
4	72	63.5	100.5	54	70	58.3	79.3	104	66	55.5	75.8
5	76	63.0	127.0	55	74	58.2	103.3	105	65	55.5	67.5
6	68	62.5	114.0	56	63	58.2	70.4	106	66	55.5	77.8
7	68	62.5	84.0	57	77	58.2	88.3	107	64	55.5	71.5
8	70	62.1	95.8	58	70	58.2	69.6	108	63	55.5	63.0
9	77	61.5	105.3	59	70	58.0	73.0	109	69	55.3	77.3
10	68	61.5	88.5	60	69	58.0	88.5	110	68	55.3	72.0
11	72	61.1	96.5	61	70	58.0	83.0	111	63	55.3	69.5
12	66	61.1	91.0	62	58	57.8	85.3	112	64	55.2	68.0
13	72	61.1	90.0	63	63	57.8	77.5	113	67	55.2	74.0
14	69	61.0	98.5	64	70	57.6	75.8	114	62	55.1	70.8
15	75	61.0	117	65	67	57.5	86.0	115	63	55.0	59.0
16	76	61.0	110.5	66	73	57.5	81.0	116	63	55.0	70.0
17	76	60.7	116.0	67	70	57.5	82.3	117	66	55.0	73.3
18	71	60.7	88.8	68	67	57.5	105.0	118	63	54.8	65.0
19	72	60.6	105.3	69	66	57.5	74.8	119	68	54.7	70.8
20	66	60.5	90.6	70	67	57.5	84.0	120	66	54.7	66.3
21	68	60.3	77.5	71	69	57.5	81.0	121	62	54.5	63.0
22	69	60.2	97.5	72	63	57.2	85.0	122	64	54.5	74.2
23	74	60.1	96.5	73	61	57.1	81.2	123	70	54.5	88.0
24	91	60.0	168.0	74	65	57.0	78.0	124	65	54.5	87.5
25	64	60.0	83.0	75	67	57.0	79.0	125	66	54.3	78.3
26	68	59.8	97.0	76	70	57.0	90.3	126	59	54.3	64.8
27	71	59.6	92.3	77	66	57.0	65.3	127	65	54.3	72.3
28	74	59.6	96.8	78	66	57.0	78.0	128	71	54.3	81.8
29	66	59.5	76.8	79	75	56.8	88.5	129	68	54.2	71.5
30	71	59.5	77.8	80	63	56.7	75.0	130	68	54.1	70.8

31	70	59.5	90.4	**81**	71	56.7	83.3	**131**	67	54.1	73.3
32	58	59.5	65.3	**82**	53	56.7	61.8	**132**	59	54.0	64.0
33	69	59.5	114.3	**83**	65	56.6	72.3	**133**	62	54.0	65.1
34	67	59.2	74.8	**84**	71	56.6	116.5	**134**	72	54.0	90.5
35	64	59.2	78.0	**85**	69	56.5	70.8	**135**	65	53.7	62.5
36	70	59.1	87.0	**86**	63	56.3	66.8	**136**	66	53.5	70.4
37	70	59.0	84.3	**87**	70	56.3	110.3	**137**	64	53.5	70.0
38	66	59.0	85.0	**88**	68	56.2	81.3	**138**	71	53.5	82.0
39	86	59.0	131.5	**89**	69	56.2	83.0	**139**	65	53.4	70.8
40	70	59.0	80.8	**90**	68	56.2	75.5	**140**	63	53.1	66.5
41	72	58.8	103.8	**91**	66	56.1	70.5	**141**	65	53.0	64.8
42	69	58.8	86.8	**92**	67	56.1	70.5	**142**	62	53.0	61.0
43	61	58.7	93.0	**93**	69	56.1	78.3	**143**	62	53.0	74.0
44	69	58.7	93.8	**94**	66	56.0	71.5	**144**	65	52.7	61.0
45	65	58.6	84.8	**95**	71	56.0	83.5	**145**	64	52.5	60.5
46	59	58.6	60.8	**96**	93	56.0	121.0	**146**	68	51.8	71.5
47	65	58.6	74.8	**97**	68	56.0	78.0	**147**	64	51.6	58.0
48	70	58.5	96.0	**98**	67	56.0	78.8	**148**	67	50.8	73.0
49	71	58.5	92.0	**99**	67	56.0	76.0	**149**	61	50.6	64.0
50	69	58.5	89.0	**100**	66	55.8	70.8	**150**	62	50.5	55.8

Treating this group as a target population, and assuming H and W are known for all elements of this population, our aim is to predict the average chest circumference for the group, based on data obtained from a sample of $n = 15$ boys from the group. Consider the following four methods of selection for this sample:

(a) We can divide the population into three equal-sized strata on the basis of H, use equal allocation, select a SRS within each stratum, and predict this mean via the stratified expansion estimator.

(b) We can adopt the same strategy as (a), but use W as our 'size' variable.

(c) We can use the sample selected in (a), but predict the mean via the separate ratio estimator based on H.

(d) We can use the sample selected in (b), but predict the mean via the separate ratio estimator based on W.

Implement each of these strategies and compute the resulting prediction of the average chest circumference, as well as a 95% prediction interval for this parameter. Which prediction do you prefer? Why? Can you suggest a more efficient sampling and prediction strategy to use with this population?

E.13 Assuming (a) in exercise E.12 is the chosen sampling strategy for the Boys population, calculate the sample size required to ensure that the stratified expansion estimator has a relative standard error of 5%.

E.14 The assumption that the variance of Y given Z increases proportionately with Z is difficult to justify when the ratio population model (5.2) is used for many economic populations, especially since all we observe is an increasing scatter in the sample data. In this context we note that we can embed (5.2) (and (5.1)) in the more general model

$$E(y_i|z_i) = \beta z_i$$
$$Var(y_i|z_i) = \sigma^2 v(z_i)$$
$$Cov(y_i, y_j|z_i, z_j) = 0 \text{ for all } i \neq j$$

where $v(z)$ is a non-decreasing function of z.

(a) Show that the ratio estimator remains unbiased for t_y under this more general model, and derive its prediction variance.

(b) Show that, if the function $v(z)$ is known, then the BLUP of t_y under this more general model is the generalised ratio estimator $\hat{t}_y^{Rv} = t_{sy} + b_v \sum_r z_i$, where $b_v = \sum_s v_i^{-1} y_i z_i / \sum_s v_i^{-1} z_i^2$. Here $v_i = v(z_i)$. You may use the fact that b_v is the BLUE of β in this case.

(c) Show that the optimal sample for this BLUP is still the extreme sample, that is the sample made up of the n population elements with largest values of Z, that was optimal for the ratio estimator under (5.2).

E.15 Verify equation (5.9). That is, show that for the prediction variance of the regression estimator $\hat{t}_y^L$ under the linear population model (5.5) is

$$Var\left(\hat{t}_y^L - t_y\right) = \frac{N^2}{n}\sigma^2\left[\left(1 - \frac{n}{N}\right) + \frac{(\bar{z} - \bar{z}_s)^2}{(1 - n^{-1})s_z^2}\right].$$

E.16 Size stratification with random sampling within the size strata is a common design strategy in business surveys. When there is a positive correlation between the size variable Z and the survey variable Y, the survey analyst also has the option of employing some form of ratio estimation to predict the population total t_y. The two most commonly used methods of ratio estimation in this situation are the separate ratio estimator and the combined ratio estimator. The former was defined in (5.13), while the latter is given by

$$\hat{t}_y^{CR} = \frac{\sum_h N_h \bar{y}_{sh}}{\sum_h N_h \bar{z}_{sh}} \sum_h t_{zh}.$$

The combined estimator above is often recommended when the sample sizes in the strata are small.

(a) Show that under the ratio population model, both the separate ratio estimator and the combined ratio estimator are unbiased, but under the separate ratio population model, only the separate ratio estimator is unbiased. Develop the prediction variances of both prediction methods under the separate ratio model. Which method do you prefer for this case?

(b) Suppose now that the (simple) ratio population model applies. Contrast prediction based on separate ratio estimator or the combined ratio estimator with each other and with prediction using the simple ratio estimator (i.e. the ratio estimator that ignores the size strata). Under what circumstances would you prefer each of these methods of prediction?

E.17 Consider a population made up of Q households. A simple random sample (denoted s) of q of these is taken and data on a survey variable Y are obtained from all individuals in each sampled household. The number of individuals in the gth household is denoted by M_g. The aim is to predict the population total t_y of Y.

There are two standard predictors typically used this situation. The first is the household level expansion estimator $\hat{t}_y^E = \frac{Q}{q}\sum_s t_{gy}$, where t_{gy} denotes the total of Y for household g. The second assumes that the total count $N = \sum_{g=1}^{Q} M_g$ of individuals in the population is known, and estimates t_y using the household level ratio estimator (often referred to as the ratio to size estimator) $\hat{t}_y^R = \frac{N}{n}\sum_s t_{gy}$.

Let y_{ig} denote the value of Y for individual i in household g. A standard model for this situation (cluster sampling) is one where it is assumed these population values follow the two level model $y_{ig} = \mu + \eta_g + \varepsilon_{ig}$, where the η_g are independently and identically distributed household level random effects with $E(\eta_g) = 0$ and $Var(\eta_g) = \omega^2$ and the ε_{ig} are independent and identically distributed individual level effects with $E(\varepsilon_{ig}) = 0$ and $Var(\varepsilon_{ig}) = \sigma^2$, distributed independently of the η_g. In what follows we shall assume this model holds.

(a) Suppose that the sizes M_g are known for all households in the population. Show that the expansion estimator $\hat{t}_y^E$ is then biased for t_y in general. Under what condition is this predictor unbiased?

(b) Under the same conditions as in (a), show that the ratio estimator $\hat{t}_y^R$ is unbiased for t_y. Under what conditions is this estimator in fact the BLUP of this population total?

(c) In many practical situations the household sizes M_g are only known for the sample households. Develop an approximation to the 'M_g known' BLUP for this situation and contrast it with the expansion estimator $\hat{t}_y^E$ above. Is this predictor unbiased in the 'M_g unknown' case?

(d) Develop an estimator for the prediction variance of the approximate BLUP suggested in (c). What are its properties?

(e) The standard estimator of the prediction variance of the expansion estimator is

$$\hat{V}(\hat{t}_y^E) = \frac{Q^2}{q}\left(1 - \frac{q}{Q}\right)\frac{1}{q-1}\sum_s (t_{gy} - \bar{t}_s)^2$$

where $\bar{t}_s = q^{-1}\sum_s t_{gy}$. Show that this estimator is approximately unbiased for the prediction variance of the expansion estimator in the 'M_g unknown' case.

E.18 Many human populations exhibit clustered structure. A common example is a population made up of individuals grouped into households. When sampling such a population it is usually necessary to account for this clustering by allowing a non-zero intracluster correlation for individual characteristics. Thus, if $y_{ig} = 1$ if the ith individual in household g has a characteristic of interest, and is zero otherwise, then a suitable model for the distribution of these values in the population is

$$Pr(y_{ig} = 1) = \pi$$
$$Pr(y_{ig} = 1, y_{jg} = 1|i \neq j) = (1 - \rho_g)\pi^2 + \rho_g \pi$$
$$Pr(y_{ig} = 1, y_{jh} = 1|g \neq h) = \pi^2.$$

Here ρ_g is the intracluster correlation of the y_{ig} in household g. Suppose that a sample survey of the households in the population is carried out, with all individuals in a selected household interviewed, and their Y-values recorded. Let Q denote the total number of households in the population, with q of these selected into the sample. Let s denote this sample of households. Let M_g denote the number of individuals in household g, so that $N = \sum_{g=1}^{Q} M_g$ denotes the total number of individuals in the population. The total number of individuals in the sample is then $n = \sum_s M_g$. The aim is to use the sample data on Y to estimate the total number t_y of people in the population with the characteristic of interest.

(a) Let $t_{gy} = \sum_{i \in g} y_{ig}$ denote the number of people in household g with the characteristic of interest. Show that under the model for y_{ig} above, the expected value of t_{gy} is proportional to M_g, and the variance of t_{gy} is proportional to $M_g[1 + (M_g - 1)\rho_g]$.

(b) Show that an unbiased predictor of t_y in this situation is then $\hat{t}_y = \sum_s t_{gy} + \hat{\pi}(N - \sum_s M_g)$, where $\hat{\pi} = \sum_s t_{gy}/\sum_s M_g$.

(c) Explain why the predictor in (b) above is not the most efficient predictor of t_y one could use in this situation.

(d) How would you estimate the prediction variance of $\hat{t}_y$?

E.19 Most large-scale government surveys are continuing surveys. That is, they are carried out on the same target population at regular intervals. An important (often primary) objective of these repeated surveys is to measure the change in the population total of a survey variable between any two time periods when the survey was carried out.

The samples for these continuing surveys typically overlap. That is, some of the population units in sample for one survey are also in sample for the second. The converse of this, of course, is that some population units in sample at the first survey are 'rotated out' of sample at the second and replaced by population units that were not selected for the first survey.

Let $u = a, b$ denote the times when the two surveys were carried out ($a < b$) and let y_{iu} denote the value of the variable of interest at time u for population unit i. To keep things simple we shall assume that the population is static (no births or deaths) and is of size N. We also assume that the sample sizes at each time are the same and equal to n, with m denoting the number of population units in sample at both times (i.e. the size of the sample overlap). Furthermore, we assume that different population units are uncorrelated with respect to their values of y_{iu}, with $E(y_{ia}) = \mu_a$, $E(y_{ib}) = \mu_b$, $Var(y_{ia}) = Var(y_{ib}) = \sigma^2$ and $Cov(y_{ia}, y_{ib}) = \rho\sigma^2$.

(a) Let $\hat{t}_{uy}$ denote the expansion estimator of the population total t_{uy} of the y_{iu}. A simple prediction of the change in this total between times a and b is then $\hat{d}_y = \hat{t}_{by} - \hat{t}_{ay}$. Show that this predictor is unbiased for the true change $d_y = t_{by} - t_{ay}$.

(b) Put $g = n^{-1}(N - n)$. Show that
$$Var(\hat{d}_y - d_y) = 2\sigma^2 \left[N(1-\rho) + n\{g^2 + 2\rho(g+1) - 1\} - m\rho(g+1)^2 \right].$$

(c) When is the variance in (b) minimised? What does this tell you about optimal sample design when measuring change?

(d) What is the BLUP of d_y given the sample data from both time periods?

(e) What is a sufficient condition for the simple difference $\hat{d}_y = \hat{t}_{by} - \hat{t}_{ay}$ to be this BLUP?

E.20 Let $X_1, X_2, \ldots X_n$ and $Y_1, Y_2, \ldots Y_n$ be two sequences of positive valued random variables satisfying $E(X_i) = \mu$, $E(Y_i) = \nu$, $Var(X_i) = \sigma^2$, $Var(Y_i) = \omega^2$, $Cov(X_i, Y_i) = \rho\sigma\omega$ and $Cov(X_i, X_j) = Cov(Y_i, Y_j) = Cov(X_i, Y_j) = 0, i \neq j$. Let $\bar{X}_n$ and $\bar{Y}_n$ denote the average values of X and Y defined by these sequences.

Exercises

(a) Show that a first order approximation to the expected value of the product of $\bar{X}_n$ and $\bar{Y}_n$ is $\mu\nu$, and a corresponding first order approximation to the variance of this product is

$$Var(\bar{X}_n \bar{Y}_n) \cong \frac{1}{n}\left(\nu^2\sigma^2 + 2\nu\mu\rho\sigma\omega + \mu^2\omega^2\right).$$

(b) Show that a first order approximation to the variance of the logarithm of $\bar{X}_n$ is

$$Var[\log(\bar{X}_n)] \cong \frac{\sigma^2}{n\mu^2}.$$

E.21 Consider the bivariate generalisation of the homogeneous strata population model (4.1) for two survey variables, Y and X:

$$E(y_i|i \in h) = \mu_{yh} \text{ and } E(x_i|i \in h) = \mu_{xh}$$
$$Var(y_i|i \in h) = \sigma_{yh}^2, \quad Var(x_i|i \in h) = \sigma_{xh}^2 \text{ and } Cov(y_i, x_i|i \in h) = \sigma_{yxh}$$
$$Cov(y_i, y_j|i \neq j) = 0, \quad Cov(x_i, x_j|i \neq j) = 0 \text{ and } Cov(y_i, x_j|i \neq j) = 0.$$

Under this model, a predictor of the ratio R of the population totals of Y and X is then

$$\hat{R} = \sum_h N_h \bar{y}_{sh} \Big/ \sum_h N_h \bar{x}_{sh}.$$

Show that an approximately unbiased large sample estimate of the prediction variance of $\hat{R}$ is

$$\hat{V}(\hat{R}) = \frac{\sum_h n_h^{-1}(1 - n_h N_h^{-1})(n_h - 1)^{-1} \sum_{sh}(y_i - \hat{R}x_i)^2}{\left(N^{-1}\sum_h N_h \bar{x}_{sh}\right)^2}.$$

E.22 Given that the matrices $\mathbf{A}$, $\mathbf{B}$, $\mathbf{C}$ and $\mathbf{D}$ are all correctly dimensioned and appropriate inverses all exist, verify that

$$(\mathbf{A} + \mathbf{CBD})^{-1} = \mathbf{A}^{-1} - \mathbf{A}^{-1}\mathbf{C}\left(\mathbf{B}^{-1} + \mathbf{DA}^{-1}\mathbf{C}\right)^{-1}\mathbf{DA}^{-1}.$$

Hence derive an expression for the weights that define the BLUP of t_y under the first order homogeneous population model (9.1).

E.23 Let ξ denote the second order homogeneous population model specified by (3.1). Suppose that this model does not hold and the first order homogeneous population model η specified by (8.1) is better suited to explaining the population. Suppose also that the size of the population is 30. The value of the σ_is are given in the following table:

i	1	2	3	4	5	6	7	8	9	10
σ_i	1	1	1	1	2	2	2	2	2	3
i	11	12	13	14	15	16	17	18	19	20
σ_i	3	3	4	4	5	8	9	9	10	10
i	21	22	23	24	25	26	27	28	29	30
σ_i	11	11	12	12	13	15	15	15	15	15

Three samples are selected from this population. These are $s_1 = \{3, 8, 13, 18, 23, 28\}$, $s_2 = \{25, 26, 27, 28, 29, 30\}$ and $s_3 = \{6, 9, 11, 12, 20, 29\}$.

(a) Can you identify the sampling design that produced these three samples?

(b) Suppose that the model η is true and we base our inference on the model ξ. What will be the safest sample? Why?

(c) Compute and compare $Var_\eta(\hat{t}_y^E - t_y)$ and $E_\eta[V(\hat{t}_y^E)]$ for s_1, s_2 and s_3.

(d) Compute the gain in precision $Var_\eta(\hat{t}_y^E - t_y) - Var_\eta(\hat{t}_{\eta y} - t_y)$ associated with each of the three samples. Here $\hat{t}_{\eta y}$ denotes the BLUP of t_y under η. Comment on these results. What is the sample with the smallest gain in precision? Is this sample a safe sample? Why?

E.24 A stratified random sample is taken from a population made up of H strata. That is, within each of H strata, indexed by $h = 1, 2, \ldots, H$, each of size N_h, we select a simple random sample without replacement of size n_h. For each sample unit we observe the value of a variable Y.

Suppose it is reasonable to assume that the population in fact is made up of G domains indexed by $g = 1, 2, \ldots, G$ that cross stratum boundaries, and are such that the values of Y are independent and identically distributed with mean μ_g and standard deviation σ_g within domain g. The total number N_g of population units making up domain g is unknown, as is the number N_{hg} of stratum h units falling in domain g. All we observe is the stratum by domain crossclassification of the sample units, with n_{hg} denoting number of sample units in stratum h and in domain g.

(a) Show that $\hat{N}_g = \sum_h \frac{N_h}{n_h} n_{hg}$ is an unbiased estimator of the total number of units in the population in domain g and so $\hat{t}_y^S = \sum_g \hat{N}_g \bar{y}_{sg}$ is an unbiased estimator of the population total t_y of Y. Here $\bar{y}_{sg}$ is the sample mean of Y for the units from domain g.

(b) Show that a first order approximation to the prediction variance of $\tilde{t}_y^S$ is

$$Var(\tilde{t}_y^S - t_y) \approx \sum_h \frac{N_h^2}{n_h}\left(1 - \frac{n_h}{N_h}\right) v_h^2(\mu) + \sum_g \left(\frac{Var(\hat{N}_g) + N_g^2}{E(n_g)}\right) \sigma_g^2$$

where

$$v_h^2(\mu) = \frac{1}{N_h}\sum_g N_{hg}\mu_g^2 - \left\{\frac{1}{N_h}\sum_g N_{hg}\mu_g\right\}^2$$

$$Var(\hat{N}_g) = \sum_h \frac{N_h^2}{n_h}\left(1 - \frac{n_h}{N_h}\right)\left(\frac{N_{hg}}{N_h}\right)\left(1 - \frac{N_{hg}}{N_h}\right)$$

and

$$E(n_g) = \sum_h n_h \left(\frac{N_{hg}}{N_h}\right).$$

(c) Write down an approximately unbiased estimator of the above approximate variance.

E.25 What is the prediction bias of the simple expansion estimator $\hat{t}_y^E$ under the following working models?

(a) The linear population model defined by (5.5). Can you suggest a method for eliminating this bias?

(b) The two factor model where $E(y_i) = \mu_{1i} + \mu_{2i}$. Here μ_{1i} represents the effect due to the level of factor 1 possessed by unit i and μ_{2i} is the corresponding effect for factor 2. Show that this bias disappears when the proportion of the sample at each level of each factor is the same as the corresponding population proportion.

E.26 A plausible alternative to the ratio population model (5.2), and one that seems to 'fit' data collected in business surveys much better, is where the regression errors are *multiplicative* rather than *additive*. That is, rather than modelling the relationship between Y and Z via (5.1), we instead assume $y_i = \beta z_i \delta_i$ where $\delta_i > 0$ with $E(\delta_i) = 1$ and $Var(\delta_i) = \omega^2$.

(a) Show that the ratio estimator remains unbiased under this alternative model. Can you suggest a more efficient estimator under this model? Would you use it?

(b) Show that if the alternative multiplicative model is in fact true, then the standard unbiased estimator (5.8) of the prediction variance of the ratio estimator is downward biased. Develop an alternative estimator for the prediction variance of the ratio estimator that is appropriate for this situation – that is one such that the estimated prediction variance of the ratio estimator is then approximately unbiased under this alternative multiplicative model.

E.27 The regression estimator $\hat{t}_y^L = N[\bar{y}_s + b(\bar{z} - \bar{z}_s)]$ of the population total t_y of a variable Y is the BLUP under the linear population model specified by (5.5). Here $\bar{y}_s$ is the sample mean of Y, $\bar{z}_s$ and $\bar{z}$ are the sample mean and population mean, respectively, of an auxiliary variable Z that is known for each unit in the population, and

$$b = \frac{\sum_s (y_i - \bar{y}_s)(z_i - \bar{z}_s)}{\sum_s (z_i - \bar{z}_s)^2}.$$

(a) Show that if $y_i \equiv z_i$ for all i then the prediction variance of $\hat{t}_y^L$ under (5.5) is zero.

(b) An alternative predictor of the population total of Y is $\hat{t}_y^D = N[\bar{z} + (\bar{y}_s - \bar{z}_s)]$. Show that under the linear population model (5.5), $\hat{t}_y^D$ is generally biased for t_y and that a sufficient condition for unbiasedness is that $\beta = 1$.

(c) Assuming that the linear population model (5.5) holds and that $\beta = 1$, which of the two predictors $\hat{t}_y^D$ or $\hat{t}_y^L$ would you prefer? Why?

(d) Suppose that you know that the value of β should be near one (rather than exactly one, as in (c) above). Describe a procedure for choosing between these two predictors in this situation. How would you implement it in practice?

E.28 Suppose our working model is the stratified ratio population model, that is one where each stratum is assumed to follow a (possibly) different version of (5.2), and we predict the population total of Y using the stratified ratio estimator $\hat{t}_y^{RS}$. Show that the robust estimator of the prediction variance of $\hat{t}_y^{RS}$ given by

$$\hat{V}_D(\hat{t}_y^{RS}) = \sum_h \left(\frac{N_h^2}{n_h}\right)\left(\frac{\bar{x}_h}{\bar{x}_{sh}} - \frac{n_h}{N_h}\right)\left(\frac{\bar{x}_h}{\bar{x}_{sh}}\right)\frac{1}{n_h}\sum_{s_h}\left\{\frac{(y_i - \hat{\beta}_h x_i)^2}{1 - (x_i/n_h \bar{x}_{sh})}\right\}$$

is approximately unbiased for the actual prediction variance of $\hat{t}_y^{RS}$ under a stratified population model where $E(y_i|z_i; i \in h) = \beta_h z_i$ and $Var(y_i|z_i; i \in h) = \sigma_i^2$, and is exactly unbiased for this prediction variance under the stratified ratio population model, that is where $Var(y_i|z_i; i \in h) = \sigma_h^2 z_i$.

E.29 Suppose the two level hierarchical population model (6.1) holds and we estimate the population total t_y of a variable Y by a cluster weighted estimator of the form $\hat{t}_{wy} = \sum_s w_g \sum_{s_g} y_i = q^{-1} \sum_s \hat{t}_{wgy}$, where $\hat{t}_{wgy} = qw_g \sum_{s_g} y_i = qw_g m_g \bar{y}_{sg}$, and its variance by the ultimate cluster variance estimator (9.13). Show that this variance estimator is generally upward biased, and derive an expression for its bias.

E.30 The ratio estimator is known to be an efficient predictor of the population total t_y of a survey variable Y when there is a strong positive correlation between this variable and a strictly positive auxiliary size variable Z. In some situations, however, the relationship between Y and Z is still strong, but the correlation is negative. It has been suggested that in these conditions it would be more efficient to use the so-called *product estimator*, defined by

$$\hat{t}_y^P = N(\bar{y}_s \bar{z}_s)/\bar{z}$$

where N denotes the number of units in the population. Two alternative models for the scenario described above are (in both cases different population units are uncorrelated)

$$\xi_1 : E(y_i|z_i) = \alpha - \beta z_i, \quad Var(y_i|z_i) = \sigma^2;$$
$$\xi_2 : E(y_i|z_i) = \gamma z_i^{-1}, \quad Var(y_i|z_i) = \eta^2 z_i^{-1}.$$

(a) Show that the product estimator is generally biased for the population total t_y of Y under ξ_1. Contrast this bias with that of the ordinary ratio estimator under ξ_1.

(b) Show that the product and ratio estimators are also generally biased for t_y under ξ_2.

(c) What is a sufficient condition for the product estimator to be unbiased under both ξ_1 and ξ_2? What about the ratio estimator?

(d) What is the BLUP of t_y under ξ_1? What is the BLUP of t_y under ξ_2? Would you recommend the product estimator if you believed either ξ_1 or ξ_2 were reasonable models for the population data? What about the ratio estimator?

E.31 Use of the ratio estimator in business surveys is usually justified on the basis of the positive correlation between the survey variable Y and the auxiliary Z. However, in reality there is often a stronger positive relationship between $\log(Y)$ and $\log(Z)$, in the sense that the model that fits the sample data best is the loglinear specification, defined by $\log(y_i) = \lambda + \gamma \log(z_i) + \zeta_i$, with $E(\zeta_i) = 0$ and $Var(\zeta_i) = \omega^2$. In what follows we will assume this model is true, with the ζ_i normally distributed. Note that if $Z \sim N(\mu, \sigma^2)$ then $E(e^Z) = e^{\mu + \sigma^2/2}$ and $Var(e^Z) = e^{2\mu + \sigma^2}(e^{\sigma^2} - 1)$.

(a) Show that if $\gamma = 1$, then the ratio estimator $\hat{t}_y^R$ remains unbiased for t_y under this loglinear model.

(b) Still assuming $\gamma = 1$, can you suggest an alternative unbiased linear predictor for t_y with smaller prediction variance?

(c) A non-linear alternative to $\hat{t}_y^R$ for arbitrary γ is the back transformed predictor $\hat{t}_y^{LL} = \sum_s y_i + \sum_r e^{\hat{\lambda} + \hat{\gamma} \log(z_i)}$ where $\hat{\lambda}$ and $\hat{\gamma}$ are the ordinary least squares estimates of λ and γ based on the sample data. Show that this estimator typically has a negative bias under the loglinear model.

(d) Show that the modified predictor $\tilde{t}_y^{LL} = \sum_s y_i + \sum_r k_i e^{\hat{\lambda} + \hat{\gamma}\log(z_i)}$ is approximately unbiased under the loglinear model. Here $k_i = e^{\hat{\omega}^2(1-g_i)/2}$ with

$$\hat{\omega}^2 = (n-2)^{-1}\sum_s \left[\log(y_i) - \hat{\lambda} - \hat{\gamma}\log(z_i)\right]^2$$

and

$$g_i = \frac{1}{n} + \frac{\left(\log(z_i) - \overline{(\log(z))}_s\right)^2}{\sum_s \left(\log(z_j) - \overline{(\log(z))}_s\right)^2}.$$

E.32 This exercise extends the results on prediction of domain totals and means that were discussed in Chapter 14 to the situation where auxiliary information on an auxiliary variable Z is available. We assume an unknown domain size N_d and replace (14.1a)–(14.1c) by

$$E(y_i|z_i, d_i = 1) = \mu(z_i; \omega_d)$$
$$Var(y_i|z_i, d_i = 1) = \sigma^2(z_i; \omega_d)$$
$$Cov(y_i, y_j|z_i, z_j, d_i, d_j) = 0$$

As in Section 14.1 we assume domain membership can be modelled as the outcome of independent and identically distributed Bernoulli trials, independently of the value of Y. However, we also assume domain membership can depend on Z, so (14.1d)–(14.1f) are replaced by

$$E(d_i|z_i) = \theta(z_i; \gamma_d)$$
$$Var(d_i|z_i) = \theta(z_i; \gamma_d)[1 - \theta(z_i; \gamma_d)]$$
$$Cov(d_i, d_j|z_i, z_j) = 0.$$

(a) Show that under the domain model defined above

$$E(d_i y_i|z_i) = \theta(z_i; \gamma_d)\mu(z_i; \omega_d)$$
$$Var(d_i y_i|z_i) = \theta(z_i; \gamma_d)\sigma^2(z_i; \omega_d) + \theta(z_i; \gamma_d)[1 - \theta(z_i; \gamma_d)]\mu^2(z_i; \omega_d)$$
$$Cov(d_i y_i, d_j y_j|z_i, z_j) = 0.$$

(b) Suppose sampling is uninformative for both Y and D, given Z, and so we estimate ω_d and γ_d by consistent estimators $\hat{\omega}_d$ and $\hat{\gamma}_d$ respectively. A plug-in model-based predictor of t_{dy} is then

$$\hat{t}_{dy} = \sum_s d_i y_i + \sum_r \mu(z_i; \hat{\omega}_d)\theta(z_i; \hat{\gamma}_d).$$

Show that the prediction variance of $\hat{t}_{dy}$ then has the consistent estimator $\hat{V}_1 + \hat{V}_2$, where

$$\hat{V}_1 = \hat{V}(\hat{\gamma}_d) \left(\sum_r \mu(z_i; \hat{\omega}_d) \frac{\partial \theta(z_i; \hat{\gamma}_d)}{\partial \hat{\gamma}_d}\right)^2 + \hat{V}(\hat{\omega}_d) \left(\sum_r \theta(z_i; \hat{\gamma}_d) \frac{\partial \mu(z_i; \hat{\omega}_d)}{\partial \hat{\omega}_d}\right)^2$$
$$+ 2\hat{C}(\hat{\gamma}_d, \hat{\omega}_d) \left(\sum_r \mu(z_i; \hat{\omega}_d) \frac{\partial \theta(z_i; \hat{\gamma}_d)}{\partial \hat{\gamma}_d}\right) \left(\sum_r \theta(z_i; \hat{\gamma}_d) \frac{\partial \mu(z_i; \hat{\omega}_d)}{\partial \hat{\omega}_d}\right).$$

and

$$\hat{V}_2 = \sum_r \{\sigma^2(z_i; \hat{\omega}_d)\theta(z_i; \hat{\gamma}_d) + \mu^2(z_i; \hat{\omega}_d)\theta(z_i; \hat{\gamma}_d)[1 - \theta(z_i; \hat{\gamma}_d)]\}.$$

Here $\hat{V}(\hat{\omega}_d)$, $\hat{V}(\hat{\gamma}_d)$ and $\hat{C}(\hat{\gamma}_d, \hat{\omega}_d)$ denote consistent estimators of $Var(\hat{\omega}_d)$, $Var(\hat{\gamma}_d)$ and $Cov(\hat{\gamma}_d, \hat{\omega}_d)$ respectively.

(c) Consider the situation where the population is stratified and the regression of Y on Z is linear and through the origin for units in the domain, but the slope of this regression line varies from stratum to stratum. In addition, the proportion of the population in the domain also varies significantly from stratum to stratum. Let θ_h be the probability that a population unit in stratum h lies in the domain and β_h be the slope of the regression line for domain units in stratum h. Show that the predictor $\hat{t}_{dy}$ in (b) above is then of the form:

$$\hat{t}_{dy} = \sum_s d_i y_i + \sum_h p_{shd} \left(N_h \bar{z}_h - n_h \bar{z}_{sh}\right) \hat{\beta}_h$$

where p_{shd} is the sample proportion of stratum h units in the domain; $\hat{\beta}_h$ is the stratum h estimate for the slope of the regression of Y on Z in the domain; $\bar{z}_h$ is the stratum h average for Z; and $\bar{z}_{sh}$ is the sample average for Z in stratum h. Consequently show that an estimator of the leading term in the prediction variance of $\hat{t}_{dy}$ is

$$\hat{V}_1 = \sum_h (N_h \bar{z}_h - n_h \bar{z}_{sh})^2 \left(\hat{V}(p_{shd})\hat{\beta}_h^2 + \hat{V}(\hat{\beta}_h) p_{shd}^2\right)$$

where $\hat{V}(p_{shd})$ is the estimated variance of p_{shd} and $\hat{V}(\hat{\beta}_h)$ is the estimated variance of $\hat{\beta}_h$.

E.33 Show that under the random means model (15.11), the BLUP of the area g effect is the so-called shrunken residual

$$\hat{u}_g^{BLUP} = \left(\frac{n_g \lambda}{1 + n_g \lambda}\right) (\bar{y}_{sg} - \hat{\mu}^{BLUE}).$$

Hence show that the BLUP of the area g mean of Y under this model is

$$\hat{\bar{y}}_g^{BLUP} = N_g^{-1} \left(n_g \bar{y}_{sg} + (N_g - n_g)(\hat{\mu}_{BLUE} + \hat{u}_g^{BLUP})\right)$$
$$= N_g^{-1} \left(n_g \bar{y}_{sg} + (N_g - n_g) \left\{\left(\frac{1}{1+n_g\lambda}\right) \hat{\mu}^{BLUE} + \left(\frac{n_g\lambda}{1+n_g\lambda}\right) \bar{y}_{sg}\right\}\right).$$

E.34 Suppose the population values of Y follow the homogeneous model (3.1). Show that

$$\hat{V}\left(\hat{F}_{Ny}(t)\right) = (n-1)^{-1}(1-n/N)\hat{F}_{Ny}(t)\left(1-\hat{F}_{Ny}(t)\right)$$

is then an unbiased estimator of the prediction variance (16.4) of the predictor $\hat{F}_{Ny}(t) = n^{-1}\sum_s d_{it}$ of the value of the population distribution function $F_{Ny}(t)$ of Y at t. Here d_{it} is the indicator for whether $y_i \leq t$.

E.35 Show that the prediction variance of the stratified distribution function predictor (16.10) is

$$Var\left(\hat{F}^S_{Ny}(t) - F(t)\right) = N^{-2}\sum_h (N_h^2/n_h)(1-n_h/N_h)F_{1hy}(t)\left(1-F_{1hy}(t)\right)$$

with unbiased estimator

$$\hat{V}\left(\hat{F}^S_{Ny}(t)\right) = N^{-2}\sum_h N_h^2(n_h-1)^{-1}(1-n_h/N_h)\hat{F}_{hy}(t)\left(1-\hat{F}_{hy}(t)\right).$$

E.36 A sample of 15 boys was drawn from the Boys Population (see exercise E.12). This sample was taken by first selecting the 5 tallest boys, then dividing the remaining 145 boys into two further subgroups, the 68 boys with heights greater than 57 inches and the remaining 77 boys. Ordered systematic samples of 5 boys each were then drawn from these subgroups, with ordering on the basis of height. Values of H, W and C for this sample are shown in the table below.

Boy	SubGrp	H	W	C
1	1	66.0	127.0	72
2	1	64.5	106.0	71
3	1	64.0	115.4	72
4	1	63.5	100.5	72
5	1	63.0	127.0	76
11	2	61.1	96.5	72
24	2	60.0	168.0	91
37	2	59.0	84.3	70
50	2	58.5	89.0	69
63	2	57.8	77.5	63
83	3	56.6	72.3	65
97	3	56.0	78.0	68
111	3	55.3	69.5	63
125	3	54.3	78.3	66
139	3	53.4	70.8	65

(a) Using these sample data, and assuming the values of the chest circumference variable (C) in the Boys Population can be modelled as a homogeneous population, see (3.1), calculate appropriate predicted values, with associated 95% prediction intervals, for the 25th, 50th and 75th percentiles of the distribution of C over the Boys Population. Comment on the quality of these predictions by comparing them with the true values deducible from the population data tabulated in exercise E.12.

(b) Repeat part (a), but this time assume a stratified working model (4.1), with strata defined by the subgroups above. How do these predictions compare with those obtained in (a)? Which ones do you prefer? Why?

(c) Finally, extend your analysis in (a) and (b) above by using the parametric predictor (16.15) to take account of the information in the auxiliary variables W and H. In this context you should use the model underpinning your choice of an efficient prediction strategy in exercise E.12.

E.37 Assuming the clustered population model (16.35) holds, show that the simple sample empirical distribution function $F_n(t) = n^{-1} \sum_s m_g \bar{d}_{sgt}^{(1)}$ is unbiased for the population distribution function $F_{Ny}(t)$. Develop an expression for the prediction variance of $F_n(t)$ under (16.35). How would you estimate this prediction variance?

Index

allocation
  compromise 36, 147, 149–50
  equal 37, 39, 45, 91
  optimal 34–7, 39, 42–3, 60, 91, 149–50
  proportional 31, 33–6
area effects 167–9, 177, 180, 184, 194
area-level model *see* population model, area level
asymptotic normality *see* central limit theorem
auxiliary information
  definition and discussion 7, 10-16
  size variables 49
  for stratification 28–9 *see also* stratified homogenous population model
  when none is available 18–19 *see also* homogenous population model
available data predictor 206–7

balanced sampling
  first-order 55–6, 89, 94, 102, 152
  overbalanced sampling 55, 91–2
  simple 91–2
  stratified 59–60, 90–1
  weighted 95
bandwidth 96–100, 115, 141, 207, 228
benchmark variables *see* auxiliary variables
best linear unbiased estimator (BLUE) 50, 73, 75, 166
best linear unbiased predictor (BLUP)
  definition 21–2
  for homogeneous population model 21–24
  for stratified homogeneous population model 30
  for gamma and ratio population models 51
  for linear population model 53
  for separate ratio population model 57
  for separate linear population model 58
  for clustered populations 63–6
  for general linear population model 73
  for correlated general linear population model 74–5
bias correction 112–13, 115, 117, 215, 217–19, 227

non-parametric 142, 150–1
robust 100, 110
binary segmentation 43–5
BLUE *see* best linear unbiased estimator
BLUP *see* best linear unbiased predictor
bootstrap 98, 118, 135–8, 180–1
business surveys 5–6, 28–9, 37, 41, 62, 95, 99, 221

calibrated weights 140–4, 151–3
central limit theorem 31, 55, 135, 158
clustered populations
  general 27, 61–3
  prediction for 63–6
  robust prediction for 93–5
  robust variance estimation for *see* ultimate cluster variance
  sample design for 66–71
combined ratio estimator *see* ratio estimator, combined
conditional mean squared error (MSE) 177–83
conditional model *see* population model, conditional
conditionality principle 10
confidence intervals
  based on Central Limit Theorem 19, 23–4, 158
  for domain estimators 158
  for the expansion estimator 23
  for small area estimators 190, 192
  for the stratified expansion estimator 31
correlated general linear population model
  best linear unbiased prediction for 74–5, 81–2
  definition 74
  sample design for 80–1
  special cases of 76–9
correlation, intra-cluster *see* intracluster correlation
cost models 68–71, 94

Dalenius–Hodges stratification *see* stratification, Dalenius–Hodges
design *see* sample design

direct estimator *see* predictors, direct
distribution function, estimation of
  for clustered populations 209–13
  definition 8, 195
  example of 204–9
  for homogeneous populations 195–7
  non-parametric regression methods for 201–4
  for stratified populations 197–8
  under a linear regression model 198–200
domains
  compared to small areas 173–6
  with known membership 158
  strata as domains 28
  with unknown membership 156–8
  weighted estimator for 159–60

EBLUP *see* empirical best linear unbiased predictor
empirical best (EB) prediction
  for clustered populations 63
  definition of 17
  for homogenous populations 21
  for populations with a regression structure 50–1, 53, 57–8
  for stratified homogenous populations 30, 50
  synthetic 187
  under a general linear population model 72–3
empirical best linear unbiased prediction (EBLUP) 65, 168–9, 171–7, 179–83
epsem *see* sample design, epsem two-stage
equal aggregate stratification *see* stratification, equal aggregate
equal allocation *see* allocation, equal
estimating equations 109, 111, 113–14, 123, 195
example populations *see* populations
expansion estimator (EE)
  robustness of optimality of 87–8
  simple 21–4
  stratified 30-1, 33–9, 42–3, 45, 60, 90, 99, 225

Fay–Herriot model *see* population model, Fay–Herriot
finite population distribution function *see* distribution function, estimation of
frame 6, 18, 51, 63, 156
full sample predictor 130, 132–3, 135

gamma population model
  definition 49
  empirical best and best linear unbiased prediction under 50–1

and overbalanced sampling 92–3
  see also ratio population model
general linear population model
  correlated *see* correlated general linear population model
  definition of 72
  empirical best prediction and best linear unbiased prediction under 72–3
  special cases of 76–9
generalised ratio estimator *see* ratio estimator, generalised

Hajek predictor 161, 173
homogenous population model
  definition of 20
  discussion and examples of 18–20
  a generalisation of 26–7
  inference and sample design for 23–6
hypergeometric distribution *see* urn model

imputation 204–7
influence function 111–15
informative sampling *see* non-informative sampling
intracluster correlation
  avoiding explicit estimation of *see* ultimate cluster variance
  options for estimating 64–6
  parameter of clustered population model 61–2

jackknife variance estimator *see* variance estimator, jackknife

kernel density estimate 126–7

labels 6-7, 20, 26, 54
large sample approximation *see* central limit theorem
linear population model
  definition 52
  empirical best and best linear unbiased prediction under 52–3
  sample design and inference under 55–6
  as a special case of the general linear population model 78
linearisation 9, 121–4, 134, 200, 225
log transformation 214–15, 218–26
logistic regression *see* population model, fixed effects logistic
LU weights 141, 143

M-estimate 111, 113
maximum likelihood (ML) 168, 171–2, 184

mean squared error (MSE), estimation
  of 169, 171, 177, 180–2, 192
  for EBLUP in small area estimation
    (conditional) 177–9
  for EBLUP in small area estimation
    (unconditional) 169–73
  for non-parametric predictors 98
median, finite population
  definition 8, 123
  estimation of 125–8
methods based on random area
    effects 164–5, 167
minimum mean squared error predictor
    (MMSEP) 16, 169, 171
mixed model see population model,
    mixed
mixture model see population model,
    mixture
model see population model
model-based direct (MBD)
    predictor 176–7, 180–3
model-based stratification see
    stratification, model-based
model calibration 215–19
MSE see mean squared error (MSE),
    estimation of
multistage sampling see sample design,
    two-stage
multipurpose surveys 139–54
multivariate stratification see
    stratification, multivariate

negative weights see weights, negative
Neyman allocation see allocation, optimal
non-informative sampling 9–11, 13, 17,
    136, 139, 157, 168
non-linear population parameters 121–2,
    124, 126, 128
non-parametric bias correction see bias
    correction, non-parametric
non-parametric prediction
  of distribution functions 201, 203
  of other quantities 95, 97–100, 141–2,
    144

optimal allocation see allocation, optimal
optimal design for clustered
    populations 66–71
ordinary least squares (OLS) 52–3, 73,
    214, 216–17, 219
outliers see robustness to outliers
overbalanced sample see balanced
    sampling, overbalanced sampling

polynomial model see population model,
    polynomial
population model

area level 162, 192
  clustered see clustered populations
  conditional 177–8
  correlated general linear see correlated
    general linear population model
  exchangeable 20–1
  Fay–Herriot 193
  fixed effects logistic 190–2
  gamma see gamma population model
  general linear see general linear
    population model
  generalised linear mixed model 184–5,
    191–2
  generalised linear model 184–5, 187–8,
    191
  homogeneous strata see stratified
    homogenous population model
  homogenous see homogenous
    population model
  linear see linear population model
  loglinear 194, 218–20, 222, 229
  marginal post-stratification 79
  mixed 65, 164, 166, 169, 175–6, 180,
    184–5, 192, 194, 214
  mixture 111–12
  polynomial 89–90, 93
  random area effects 164–5, 167, 174,
    184, 190–1
  random effects logistic 190–1
  ratio see ratio population model
  separate ratio see ratio population
    model, separate
  two-level 105
  urn see urn model
  working see working model
populations
  AAGIS (Australian Agricultural and
    Grazing Industries Survey) 181–3
  Albanian 181–2
  beef farms 116–18
  cities 32–5, 39–40, 43, 52–3, 108, 126
  farms 50, 57, 145, 147, 180–1
post-stratification 31, 79, 108, 110–11,
    148, 150, 181
PPZ sampling see sample design,
    probability proportional to size
prediction intervals
  definition 24
  robust 83, 88, 101, 117
  via subsampling 117, 135, 137
prediction mean squared error 98, 169,
    171–3, 178
prediction variance
  defined 16
  estimation of 102, 105, 125, 128, 172,
    177, 221–2
  of expansion estimator (EE) 23
  of non-parametric predictors 98

prediction variance (*cont.*)
  of ratio estimator 54
  of stratified expansion estimator 30
  of synthetic EB predictor 187
predictors
  best linear unbiased *see* best linear unbiased predictor
  bias corrected 112, 115, 117, 215
  calibrated *see* calibrated weights
  direct 161, 173, 175, 182–3, 187
  expansion *see* expansion estimator
  indirect 174–5
  mixed-synthetic 169
  ratio *see* ratio estimator
  regression *see* regression estimator
  regression-adjusted 216–17
  transformed 214–15
primary sampling units *see* PSUs
proportional allocation *see* allocation, proportional
proportions
  allocation for 35
  estimation of 19, 36
PSUs (primary sampling units) 62, 130, 133

quantiles (*see also* distribution function, estimation of)
  definition and discussion 8, 195
  estimation for homogenous populations 195–7
  estimation for stratified populations 197–8
  example of 204–9
  Woodruff method for prediction intervals for 197–8, 204

random effects *see* methods based on random area effects
random selection procedure 19–20
ratio, estimating a 24, 122
ratio-adjusted predictor 216, 219, 223, 225–7
ratio estimator 51–52
  combined 227–8, 232
  generalised 99–100
  separate 57
ratio population model
  definition 49
  empirical best and best linear unbiased prediction for 50–1
  robustness and 88–93
  sample design and inference under 53–5
  separate 56–60, 78
  as a special case of the general linear population model 78
  see also ratio estimator

regression estimator 53, 55–6
  separate 58–60
resampling methods 138
ridge regression 143, 148–50
robust bias correction *see* bias correction, robust
robust predictors 84–100
robust variance estimation 83–4, 101–7
robustness to outliers 83–4, 108–18

sample
  balanced *see* balanced sampling
  compared to census 4, 6
  definition 4, 6–7
  non-informative *see* non-informative sampling
  overbalanced *see* balanced sampling, overbalanced sampling
sample design
  epsem two-stage 95
  probability proportional to size 92, 95, 230–1
  simple random sampling 11, 19, 24–6, 86
  stratified 31–48
  two-stage 61–71
sample labels 10, 19
sample size, determining 25, 36–7, 60
separate ratio estimator *see* ratio estimator, separate
separate regression estimator *see* regression estimator, separate
simple random sampling *see* sample design, simple random sampling
size strata 32–4, 37–43, 47–8, 59, 147, 221
small area estimation 161–94
smearing 218–21, 223, 225–8
smoothing 96–9, 148, 196, 201
stratification
  equal aggregate 42–3, 45–7, 147, 221
  model-based 40–2
  multivariate 43–5
  Dalenius–Hodges 38–40
stratified expansion estimator *see* expansion estimator (EE), stratified
stratified homogenous population model
  allocation of sample to strata 31–7
  defining stratum boundaries 37–43
  discussion and definition 28–30
  empirical best and best linear unbiased prediction under 30–1
  how many strata 45–8
  multivariate stratification 43–5
  sample design under 31–48
  as a special case of the general linear population model 78
stratified ratio and regression population models 56–60

superpopulation model *see* population model
synthetic predictors 162–4, 169
systematic sampling 87

Taylor series *see* linearisation
transformation 42, 214–16, 218–22, 224–30, 232
two-stage sampling *see* sample design, two-stage

UK Labour Force Survey 185, 187–8
ultimate cluster variance 105–7
units 5, 10
urn model 19–20

variance
  for clustered populations 66
  of expansion estimator (EE) 23
  of ratio and regression estimators 54–5, 91
  of small area estimators 169, 171–2
  of stratified expansion estimator 75, 152
  under balanced sampling 56

variance components 75, 165–6, 168–73, 177–9
variance estimation
  for clustered populations *see* ultimate cluster variance
  for expansion estimator (EE) 23–4
  jackknife 131–5, 221, 225, 228
  for ratio and regression estimators 55, 57
  for stratified expansion estimator 30–1
  under the correlated general linear population model 75
  under the general linear population model 73

weights
  calibrated *see* calibrated weights
  negative 143, 148–9, 182
  non-parametric *see* non-parametric prediction
  ridged 143–5, 149–50
Woodruff method for quantiles *see* quantiles, Woodruff method for prediction intervals
working model 83–95, 101, 108–11, 156, 162, 216–17